D0991129

Manned Control Centre, built around External Tank, now being studied by
NASA. Called Solar Power Satellite, it would be built by 2 Teleoperator Space
Spiders (seen working on rims) from rolls of prefabricated materials delivered
by Shuttle in one integrated operation. It could provide living quarters for
Shuttle astronauts and focal point for varied space missions

17	Jun	Spacelab-3, multi-customer flight to test science equipment. ? 30 days. 3 crew + 2 payload specialists
18–22		Scheduled for completion by NASA Financial Yr ending Oct 1981
23–43		21 Flights Scheduled for completion by NASA Financial Yr ending Oct 1982
44–76		33 Flights by Oct 1983
77–121		45 Flights by Oct 1984

Notes Original plan for 725 flights in 1980–91 has been cut to 514, largely
because Defense Dept has reduced 224 planned flights to 106, or 9 per year.
ESA expects 'several' Spacelab flights by 1983, and 5 Spacelabs with
European payloads during 1983–85. Early launches all from Kennedy

THE OBSERVER'S
SPACEFLIGHT DIRECTORY

THE OBSERVER'S
SPACEFLIGHT
DIRECTORY

Reginald Turnill

with Foreword by George Low

FREDERICK WARNE
LONDON

First published by
Frederick Warne (Publishers) Ltd, London 1978

© Reginald Turnill 1978

178083

ISBN 0 7232 2051 4

Filmset and printed in Great Britain by
BAS Printers Limited, Over Wallop, Hampshire
2638·977

CONTENTS

LIST OF COLOUR PHOTOGRAPHS

FOREWORD BY GEORGE LOW

NASA Deputy Administrator 1969–76. During 27 years with NASA he had overall responsibility for the Gemini and Apollo programmes, and has also led the planning of future space projects. Mr Low is now President of the Rensselaer Polytechnic Institute at Troy, New York.

The Observer's Spaceflight Directory is graphic proof that space is here as part of our everyday lives. Looking through this compilation of spacecraft it is difficult to realize that just over 20 years ago no man-made object had orbited our planet and no man had penetrated even to the upper limits of our atmosphere.

Since the orbiting of history's first artificial Earth satellite by the Soviet Union in October 1957, there have been nearly 2000 payloads successfully placed in space, and ships from Earth have explored the Moon and other planets, and 2 are en route between Jupiter and Saturn. In manned missions 12 astronauts have visited the Moon's surface, 43 astronauts and 34 cosmonauts have spent nearly 33,000 man-hours in space, and have demonstrated that humans can live and work in space for periods of up to 3 months. The Apollo-Soyuz flight in 1975 paved the way toward more ambitious co-operative flights in the future.

Further proof that space is a part of our everyday life is the continuous operation of a series of commercial communications satellites that, among other things, connect the continents with routine channels of communications such as telephone service and television broadcasts. Earth observation satellites and weather satellites have also contributed to our well-being on Earth.

We indeed have come a long way and we are fortunate that there have been men like Reg Turnill on hand to document our progress.

George Low

AUTHOR'S INTRODUCTION

The Observer's Spaceflight Directory can claim to be the first comprehensive directory covering world-wide space activities to be published. As George Low of NASA generously acknowledges in his Foreword, it is the result of 22 years of first-hand space coverage by the author. The aim has been to expand and update all the information already supplied in the smaller Observer's books of *Manned* and *Unmanned Spaceflight*. The larger, more comprehensive format makes it easier to appreciate how, in the pioneering days, unmanned robot spacecraft made it possible for men to reach the Moon; now too, the unmanned planetary explorers are gathering the knowledge and developing the techniques which will soon enable men to move freely around the planets of our Solar System.

Already near-Earth space is humming with man-made activity. Only the most ignorant of the 'What's-in-it-for-me?' sceptics persist with their opposition in the face of the vast contributions to our everyday life already provided by the rival Soviet and US space programmes. Increasingly, fresh contributions are being added by the developing European Space Agency, and by the national programmes of Japan, China and other nations.

Manned flights have yielded astonishing medical benefits. They range from the use of space helmets for saving the lives of premature babies, and spacesuit techniques for treating leukaemia, scalds and burns, to the development of tiny space-developed cardiac 'pacemakers' being fitted to heart-sufferers at the rate of 60,000 a year. **Communications** satellites, now taken for granted, provide world-wide TV, telephone, telex and data links; nowadays city-dwellers in every country in the world, expect to be able to watch things like the Olympic games as they happen; but in remote places like Alaska and India, village-dwellers are still astonished and delighted when ATS-6 brings them educational TV and medical treatment by a doctor hundreds of miles away who is able to look at an accident case by means of a satellite picture. **Meteorological** satellites help ships to make faster journeys by telling them where to find favourable currents, and guide them safely through moving ice-packs. **Applications** satellites, like the US Landsat (formerly known as ERTS), help us to foresee and avoid disasters such as flood, drought and earthquake.

The US Space Shuttle, the world's first spaceplane, will speed up and extend all these simultaneous developments. With a flight a week, we shall see over 300 men and vast quantities of equipment being placed in orbit annually by the combined efforts of America and Europe. Russia's exciting new Space Stations—and other projects not yet revealed—will match and perhaps surpass these efforts. Both East and West will be devoting much time and effort to using zero-gravity conditions for creating new metals and manufacturing techniques, with the long-term creation of space colonies—millions of people living in near-perfect conditions in man-made planets—very much in mind. Such colonies are no more far-fetched than were the prospects 20 years ago of landing men on the Moon. And the next generation of 'Who's-going-to-pay?' experts can be reassured that such colonies will pay their own economic way by gathering solar energy and selling it to those still on Earth; by exporting their unique high-quality components; and no doubt by providing holiday accommodation and Zero-G hospitals for patients suffering from heart attacks etc. These prospects are not just the author's private flight of fancy; the detailed plans for them are to be found in NASA's own *Outlook for Space*, which aims to give back to youth its opportunities for pioneering adventure, and its hopes that man's timeless dreams of Utopia can be fulfilled. At the time of writing, America's search for life on Mars with the Viking spacecraft is still underway; serious long-term studies are also under way about the possibilities of creating—or re-creating—a Martian atmosphere by exploding nuclear devices on its frozen Poles. The reverse process of reducing the incredibly dense Venusian atmosphere could be just one step beyond that.

Manned flights to the planets are being brought steadily nearer by Russian work in Salyut space stations. Already cosmonauts have learned to re-cycle the water they drink—even recovering their own perspiration from the atmosphere—so that it can be used over and over again. These are the methods, together with such devices as on-board greenhouses producing food, which in the near future will make possible man's first hazardous expeditions to the planets. Improved propulsion systems, making the round trip less than the 18 months needed at present, will surely follow.

All these exciting achievements and prospects are contained in the facts and figures chronicled in this Space Directory. The second half of the 20th century is undoubtedly the most-exciting, fastest-moving period in the development of mankind.

ACKNOWLEDGEMENTS

The Observer's Spaceflight Directory includes all the contents, which have been revised and updated, of the smaller Observer's books of *Manned* and *Unmanned Spaceflight*, together with much additional material. The thanks of the publishers and author must go first to NASA (the US National Aeronautics and Space Administration). All their officials, from the Administrator downwards, have been unfailingly helpful with information and pictures over a period of many years; somewhat invidiously, the author would like to single out Dr Wernher von Braun, who provided interviews in many parts of the world; and William J. O'Donnell, Public Affairs Officer, whose quiet and efficient generosity is typical of NASA News departments at Washington, Cape Canaveral, Houston, Huntsville and Pasadena. NASA Press Kits and News Reference books, unrivalled by any other international organization, are provided jointly by them and the space contractors for individual missions. These include Rockwell International (formerly North American Rockwell), prime contractors on Apollo and the Space Shuttle; and Grumman (prime contractors for the LM). Equally important among the major contractors are Boeing, Chrysler, Hughes, IBM, Martin Marietta and TRW (who also publish an invaluable Space Log). It is difficult to imagine life without *Aviation Week & Space Technology* and its helpful Editor, Robert Hotz. Thanks too to the increasingly important European Space Agency, with headquarters in Paris; to Aerospatiale, the British Aircraft Corporation and Hawker Siddeley. The Royal Aircraft Establishment's *Table of Earth Satellites* is another essential source; thanks are due to Dr D. G. King-Hele for permission to reproduce its latest census of space vehicles; and equally to Dr C. S. Sheldon, Chief, US Science Policy Research Division, for permission to use tables from his annual surveys of US and Soviet progress; and to Frederick C. Durant III, Asst Director, Astronautics at Washington's new National Air and Space Museum. Novosti, Tass and Graham Turnill have provided some Soviet pictures and Associated Press and Popperfotos some other pictures. Chartwell Illustrators, C. P. Vick and D. R. Woods are responsible for the line drawings. Grateful thanks, too, to Margaret Turnill, whose research, typing and checking have matched the author's output for many years, to Anne Emerson for the final editing and to John Warren for his help in selecting the photographs.

METRIC SYSTEM

The increasing, but not yet universal, use of the Metric System, together with wildly fluctuating currency rates, have made it increasingly difficult to render weights, measures and costs in a form which would be both accurate and acceptable to readers. The swing to metrication has been recognized in recent Observer books by giving metric figures first, and then Imperial measures in brackets, instead of the other way round. Now the author and publishers have come to the conclusion that it makes a more readable book to use only metric measurements; but since so many people now have pocket calculators, to help those who prefer miles to metres, or to express thrust in pounds or newtons rather than kilogrammes, the Conversion Table used by NASA is reproduced on page 12.

So far as currency conversions are concerned, where it is possible with completed projects, their cost has been given, for instance, in dollars and pounds. In continuing projects it is more practical just to give dollar costs, and not to attempt conversions to pounds which constantly alter; the same applies, of course, to other currencies.

Conversion Table

	Multiply	By	To Obtain
Distance	feet	0·3048	metres
	metres	3·281	feet
	kilometres	3281	feet
	kilometres	0·6214	statute miles
	statute miles	1·609	kilometres
	nautical miles	1·852	kilometres
	nautical miles	1·1508	statute miles
	statute miles	0·8689	nautical miles
	statute miles	1760	yards
Velocity	feet/sec	0·3048	metres/sec
	metres/sec	3·281	feet/sec
	metres/sec	2·237	statute mph
	feet/sec	0·6818	statute miles/hr
	feet/sec	0·5925	nautical miles/hr
	statute miles/hr	1·609	km/hr
	nautical miles/hr (knots)	1·852	km/hr
	km/hr	0·6214	statute miles/hr
Liquid measure, weight	gallons	3·785	litres
	litres	0·2642	gallons
	pounds	0·4536	kilogrammes
	kilogrammes	2·205	pounds
	metric tonne	1000	kilogrammes
	short ton	907·2	kilogrammes
Volume	cubic feet	0·02832	cubic metres
Pressure	pounds/sq in	70·31	grammes/sq cm
Thrust	pounds	4·448	newtons
	newtons	0·225	pounds
Temperature	Centigrade	1·8 add 32	Fahrenheit

MONITORING SPACE

Satellite Situation Report (To Dec 31 1976)

	Objects in orbit	Decayed Objects
Australia	1	1
Canada	8	0
ESA	1	0
ESRO	1	9
France	55	24
France/FRG	2	0
FRG	9	3
India	1	0
Indonesia	1	0
Italy	0	4
Japan	16	0
NATO	3	0
Netherlands	1	3
PRC	7	13
Spain	1	0
UK	11	4
US	2746	1373
USSR	1277	4070
TOTAL	4141	5504

The table above, compiled and continuously updated by NASA's Goddard Space Flight Center at Greenbelt, Maryland, shows that since the launch of Sputnik 1 on Oct 4 1957, around 10,000 man-made objects—satellites, rocket casings, ejected panels and other debris—have been placed in orbit. The progress of every one of these objects is continuously monitored by America's Space Detection and Tracking System (SPADATS); and no doubt by similar Soviet equipment as well. On a typical day, more than 20,000 observations of space objects are processed by the computers; an example in 1976 was 888 satellites with payloads (USSR 404, USA 429, Others 55). The total of 4107 objects, many, of course, seen up to 18 times a day, was made up by 3219 pieces of debris.

Enormous effort and expense is put into this unrelenting space watch, for two reasons. First, because re-entering space debris falling back to Earth could trigger a false alarm in the Western World's missile alarm system, giving the impression that Russia had launched an attack with nuclear-tipped missiles; and secondly because, under the 1967 United Nations Space Treaty, each country is responsible for any damage caused by its returning space debris. As the number of large objects in Earth orbit steadily increases—easily the largest at present is the 90-tonne Skylab—the danger that some of it will survive its fiery re-entry through the Earth's atmosphere, and fall on populated areas, steadily increases.

America's space watch, operated by the US Air Force, has been steadily developed since the late 1950s, and covers both missile firings and satellite launches. The radar warning networks include the Ballistic Missile Early Warning System (BMEWS), the Sea-Launched Ballistic Missile Detecting and Warning System (SLBM), the Over-the-Horizon Radar (OTHR), and its most recent addition, SPADATS. Information from this warning network is fed into NORAD's (North American Defense) Command and Control system,

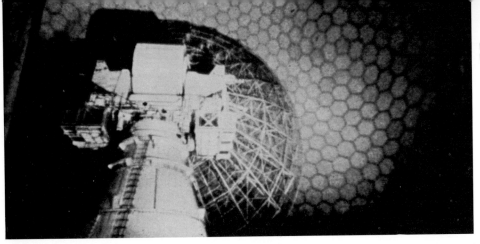

BMEWS Tracker at
Fylingdales, Yorks

buried deep inside Cheyenne Mountain, Colorado, to protect it from possible nuclear attack.

It is wholly a military system, of course, developed as a response to the possibility of attack by ICBM's (Intercontinental Ballistic Missiles). But since an understanding of its detection capabilities sheds much light on spaceflight generally, the system is summarized here.

BMEWS was developed in the late 1950s to extend the existing range of tracking radars from a few hundred km to about 5000 km. The most probable flight-path of a Soviet-launched ICBM attack was then over the Arctic; so 3 BMEWS radar stations were built at Thule, in N Greenland, at Clear in Alaska and at Fylingdales, England. Each has antennas the size of a football field able to detect a missile as it rises above the horizon; trackers and computers then calculate its trajectory, and indicate its impact point. A single missile suggests a satellite launch or test-firing, and is regarded as having a low 'threat value'; in military terms, multiple launches would obviously have a high-threat value. BMEWS trackers, even when they are satisfied that a Soviet-launch is not hostile, but a routine operation, still automatically feed the details to the Space Defense Center in Cheyenne.

Many experts thought that, by the early 70s, BMEWS would be completely out of date, its operations outmoded by later techniques. In fact, valuable new roles have been found for it. The trackers, particularly in England, compute the precise location of any re-entering space debris; the result is that America especially has been able to learn a great deal about Soviet metallurgy and space techniques by recovering and analysing sections of rockets and spacecraft which reach the surface. This has become a busy activity in both East and West—and led both sides to fit their 'spy' satellites with self-destruct mechanisms, so that they can be blown up or de-orbited at the end of these missions in such a way that any debris falls on 'home' territory. It has been discovered too that by rolling the 3 Fylingdales trackers on their backs, they can look south, instead of north, and if necessary provide some cover against missile attacks via the 'back-door'—the South Pole. Astronomers and space scientists also look forward to the day when Fylingdales' military role does come to an end; it has endless possibilities as a space observatory.

SLBM Detection and Warning, developed since 1968, is based on 8 radar sites, intended as an interim system, located in the Atlantic, Pacific and Gulf Coasts. They can both detect the launch of a missile from a submarine, and then track its course. They are being replaced with two more advanced 'phased-array' radars, one each on the US East and West coasts. To detect submarine-launched missiles beyond the range of ground-based radars, a system of SAMOS (Satellite and Missile Observation System) satellites is kept in synchronous orbits. Their infrared sensors can detect a submarine-launched missile by comparing the missile's hot plume with the cool ocean background. But SAMOS satellites have their limitations; sun shining off the clouds can sometimes look like the infrared 'signature' of a missile launch; and they do not cover all possible Soviet launch sites, nor provide accurate impact-point predictions.

OTHR (Over the Horizon Radar) was developed in the 1960s and went into operation in Mar 1968, to meet the threat of more advanced Soviet missiles which could be launched into orbit via the South Pole, travelling three-quarters of the way round the world before re-entering to strike their target. These missiles are called FOBS (Fractional Orbital Bombardment Missiles). Over-the-horizon radar is not really radar at all. It consists of a series of transmitters and receivers located throughout the Pacific and Europe; one of them was at Orfordness on England's east coast for several years until it was closed down in 1974. A powerful, high-frequency transmitter bounces a continuous signal off the ionosphere, an outer layer of the Earth's atmosphere which reflects radio waves. The signal bounces back and forth repeatedly between the ionosphere and the Earth's surface until being received at another station several thousand miles away. If a missile penetrates the ionosphere, as it must do during the initial rocket burn, it disturbs the bouncing radio signal, and is thus detected.

Finally came SPADATS, which includes the US Navy's Space Surveillance System, made up of a line of radio transmitters and receivers strung across the US; Canada's Satellite Tracking Unit, which operates a telescopic camera able to photograph satellites; and SPACETRACK, operated by the US Air Force. Each day this system's world-wide network of cameras and radars makes 18,000 observations of the total of over 3000 satellites now in space. As they gradually slow down, or 'decay', the time when each satellite will fall back to Earth is predicted, and its actual descent tracked and logged, so that it will not be mistaken for an attacking missile. It is the latest phased-array radar, the AN/FPS-85, already operational at Eglin Air Force Base, Florida, that is making the briefly used OTHR out of date. These new generation radars will play an ever-increasing role in space surveillance. The Eglin radar has no moving parts, but consists of 2 fixed electronic arrays set in a 10-storey high bank of concrete built at an angle of 45° to the horizon. One array, consisting of 5184 transmitter modules, sends out a beam, aimed and steered electronically, which sweeps the area of coverage in milliseconds; the other array, 4660 modules in a hexagonal shape, receive the radar signal as it is bounced back from an orbiting satellite. The Florida base was chosen for the first of these new arrays because most space objects pass within its coverage every day from there.

SPACETRACK radar sensors can also be used to reconstruct the exact shape and size of a space object. When the Skylab space station was damaged during launch in 1973, correspondents like myself were unable to obtain confirmation of our suspicions that NASA had obtained a top-secret military 'picture' of Skylab's radar 'signature', confirming that a solar panel had failed to deploy. Over a year later this was officially admitted. That incident demonstrates how America can ascertain, within minutes of its launch, the size, shape and purpose of every Soviet spacecraft, whether its job is military or civil, and whether it is operational or damaged. If Russia does not yet have a similar capability to monitor American launches, she is no doubt working hard to obtain it.

More information about the East-West electronic space 'war' can be found in the **Military Satellites** entry under US Space Projects, and under **Cosmos** in the Soviet section.

Note: The log opposite includes manned flights. It is compiled by the Royal Aircraft Establishment, Farnborough, England, and reproduced with their permission.

16

SPACE LOGS

Satellites & Space Vehicles 1957–76

Country of origin	Year of launch					Total 1957–76
	1957–68	1969–73	1974	1975	1976	
USSR	314	381	79	85	97	956
USA	432	126	13	23	21	615
France	4	2	0	3	0	9
Japan	—	4	1	2	1	8
China	—	2	0	3	2	7
UK	—	1	0	0	0	1
USA/Intelsat	6	11	1	2	1	21
USSR/Intercosmos	—	10	2	2	2	16
USA/UK	3	3	4	0	0	10
USA/ESRO-ESA	3	4	0	1	0	8
USA/Canada	2	4	0	1	1	8
USA/France	1	1	1	1	0	4
USSR/France	—	3	0	1	0	4
USA/FRG	—	2	2	0	1	5
USA/Indonesia	—	—	—	—	1	1
USA/Italy	2	1	1	0	0	4
USA/NATO	—	2	0	0	1	3
USA/Australia	1	1	0	0	0	2
France/FRG	—	1	0	0	0	1
USA/Netherlands	—	—	1	0	0	1
USA/Spain	—	—	1	0	0	1
USSR/India	—	—	—	1	0	1
Total launches	768	559	106	125	128	1686

US Manned Spaceflights

Spacecraft	Launch Date	Astronauts	Revs	Flt Time Days	Hr	Min	Highlights
Mercury 3	5. 5.61	Alan Shepard	Sub-orbital	00.	00.	15	1st American in space
Mercury 4	21. 7.61	Virgil Grissom	Sub-orbital	00.	00.	16	Capsule sank
Mercury 6	20. 2.62	John Glenn	3	00.	04.	55	1st American in orbit
Mercury 7	24. 5.62	M. Scott Carpenter	3	00.	04.	56	Landed 402 km from target
Mercury 8	3.10.62	Walter Schirra	6	00.	09.	13	Landed 8 km from target
Mercury 9	15. 5.63	L. Gordon Cooper	22	01.	10.	20	1st long flight by an American
Gemini 3	23. 3.65	Virgil Grissom John Young	3	00.	04.	53	1st manned orbital manoeuvres
Gemini 4	3. 6.65	James McDivitt Edward White	62	04.	01.	56	21-min 'spacewalk' (White)
Gemini 5	21. 8.65	L. Gordon Cooper Charles Conrad	120	07.	22.	56	1st extended manned flight
Gemini 7	4.12.65	Frank Borman James Lovell	206	13.	18.	35	Longest US flight for 8 years
Gemini 6	15.12.65	Walter Schirra Thomas Stafford	16	01.	01.	51	RV to 1.8 m of Gemini 7
Gemini 8	16. 3.66	Neil Armstrong David Scott	6	00.	10.	41	1st docking; emergency splashdown
Gemini 9	3. 6.66	Thomas Stafford Eugene Cernan	45	03.	00.	21	2-hr spacewalk (Cernan)
Gemini 10	18. 7.66	John Young Michael Collins	43	02.	22.	47	RV with 2 targets; Agena package retrieved
Gemini 11	12. 9.66	Charles Conrad Richard Gordon	44	02.	23.	17	RV and docking
Gemini 12	11.11.66	James Lovell Edwin Aldrin	59	03.	22.	34	Dockings; 3 spacewalks
Apollo 7	11.10.68	Walter Schirra Donn Eisele Walter Cunningham	163	10.	20.	09	1st manned Apollo flight

Spacecraft	Launch Date	Astronauts	Revs	Flt Time			Highlights
				Days	Hr	Min	
Apollo 8	21.12.68	Frank Borman James Lovell William Anders	Lunar revs 10	06.	03.	00	1st manned flight around Moon
Apollo 9	3. 3.69	James McDivitt David Scott Russell Schweickart	151	10.	01.	01	Docking with Lunar Module
Apollo 10	18. 5.69	Thomas Stafford Eugene Cernan John Young	Lunar revs 31	08.	00.	03	Descent to within 14 km of Moon
Apollo 11	16. 7.69	Neil Armstrong Edward Aldrin Michael Collins	CM L. revs 31	08.	03.	18	Armstrong and Aldrin land on Moon; 20 kg samples
Apollo 12	14.11.69	Charles Conrad Richard Gordon Alan Bean	CM L. revs 45	10.	04.	36	2 EVAs, total 7 hr 39 min; 34 kg samples
Apollo 13	11. 4.70	James Lovell John Swigert Fred Haise		05.	22.	55	Mission aborted following oxygen tank explosion
Apollo 14	31. 1.71	Alan Shepard Stuart Roosa Edgar Mitchell	CML. revs 34	09.	00.	42	2 EVAs, total 9 hr 25 min; 44 kg samples
Apollo 15	26. 7.71	David Scott James Irwin Alfred Worden	CM L. revs 74	12.	07.	12	3 EVAs, total 18 hr 36 min; 78 kg samples
Apollo 16	16. 4.72	John Young Thos Mattingly Charles Duke	64	11.	01.	51	3 EVAs, total 20 hr 14 min; 97.5 kg samples
Apollo 17	6.12.72	Eugene Cernan Ronald Evans Harrison Schmitt	CM L. revs 75	12.	13.	51	3 EVAs, total 22 hr 06 min; 113 kg samples
Skylab 1	14. 5.73	Unmanned					Vibration damage during lift-off

Spacecraft	Launch Date	Astronauts	Revs	Flt Time			Highlights
				Days	Hr	Min	
Skylab 2	25. 5.73	Charles Conrad Joseph Kerwin Paul Weitz	404	28.	00.	50	Exceeded Soviet duration record EVAs repaired damage
Skylab 3	28. 7.73	Alan Bean Owen Garriott Jack Lousma	858	59.	11.	09	Rescue mission prepared but not needed
Skylab 4	16.11.73	Gerald Carr Edward Gibson William Pogue	1214	84.	01.	15	Total Skylab EVAs 3 days, 5 hr 48 min
ASTP	15. 7.75	Thomas Stafford Vance Brand Donald Slayton	138	09.	01.	28	1st US–Soviet Joint Flight; 2 days docked activities

USSR Manned Flights

Spacecraft	Launch Date	Cosmonauts	Revs	Flt Time			Highlights
				Days	Hr	Min	
Vostok 1	12. 4.61	Yuri Gagarin	1	00.	01.	48	1st man in space
Vostok 2	6. 8.61	Herman Titov	17	01.	01.	18	1st day in space
Vostok 3	11. 8.62	Andrian Nikolayev	64	03.	22.	27 }	1st double flight
Vostok 4	12. 8.62	Pavel Popovich	48	02.	22.	29 }	
Vostok 5	14. 6.63	Valeri Bykovsky	81	04.	23.	06 }	2nd group flight
Vostok 6	16. 6.63	Valentina Tereshkova	48	02.	22.	50 }	1st woman in space
Voskhod 1	12.10.64	Vladimir Komarov Konstantin Feoktistov Boris Yegorov	16	01.	00.	17	1st 3-man craft
Voskhod 2	18. 3.65	Pavel Belyayev Alexei Leonov	17	01.	02.	02	1st spacewalk (10 min) by Leonov

Spacecraft	Launch Date	Cosmonauts	Revs	Flt Time			Highlights
				Days	Hr	Min	
Soyuz 1	23. 4.67	Vladimir Komarov	18	01.	02.	45	Re-entry parachute snarled; Komarov killed
Soyuz 3	26.10.68	Giorgi Beregovoi	64	03.	22.	51	RV practice with unmanned Soyuz 2
Soyuz 4	14. 1.69	Vladimir Shatalov	48	02.	23.	14	1st docking of 2 manned craft; Khrunov and Yeliseyev returned in Soyuz 4
Soyuz 5	15. 1.69	Boris Volynov Yevgeny Khrunov Alexei Yeliseyev	50	03.	00.	46	
Soyuz 6	11.10.69	Georgi Shonin Valeri Kubasov	80	04.	22.	42	30 manual control manoeuvres; 1st space welding
Soyuz 7	12.10.69	Anatoli Filipchenko Vladislav Volkov Viktor Gorbatko	80	04.	22.	41	
Soyuz 8	13.10.69	Vladimir Shatalov Alexei Yeliseyev	80	04.	22.	41	
Soyuz 9	1. 6.70	Andrian Nikolayev Vitali Sevastyanov	285	17.	16.	59	Endurance record (nearly 18 days); mileage was 7,387,418
Salyut 1	19. 4.71	Orbital Science Station					Destroyed Oct 11 1971 after approx. 2800 orbits
Soyuz 10	23. 4.71	Vladimir Shatalov Alexei Yeliseyev Nikolai Rukavishnikov	30	02.	00.	45	Docked with Salyut but no entry made
Soyuz 11	6. 6.71	Georgi Dobrovolsky Vladislav Volkov Viktor Patsayev	380	23.	17.	40	23 days spent in Salyut; crew killed on re-entry
Soyuz 12	27. 9.73	Vasily Lazarev Oleg Mararov	32	00.	47.	16	Tested post-disaster modifications
Soyuz 13	18.12.73	Pyotr Klimuk Valentin Lebedev	128	07.	20.	55	Soviet and US spacemen in orbit together for 1st time
Salyut 3	25. 6.74	Unmanned					Destroyed 24.1.75
Soyuz 14	3. 7.74	Papel Popovich Yuri Artyukhin	252	15.	17.	30	Docked with Salyut 3 Jul 4

Spacecraft	Launch Date	Cosmonauts	Revs	Flt Time Days	Hr	Min	Highlights
Soyuz 15	26. 8.74	Gennady Sarafanov Lev Demin	32	02.	00.	12	Emergency return after failing to dock with Salyut 3
Soyuz 16	2.12.74	Anatoli Filipchenko Nikolai Rukavishnikov	96	05.	22.	24	Successful ASTP rehearsal
Salyut 4	26.12.74	Unmanned					Destroyed Feb. 3 1977
Soyuz 17	11. 1.75	Alexei Gubarev Georgi Grechko	467	29.	14.	40	New Soviet duration record in Salyut 4
Soyuz 00	5. 4.75	Vasily Lazarev Oleg Makarov	00	00.	00.	00	1st abort between launch and insertion
Soyuz 18	24. 5.75	Pyotr Klimuk Vitali Sevastyanov	993	62.	23.	20	Soviet duration record increased
Soyuz 19 (ASTP)	15. 7.75	Alexei Leonov Valeri Kubasov	96	05.	22.	31	1st US/USSR Joint Flight
Soyuz 20	17.11.75	Unmanned					Re-supply rehearsal to Salyut 4
Salyut 5	22. 6.76	Unmanned					
Soyuz 21	6. 7.76	Boris Volynov Vitaly Zholobov	789	49.	05.	24	48 days in Salyut 5
Soyuz 22	15. 9.76	Valery Bykovsky Vladimir Aksenov	127	07.	21.	54	Probably 1st manned spy satellite
Soyuz 23	14.10.76	Vyacheslav Zudov Valery Rozhdestvensky	32	02.	00.	06	1st Soviet splashdown after emergency return
Soyuz 24	7. 2.77	Victor Gorbatko Yury Glazkov	286	17.	16.	08	Docked with Salyut 5

Note: Even NASA's own official figures frequently vary, leading to differences in cumulative totals. Earth Revolution figures are very approximate, since they are inevitably a mixture of orbit and revolution totals, and a revolution is 6 min longer than an orbit.

USSR totals are based on official figures when available, and estimates when they are not. The US has been credited with two revolutions for each Apollo Moonflight.

MAJOR UNMANNED FLIGHTS

Date	Craft	Description
Oct 4 1957	Sputnik 1	1st artificial satellite
Jan 31 1958	Explorer 1	Discovered Van Allen radiation belts
Mar 17 1958	Vanguard 1	Measured Earth's shape
Jan 2 1959	Luna 1	Earth escape spacecraft
Aug 7 1959	Explorer 6	TV pictures from space
Sep 12 1959	Luna 2	Lunar impact
Oct 4 1959	Luna 3	Farside lunar picture
Apr 1 1960	Tiros 1	Weather satellite
May 24 1960	Midas 2	Missile detection satellite
Jun 22 1960	Transit/Solrad	1st multiple payload
Aug 10 1960	Discoverer 13	Payload recovered
Feb 12 1961	Sputnik 8	Orbital platform launch
Feb 12 1961	Venus 1	Venus fly-by launched by Sputnik 8
Aug 27 1962	Mariner 2	Data from Venus
Nov 1 1962	Mars 1	Mars fly-by
Jul 26 1963	Syncom 2	Synchronous satellite
Oct 17 1963	Vela 1	Satellite to detect nuclear explosions
Jul 28 1964	Ranger 7	1st close-up Moon pictures
Nov 28 1964	Mariner 4	Mars pictures
Apr 3 1965	Snapshot 1	Nuclear reactor in orbit
Apr 6 1965	Early Bird (Intelsat 1)	Commercial TV communications
Jul 16 1965	Proton 1	Cosmic ray measurements
Nov 16 1965	Venus 3	Venus impact
Jan 31 1966	Luna 9	Soft-landed on Moon; pictures from surface
Mar 31 1966	Luna 10	Lunar orbiter
Aug 10 1966	Orbiter 1	Pictures from lunar orbit
Dec 21 1966	Luna 13	Tested denisty of lunar surface
Jan 25 1967	Cosmos 139	FOBS—Fractional Orbit Bombardment Satellite
Apr 17 1967	Surveyor 3	Dug lunar trench
Jun 12 1967	Venus 4	Investigated Venusian atmosphere
Jul 1 1967	Dodge 1	1st full-face colour picture of Earth
Sep 8 1967	Surveyor 5	Chemical analysis of lunar soil
Oct 30 1967	Cosmos 186/188	Automatic docking
Sep 21 1968	Zond 5	Circumlunar flight and recovery
Oct 20 1968	Cosmos 249	Inspection/destructor satellite
Nov 10 1968	Zond 6	Skip re-entry after lunar flight
Aug 17 1970	Venus 7	Soft-landed on Venus
Sep 12 1970	Luna 16	Automatic return of lunar soil
Nov 10 1970	Luna 17	Lunar robot
May 19 1971	Mars 2	Impacted on Mars
May 28 1971	Mars 3	Soft-landed on Mars
May 30 1971	Mariner 9	Orbited Mars
Mar 3 1972	Pioneer 10	Launched to Jupiter; 300 pictures of Jupiter, Ganymede, Callisto and Europa
Mar 27 1972	Venus 8	Analysed Venusian soil
Jul 23 1972	Landsat 1	1st Earth Resources Satellite
Jun 8 1975	Venus 9	Venus Surface pictures
Aug 20 1975	Viking 1	Mars Surface pictures

PLANETARY FLIGHTS

Mars

1	Oct	10 1960	USSR (U)	Failed to reach Earth orbit
2	Oct	14 1960	USSR (U)	Failed to reach Earth orbit
3	Oct	24 1962	USSR (U)	Exploded in Earth orbit
4	Nov	1 1962	Mars 1	Passed Mars at 200,000 km after communications failed
5	Nov	4 1962	USSR (U)	Failed to leave Earth orbit
6	Nov	5 1964	Mariner 3	No communications when shroud failed to jettison, throwing it off course
7	Nov	28 1964	Mariner 4	Returned Mars pictures and data, passing at 9844 km. Still active
8	Nov	30 1964	Zond 2	Communications failed, but passed Mars at 1500 km
9	Jul	18 1965	Zond 3	Engineering test towards orbit of Mars; returned lunar farside pictures en route
10	Feb	24 1969	Mariner 6	Returned Mars pictures; passed planet on Jul 31 1969 at 3215 km
11	Mar	27 1969	Mariner 7	Returned Mars pictures and data; passed planet on Aug 5 1969 at 3516 km
12	May	8 1971	Mariner 8	Failed to achieve orbit
13	May	10 1971	Cosmos 419	Failed to leave Earth orbit
14	May	19 1971	Mars 2	Gathered data in Martian orbit; lander, carrying USSR coat-of-arms, the hammer-and-sickle emblem, crashed on surface Nov 27 1971
15	May	28 1971	Mars 3	Gathered data in Martian orbit; lander survived, touched down, but TV transmissions from surface on Dec 2 1971 ended after 20 sec
16	May	30 1971	Mariner 9	Entered Martian orbit on Nov 13 1971; sent over 7000 TV pictures of planet and its moons
17	Jul	21 1973	Mars 4	Failed to orbit
18	Jul	25 1973	Mars 5	Operated in Mars orbit for a few days and sent some good pictures of S Hemisphere
19	Aug	5 1973	Mars 6	Lander transmitted 1st atmospheric data but crashed on surface
20	Aug	9 1973	Mars 7	Lander missed Mars
21	Aug	20 1975	Viking 1	Successfully orbited; after 2 weeks delay lander sent back 1st Mars surface pictures
22	Sep	9 1975	Viking 2	2nd successful orbiter and lander

Venus

1	Feb	4 1961	Sputnik 7	Earth-orbiting platform failed to launch probe
2	Feb	12 1961	Venus 1	Communications failed, but passed Venus at 100,000 km
3	Jul	22 1962	Mariner 1	Destroyed by Range Safety Officer at 161 km altitude
4	Aug	25 1962	USSR (U)	Failed to leave Earth orbit
5	Aug	27 1962	Mariner 2	Returned data, passing Venus at 34,830 km
6	Sep	1 1962	USSR (U)	Failed to leave Earth orbit
7	Sep	12 1962	USSR (U)	Failed to leave Earth orbit
8	Nov	11 1963	Cosmos 21	Engineering test only; failed to leave Earth orbit
9	Mar	27 1964	Cosmos 27	Failed to leave Earth orbit
10	Apr	2 1964	Zond 1	Communications failed; passed Venus at 100,000 km
11	Nov	12 1965	Venus 2	Communications failed; passed Venus at 23,810 km
12	Nov	16 1965	Venus 3	Communications failed; impacted on Venus
13	Nov	23 1965	Cosmos 96	Failed to leave Earth orbit
14	Jun	12 1967	Venus 4	Returned atmospheric data on Oct 18 1967 during descent and impact
15	Jun	14 1967	Mariner 5	Returned data during fly-by at 3990 km on Oct 19 1967
16	Jun	17 1967	Cosmos 167	Failed to leave Earth orbit
17	Jan	5 1969	Venus 5	Returned atmospheric data during descent and impact on May 16 1969
18	Jan	10 1969	Venus 6	Returned atmospheric data during descent and impact on May 17 1969
19	Aug	17 1970	Venus 7	Reached surface and returned data for 23 min on Dec 15 1970
20	Aug	22 1970	Cosmos 359	Insufficient velocity to leave Earth orbit
21	Mar	27 1972	Venus 8	Reached Venusian surface and returned data for 50 min on Jul 22 1972
22	Mar	31 1972	Cosmos 482	Failed to leave Earth orbit
23	Nov	3 1973	Mariner 10	Venus/Mercury fly-by. 6800 pictures of Venus, Mercury, Earth and Moon; 1st pictures of Venus; 1st mission to Mercury
24	Jun	8 1975	Venus 9	1st pictures from Venus' surface; operated 53 min after landing
25	Jun	14 1975	Venus 10	Lander survived 65 min on surface; sent more pictures

Jupiter

1 Mar 3 1972 Pioneer 10 — Fastest man-made object; passed Jupiter at distance of 130,300 km on Dec 4 1973

2 Apr 5 1973 Pioneer 11 — Passed Jupiter at 42,800 km on Dec 2 1974. Will be 1st craft to reach Saturn

3 Sep 1 1977 Voyager 1 — 1st pictures due Dec 15 1978; closest approach (278,000 km) Mar 5 1979

4 Aug 20 1977 Voyager 2 — 1st pictures Apr 20 1979; closest approach (643,000 km) Jul 10 1979

Saturn

1 Apr 5 1973 Pioneer 11 — Due to reach Saturn and possibly fly between rings in Sep 1979

2 Sep 1 1977 Voyager 1 — 1st pictures due Aug 24 1980; closest approach (138,000 km) Nov 12 1980

3 Aug 20 1977 Voyager 2 — 1st pictures due Jun 8 1981; closest approach (1,020,000 km) Aug 27 1981

Mercury

1 Nov 3 1973 Mariner 10 — Passed Mercury at 271 km on Mar 29 1974, sent 2300 TV pictures; 2nd and 3rd passes in Sep 74 and Mar 75; 10,000 pictures total

Uranus

1 Aug 20 1977 Voyager 2 — Pictures due about Jan 30 1986

Tabulation of USSR Spaceflights
(Payloads to Dec 31 1976)

	1957	58	59	60	61	62	63	64	65	66	67	68	69	70	71	72	73	74	75	76	TOTAL
Sputnik	2	1	—	3	4	—	—	—	—	—	—	—	—	—	—	—	—	—	—	—	10
Luna (Lunik)	—	—	3	—	—	—	2*	—	4	5	—	1	1	2	2	1	1	2	—	1	25
Vostok, Voskhod	—	—	—	—	2	2	2	1	1	—	—	—	—	—	—	—	—	—	—	—	8
Cosmos	—	—	—	—	—	12	12	27	52	34	61	64	55	72	81	72	85	74	85	101	887
Venus (Venik)	—	—	—	—	—	3*	—	—	2	—	1	—	2	1	—	1	—	—	2	—	12
Mars	—	—	—	—	—	3*	—	—	—	—	—	—	—	2	—	—	4	—	—	—	9
Polyot	—	—	—	—	—	—	1	1	—	—	—	—	—	—	—	—	—	—	—	—	2
Electron	—	—	—	—	—	—	—	4	—	—	—	—	—	—	—	—	—	—	—	—	4
Zond	—	—	—	—	—	—	—	2	1	—	—	3	1	1	—	—	—	—	—	—	8
Molniya	—	—	—	—	—	—	—	—	2	2	3	3	2	5	3	6	8	7	10	7	58
Proton	—	—	—	—	—	—	—	—	2	1	—	1	—	—	—	—	—	—	—	—	4
Soyuz (Union)	—	—	—	—	—	—	—	—	—	—	1	2	5	1	2	—	2	3	4	3	23
Meteor	—	—	—	—	—	—	—	—	—	—	—	—	2	4	4	3	2	5	4	3	27
Intercosmos	—	—	—	—	—	—	—	—	—	—	—	—	2	2	1	3	2	2	2	2	16
No Designation	—	—	—	—	—	—	—	—	—	—	2	—	—	—	—	—	—	—	—	—	2
Salyut-1	—	—	—	—	—	—	—	—	—	—	—	—	—	—	1	—	1	2	—	1	5
Oreol-1	—	—	—	—	—	—	—	—	—	—	—	—	—	—	1	—	1	—	—	—	2
PROGNOZ	—	—	—	—	—	—	—	—	—	—	—	—	—	—	—	2	1	—	1	1	5
USSR International Cooperatives	—	—	—	—	—	—	—	—	—	—	—	—	—	—	—	1	—	2	—	—	3
Radoga	—	—	—	—	—	—	—	—	—	—	—	—	—	—	—	—	—	—	1	1	2
Ekran	—	—	—	—	—	—	—	—	—	—	—	—	—	—	—	—	—	—	—	1	1
Totals to Date	2	1	3	3	6	20	17	35	64	44	66	74	70	88	97	89	107	95	111	121	1113

*Includes launches identified by the US but not announced by USSR

Space Programmes

AUSTRALIA

History Australia's space activities have diminished with the run-down of the Woomera rocket range, set up jointly with Britain at the end of the war to test military missiles, and the Blue Streak, Black Knight and Black Arrow launchers. Later, the provision of the Woomera facilities enabled Australia to become a member of the now defunct European space club, ELDO. The space logs credit Australia with the 2 small test satellites detailed below:

Wresat L Nov 29 1967 by Sparta from Woomera. Wt 45 kg. Orbit 106 × 777 km. Incl 83°. Australia's first satellite was launched by a modified American Redstone rocket with additional 2nd and 3rd stage solid-propellant motors. Wresat (from *Weapons Research Establishment*, Australia, which developed it) transmitted data for 5 days and re-entered after 42 days.

Oscar 5 L Jan 23 1970 by Thor Delta from Vandenberg. Wt 18 kg. Launched piggyback on ITOS-1 into a 1435 × 1481 km orbit, with 102° incl, this was the 5th amateur radio satellite and transmitted for 46 days. Orbital life 10,000 yr. (*See* US—Oscar.)

BRAZIL

Brazil took the lead among developing nations in making use of satellites for both economic and educational purposes. She also has a long-term programme to develop her own launch system.

Since Apr 1973, a tracking and data processing station has been in operation at Sao Jose dos Campos near the centre of the country, gathering information from LANDSAT-1. Dr Fernando de Mendonca, director of the Institute for Space Research (Instituto de Pesquisas Espaciais) said that for internal development 'LANDSAT is a better form of aid than dollars'. A remarkable example is the use of its data to identify areas favourable for the breeding of a snail identified as the carrier of an intestinal disease germ which affects 10% of the Brazilian population. Once identified, teams will be sent out to eradicate the snails. LANDSAT maps of Brazil are also being prepared, showing forest areas, geological structures, inland lakes and waterways; settlers can also be watched by this means to ensure they clear only the agreed one-third of their areas for agricultural use, preserving the remainder as forest. In co-operation with NASA, plans were drawn up (in 1974) to use ATS-6 for experimental educational TV relays to 500 schools and 15,000 students in NE Brazil early in 1975. NASA agreed to reorient ATS-6 to provide Brazil with a 30min transmission each day for two months. Depending on the success of this experiment, Brazil planned to establish a 3-satellite domestic communications network in the late 1970s, for the sole use of educational TV channels. Launches would be under NASA contract from Cape Canaveral.

Brazil's Space Institute, now accelerating its rocket design and launch programme, has already produced three sounding rockets: Sonda 1, a single-stage solid propellant rocket, has carried a 5kg payload to altitudes of between 70 and 120 km more than 200 times. Sonda 2, also single-stage, has carried 59kg payloads to heights of 80 km; on about 30 occasions. Sonda 3, 2-stage, is designed to carry 50 kg to heights of 500 km. Further developments are intended to lead to a 3-stage rocket capable of use as a satellite launcher.

CANADA

History   15 yr of close co-operation between Canada and NASA culminated in the launch of CTS (Communications Technology Satellite) on Jan 16 1976. It was Canada's 8th satellite—all launched by NASA. The Anik satellites described below are also known as 'Telesata', since this domestic commercial system is operated by Telesat Canada. Having become an 'observer' state of the European Space Agency, Canada is participating in the ESA/US Aerosat programme, and is also undertaking development of the manipulator arm for use on the Space Shuttle Orbiter when placing and recovering satellites and payloads in orbit.

Alouette   Canada's first satellite, Alouette 1, L Sep 29 1962 by Thor-Agena B from Vandenberg into a 997×1026 km orbit, with 145 kg wt and $80°$ incl, was the first satellite to return useful data—on the ionosphere—for more than 6 yr. It has a 2000-yr life. Alouette 2, L Nov 29 1965 (orbit 505×2987 km; incl $80°$, orbital life 500 yr) was the first in a series of 3 craft in a joint Canadian-NASA International Satellites for Ionospheric Studies (ISIS) programme. This also transmitted useful data for over 6 yr.

ISIS   A continuation of Alouette studies, ISIS 1, L Jan 30 1969 by TAID (Thor-Delta) from Vandenberg, with 578×3526 km orbit, 241 kg wt and $88°$ incl. It carried 8 Canadian and 4 US experiments, to measure the daily and seasonal fluctuations in the electron density of the upper atmosphere, to study radio and cosmic noise emissions, and measure energetic particles interacting with the ionosphere. Orbital life 250 yr. ISIS 2, L Apr 1 1971 by Thor Delta from

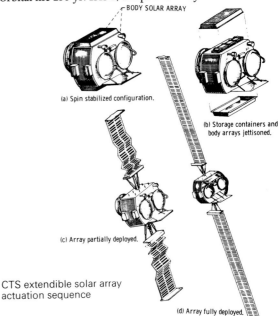

(a) Spin stabilized configuration.

(b) Storage containers and body arrays jettisoned.

(c) Array partially deployed.

(d) Array fully deployed.

CTS extendible solar array actuation sequence

Vandenberg, with 1358×1429 km orbit, 88° incl and 264 kg wt, and 8000 yr orbital life, continued the studies. A 3rd ISIS is planned.

Anik This is the Eskimo word for 'brother'; it was the world's first domestic commercial communications system using a stationary orbit. Anik 1, L Nov 10 1972 by Thor-Delta from Cape Kennedy (orbit $35,822 \times 36,508$ km; and placed at 114°W) provided commercial communications over Canada's sprawling 10M sq km. Some circuits were leased to US carriers for private voice, data and video transmissions to Alaska etc. It was followed by Anik 2 on Apr 20 1973, at 109°W, with Anik 3 held in reserve. With launch wt of 562 kg and orbital wt of 295 kg, the craft were spin-stabilized with despun antenna and feeds, and 7-yr station-keeping capability. Each could carry 1 colour TV picture or 960 one-way phone conversations. Anik 3, L May 7 1975, was placed at 119°W (due S of Los Angeles), with 10 colour TV channels or up to 9600 phone circuits.

CTS L Jan 16 1976 by Delta from Cape Canaveral and placed in synchronous orbit at 116°W (W of S America), the first Communications Technology Satellite was even more powerful than America's ATS-6. With a 2-yr experimental life this joint Canadian/NASA satellite was able to transmit power levels 10–20 times higher than current commercial satellites. The object was to provide TV and other services to small, low-cost ground terminals for the benefit of remote areas. With a launch wt of 675 kg, CTS-1 achieves its high radiated power (59 DBW compared with 53 DBW on ATS-6) by means of a pair of solar 'sails', or arrays, with a total length of 16·5 m. They were contributed by the European Space Agency (*qv*); cost to NASA alone of this shared project totalled $22·2M for launch and spacecraft.

CHINA

History By the end of 1976 China had launched 7 satellites, from her Shuang-cheng-tzu launch site in Central China approx 1600 km W of Peking, and near Lop Nor, the nuclear test base. About the same weight as early US communications satellites, China's early satellites were believed to have been orbited by a booster based on Soviet technology—a 2-stage, liquid-propelled SS-2 Sandal intermediate range ballistic missile (IRBM), capable of placing a 136–272kg satellite in Earth orbit, depending on launch latitude and inclination. By the 4th mission, the CSS-X-3 (based on the 5600km range ICBM) was being used to orbit up to 4500 kg. With the 7th mission it was clear that China was nearing the point when she would be able to orbit her first men.

Development of China's space boosters and long-range military rockets is largely credited to Dr Tsien Wei-Ch'ang, who was expelled from the US in 1949 as a result of the late Senator Joseph McCarthy's anti-Communist campaigns. Tsien had taken a PhD at Toronto University and then worked as a research engineer at the Jet Propulsion Laboratory, Pasadena, until ordered to return to mainland China. He came back to the US as leader of 7 Chinese scientists in Nov 1972; he said China was developing domestic communications satellites, and would launch one 'in the very near future'.

There has been speculation that the gap of over 4 yr between the launch of China 2 and 3 may be the result of the Chinese Government deciding to dismantle their Inner Mongolian test sites, and transfer them much further from the Soviet border. This would account for 3 quick launches following the gap. Evidence for this, however, remains inconclusive.

China 1   L Apr 24 1970 by SS-2 from Shuang-cheng-tzu. Wt ? 172 kg. Orbit 441 × 2386 km. Incl 68·4°. China's first satellite made her the 5th space country. It carried a transmitter which broadcast 'The East is Red', a Chinese song paying tribute to Chairman Mao, and announcing the times it passed over various parts of the world. Probably spheroid, with 1 m dia, it transmitted until Jun 1970; orbital life 100 yr.

China 2   L Mar 3 1971 by SS-2 from Shuang-cheng-tzu. Wt 221 kg. Orbit 265 × 1825 km. Incl 69·9°. Powered by solar cells, transmitted data for 12 days, when the batteries apparently failed. Orbital life 5 yr.

China 3   L Jul 26 1975 by CSS-X-3 (ICBM derivative) from Shuang-cheng-tzu. Wt ? Orbit 184 × 461 km. Incl 69°. During its 50 days' life, this passed regularly over Russia, the US and Eastern Europe. No details of its weight and capability, or how long transmissions lasted, were given. Decayed Sep 14 1975.

China 4   L Nov 26 1975 by CSS-X-3 from Shuang-cheng-tzu. Wt ? 2700–4500 kg. Orbit 173 × 483 km. Incl 63°. Part of this satellite was apparently successfully recovered on Dec 2, after 7 days in orbit—a major step forward for China's space programme, indicating the ability of her rockets to place craft in precise orbits from which recovery within Chinese territory is possible. The satellite's recovery was front-paged on Chinese newspapers alongside reports of the first meeting in Peking between President Ford and Chairman Mao; but no details were given as to whether recovery was made by aircraft snatch during descent, or on land or

Landsat 1 view of China's Lop Nor launch area. Circular feature is dry lake bed

sea—nor whether the contents were undamaged. The remainder of the satellite continued to transmit from a 399 × 175 km orbit, until re-entry on Dec 29 1975.

China 5   L Dec 16 1975 by CSS-X-3 from Shuang-Cheng-tzu. Wt ? 4500 kg. Orbit 186 × 387 km. Incl 69°. With a similar orbit to China 3, this test vehicle decayed in 42 days.

China 6   L Aug 30 1976 by ? SS-2 from Shuang-Cheng-tzu. Wt ? 270 kg. Orbit 195 × 2145 km. Incl 69°. Similar to earlier small satellites, with an orbital life of 6 yr, and was probably still transmitting at the end of 1976.

China 7   L Dec 7 1976 by CSS-X-3 from Shuang-Cheng-tzu. Wt ? 3600 kg. Orbit 172 × 479 km. Incl 59°. The spacecraft is believed to have been recovered about Dec 10, leaving a residual object, possibly the booster, weighing 1200 kg, in orbit for a further 23 days. A fragment, however, which decayed after only 2 days might have been a spacecraft.

EUROPEAN SPACE AGENCY

History Now well-established, with 8 successful scientific satellites and 200 sounding rockets behind it, and with Spacelab ahead to provide European participation in future manned spaceflights, ESA promises at last to provide an independent challenge to the massive American and Russian space programmes. The Ariane rocket is developing so well that ESA is urging its use (in competition with the US Shuttle) for launching future Intelsat and other internationally financed payloads, as well as ESA's own. Although small compared with the Soviet and US programmes, ESA is already adding to its scientific satellites studying Sun-Earth relations, 'black holes', etc, a series of applications satellites covering international telephone and TV communications, weather forecasting, and both maritime and aviation systems. Formation of a European Space Agency was agreed upon by 11 countries at a meeting in Brussels on Jul 31 1973, and it finally came into operation in May 1975. The widely differing views of the biggest participating countries were met by agreement that Germany would be the largest contributor to Spacelab; France to production of the Ariane launcher, because she was the only country insisting on an independent European launch capability; and Britain to production of Marots, a maritime communications satellite.

Because every member has a different currency, a new, universal currency of 'Accounting Units' (AU) was created, with a value at that time of AU1 = $1·3. In 1976, about one-third of ESA's budget of AU429M ($540M) was devoted to its scientific activities. The member states, and their percentage contributions—in effect their membership fees— were as follows:

Belgium	3·72%	Netherlands	4·88%
Denmark	2·19%	Spain	5·14%
France	20·43%	Sweden	4·61%
W. Germany	25·00%	Switzerland	3·30%
Ireland		United Kingdom	17·15%
Italy	13·58%		

Observer states, who occasionally also participate in experiments are Australia, Canada and Norway; in addition there are co-operation agreements with India, Japan, the US and USSR.

ESA was formed by merging ELDO (European Launcher Development Organization) and ESRO (European Space Research Organization). In 1976 it had a total staff of 1440, of whom 600 were engineers, 120 scientists and 220 technicians. Its largest establishment, ESTEC (the European Space Research and Technology Centre) with 841 staff responsible for the study, design, development and testing of both space components and complete space vehicles, is at Noordwijk, near Amsterdam, Netherlands, ESOC (European Space Operations Centre), with 257 staff responsible for launch control and satellite tracking and data processing, is at Darmstadt, near Frankfurt, Germany.

Like NASA, ESA's activities are entirely non-military; its main aims are to develop a long-term European space programme, covering research, technology and applications. To achieve this, it must progressively 'Europeanize' the parallel national space programmes on which many European nations still spend much of their meagre space budgets. In the early days of ELDO and ESRO (detailed below), there was much national conflict about the need to build European rockets capable of orbiting large satellites independently of either Russia or America. The scientists mostly preferrred to use their limited money and resources to develop satellites, and to pay America for rockets and launch facilities. Technicians and some politicians took the view that America would ultimately be reluctant to provide launching facilities for TV, telephone and communication satellites which would conflict with her own systems— particularly INTELSAT, being developed for international use under America's space 'umbrella'. When the British Government finally ceased to support Blue Streak, originally developed as a British ICBM, ELDO withered away; but ESRO survived. Between 1968 and 1972, 8 scientific satellites were launched on its behalf by America—7 successfully. Their major theme was the study of Sun-Earth relationships, and their work is remarkably illustrated in the diagram showing the effects of the solar wind on Earth's magnetic field as it speeds around the Sun.

HEOS 1 and 2 (Highly Eccentric Orbit Satellites), drum-shaped but with 16 sides with antenna and booms extending from the top, are models of what can be achieved by genuine European collaboration. Their experiments were contributed by Imperial College, London; Rome University; the Danish Space Research Institute; the Netherlands' European Space Research and Technology Centre; the Centre D'Etudes Nucleaires, Saclay; Milan University; and the Max Planck Institutes in Garching and Heidelberg. They showed how the Sun continuously blows out a stream of plasma (the solar wind) at speeds around 400 km per sec; when this meets Earth's shock-wave as it travels through space, Earth's magnetic field is squeezed into the sort of shape shown in the figure on page 33.

HEOS 1 before launch from Cape Canaveral

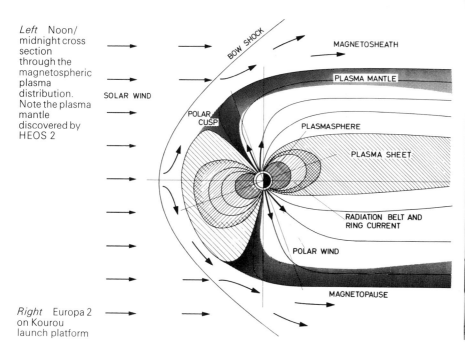

Left Noon/midnight cross section through the magnetospheric plasma distribution. Note the plasma mantle discovered by HEOS 2

SOLAR WIND

BOW SHOCK

MAGNETOSHEATH

PLASMA MANTLE

POLAR CUSP

PLASMASPHERE

PLASMA SHEET

RADIATION BELT AND RING CURRENT

POLAR WIND

MAGNETOPAUSE

Right Europa 2 on Kourou launch platform

ELDO The 7-nation European Launcher Development Organization was set up on Feb 29 1964. The initiative came from Britain, anxious to find a use for Blue Streak, originally developed as a strategic missile, as the first stage of an independent European launcher system. Member countries were Belgium, France, W Germany, Italy, Netherlands, and the United Kingdom, plus Australia; her Government was equally anxious to keep the Woomera Test Range, North of Adelaide, in business. With France providing the 2nd stage, Coralie; Germany the 3rd stage, Astris; Italy the satellite test vehicles, Belgium the ground guidance station; and Netherlands, telemetry links and other equipment, Europa 1 was built. By Jun 12 1970 10 test firings had been carried out at Woomera; despite a series of earlier failures, there was confidence that the last, F-9, would place a test satellite in orbit. But loss of thrust on the 3rd stage, and failure to jettison the nose cone, prevented this. Despite growing doubts about the project, and Britain's decision to withdraw, announced in 1970, ELDO pressed ahead with development of the more advanced Europa 2 vehicle, which had the addition of a French-built, 4th stage perigee motor, able to place a satellite of up to 200 kg in stationary orbit. The objective was to launch 2 Franco-German Symphonie telecommunications satellites in 1973 and 1974. Unfortunately the first test-firing of Europa 2, from Kourou on Nov 5 1971, with the world's space correspondents observing, ended in disaster. After $2\frac{1}{2}$ min, it deviated from its flight trajectory, became overstressed, and exploded. Amid much bitter argument, plans continued for the next test launch, F-12, and the vehicle was aboard a French ship en route to Kourou, when on Apr 27 1973 the project was finally abandoned in favour of creating a European Space Agency, and building Spacelab for America's Space Shuttle system. Up to that time, Europe had spent about £310M ($745M) on ELDO.

ESRO The 10-nation European Space Research Organization was set up in 1964 to promote European collaboration in space research and technology. Member states were Belgium, Denmark, France, W Germany, Italy, Netherlands, Spain, Sweden, Switzerland, and Great Britain. With headquarters in Paris, and pay-roll of 1100, its 1973 income was £53M ($127·5M).

ESRO/ESA Satellites

Note: The early launches were out of sequence because of rocket faults and technical delays with the satellites themselves.

ESRO 2A L May 29 1967 by Scout from Vandenberg. Wt 74 kg. This should have been ESRO's first satellite, to study solar astronomy and cosmic rays. It failed to orbit because of 3rd stage failure of the Scout rocket.

ESRO 2B (Iris) L May 17 1968 by Scout from Vandenberg. Wt 164 kg. Orbit 325 × 957 km. Incl 98·90°. Returned solar and cosmic radiation data. With ESRO 1A, as it passed over the Polar cap close to Earth, it recorded how and when the particles already measured by HEOS 1 further out, arrived in that area. Dec. May 8 1971.

Right ESRO 2B (Iris)

ESRO 1A

ESRO 1A (Aurorae) L Oct 3 1968 by Scout from Vandenberg. Wt 84 kg. Orbit 252 × 1231 km. Incl 93·7°. Carried 8 experiments to measure the polar ionosphere during magnetic storms. During 3504 passes through the Van Allen radiation belts, it established the position of their 'poleward' boundary. Dec. Jun 26 1970.

HEOS 1 L Dec 5 1968 by TAID from Cape Canaveral. Wt 108 kg. Orbit 418 × 223,440 km. Incl 28°. Re-entered on Oct 28 1975, having operated perfectly for 7 yr—6 yr longer than planned. The orbit, taking it two-thirds of the way to the Moon was to study interplanetary magnetic field conditions. Combined with Heos 2, it enabled this study to cover 7 of the 11 yr of a solar cycle. It not only measured solar particles in interplanetary space, but established in what strength and from which direction they arrived. By the time it re-entered, after 542 revs, 90 scientific publications had resulted from correlating its results with data from other ESRO, European, American and Russian spacecraft.

ESRO 1B (Boreas) L Oct 1 1969 by Scout from Vandenberg. Wt 80 kg. Orbit 291 × 389 km. Incl 85°. Short life of 52 days was due to booster malfunction. It was a duplicate of ESRO 1A, and carried the same 8 experiments.

HEOS 2 L Jan 31 1972 by Delta from Cape Kennedy. Wt 117 kg. Orbit 405 × 240,164 (raised in Jul 1973 to 5442 × 235,589 km). Incl 89·9° (later 88°). With HEOS 1 this discovered the 'plasma mantle' shown on page 33. It also found that the very high energy particles emitted by the Sun, which can travel to Earth within a few minutes, travel through 3-dimensional, snake-like 'tubes' in interplanetary space.

TD-1A L Mar 12 1972 by Thrust Augmented Delta from

TD-1A

Cape Kennedy. Wt 472 kg. Orbit 524 × 551 km. Incl 97·55°. At the time Europe's largest scientific spacecraft, built by 5 European firms in W Germany, Sweden, Britain and Italy, with France's Engins Matra as prime contractor. Despite the failure of its tape recorder, a British/Belgian experiment surveyed 95% of the sky and examined 15,000 stars as seen through 'wide-open' UV (ultraviolet) eyes. From it a Bright Star Catalogue is being published. Dutch astronomers also used it to study individual stars in great detail, and found one, called Cygni, which was losing mass (its atmosphere) at a very fast rate. Orbital life, 20 yr. (A proposed TD-2 was abandoned.)

ESRO 4   L Nov 22 1972 by Scout from Cape Kennedy. Wt 130 kg. Orbit 245 × 1173 km. Incl 91°. Its elliptical, near-Polar orbit, enabled the 5 experiments provided by Netherlands, Sweden, Britain and W Germany to measure neutral gas densities above 250 km and their relationship with the ionosphere. It found that the temperature of the upper atmosphere over the Poles was higher, instead of being lower, than that over the equator. Dec. Apr 15 1974.

ESRO 4

COS-B

COS-B   L Aug 9 1975 by Delta from Vandenberg. Wt 275 kg. Orbit 342 × 99,873 km. Incl 90°. Constructed by 7 ESA members (Belgium, Denmark, France, W Germany, Italy, Spain and Britain), COS-B carried a single gamma-ray experiment provided by institutes in France, W Germany, Italy and Netherlands. Intended to study the unexplained gamma-ray bursts detected earlier by US satellites, soon

Left COS-B ready for launch by Thrust-Augmented Delta

after launch it was confirming X-ray and X- and gamma-radiation from the Crab Nebula. Orbital life 5 yr; operational life 2 yr.

GEOS L Apr 20 1977 by Delta from Cape Canaveral. Wt 574 kg (incl 305 kg apogee motor). Owing to a launcher fault between 2nd and 3rd stage, geostationary orbit was not achieved; on Apr 25, ESOC fired the apogee motor to place it in a 12-hr, 38,498 × 213km orbit with rotation to 35°E. This gave ESA's specially built Odenwald ground station a 2–7-hr visibility daily instead of the planned 24 hr. GEOS, the world's first purely scientific satellite, intended to be the reference spacecraft for the International Magnetosphere Study (IMS), had been dogged by 'severe delays' due to 'late arrival and poor quality of high reliability components'. It was originally intended to be launched from

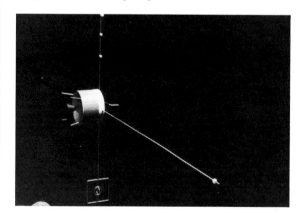

Right GEOS

Kourou by Europe's now-abandoned Europa rocket. Failure of the NASA launch, which had to be paid for even though it was a failure, was a major blow to ESA. Its development, with British Aircraft Corporation as the main contractor, involved 15 companies in 10 countries. It carries 7 experiments provided by ESTEC, Denmark, France, Italy, Germany, Sweden, Switzerland, and the UK, and cost AU42·5M ($54M); ESA had to face the difficult decision of whether to launch the back-up spacecraft.

Projects

These fall into 4 groups. The Scientific Satellites, as already explained, are part of the 'obligatory programme'. The Applications Satellites are mostly funded by the countries most interested in them. Because each country expects to receive work in proportion to the amount it pays, usually funds cannot be switched from one project to another, which provides ESA administration with a difficult task.

Scientific Satellites

IUE Due for launch in mid-1977 with a 2-yr life, the International Ultra-violet Explorer is a joint NASA, ESA and UK Science Research Council project. ESA is designing and constructing a ground station near Madrid for use by astronomers from NASA, SRC and ESA in the ratio 4:1:1.

ISEE Due for launch in late 1977, with a 3-yr life, the International Sun-Earth Explorer is part of a 3-satellite project. NASA is developing ISEE-A and C, and will launch A and B on the same rocket into a 282 × 140,000km orbit. It

Above ISEE-B

Below EXOSAT

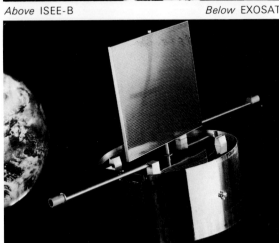

will continue the study of the magnetosphere, Earth's interaction with the solar wind. B will carry 8 experiments provided by universities and scientific institutions in France, Italy, W Germany and US.

EXOSAT Due for launch by Thor-Delta in Sep 1980 with a 2-yr life, and 340 kg wt. It will be placed in a near-Polar orbit, with apogee of 200,000 km in the N Hemisphere. With X-ray detectors it will study neutron stars and 'black holes'. Formerly known as HELOS (Highly Eccentric Lunar Occultation Satellite). Data will be collected by ESA's Madrid station.

Applications Satellites

OTS Due to be launched by Delta from Cape Canaveral in Jun 1977. The Orbital Test Satellite (wt in orbit 444 kg) is

Left OTS

Right OTS details

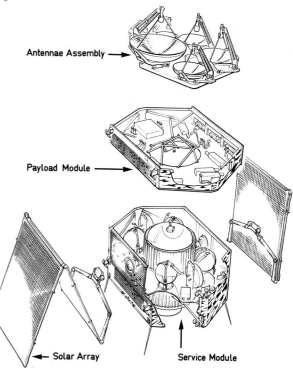

Antennae Assembly ⟶

Payload Module ⟶

◄— Solar Array

Service Module

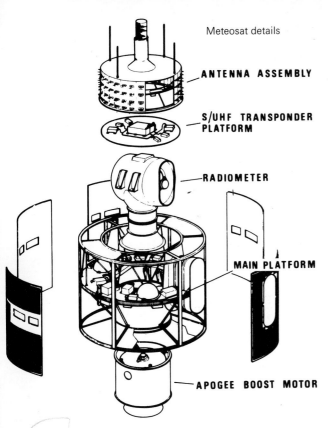

Meteosat details

ANTENNA ASSEMBLY

S/UHF TRANSPONDER PLATFORM

RADIOMETER

MAIN PLATFORM

APOGEE BOOST MOTOR

intended to be an experimental start to the European Communications Satellites (ECS) programme. This is designed to meet the needs of the increasing volume of intra-European telephone telegraph and telex traffic, as well as of the European Broadcasting Union. Other proposed OTS uses include off-shore oil rig communications and educational TV. It will be placed in geostationary orbit at 10°E Long. Prime contractor is Hawker Siddeley Dynamics (Britain); coverage should include W Europe, Middle East, N Africa, Azores, Canary isles, Madeira and Iceland.

Meteosat   Due for launch by Thor-Delta in Jul 1977, with a 3-yr life, this is intended to be Europe's contribution to the Global Atmospheric Research Programme (GARP) in 1978–9, and later to the World Weather Watch (WWW) of the World Meteorological Organization. It will be one of 5 geostationary meteorological satellites; 2 will be provided by US, 1 by Japan and 1 by USSR. Between them they should provide more accurate weather forecasting. Launch wt of 700 kg will include the ejectable apogee boost motor. Prime contractor is SNIAS (France); other ESA participants are Belgium, Denmark, W Germany, Italy, Sweden, Switzerland and UK.

Marots   Due for launch late 1977 with 7-yr life, a maritime communication satellite, based on OTS experience, aims to provide the service required by the 43-nation INMARSAT. This organization, now being formed, is expected to include America and Russia as well as European and other nations, who own between them much of the world's shipping. Main contractors are Hawker Siddeley Dynamics and Marconi Space Systems, since Britain is the largest contributor; other ESA participants are Belgium,

France, W Germany, Italy, Netherlands, Norway, Spain and Sweden. With 466 kg max wt, Marots will be stationed about 40°E over the Indian Ocean.

Aerosat Due to be launched mid 1979, with a 7-yr life, this will be one of 2 satellites launched jointly by ESA, Canada and Comsat General (see INTELSAT in the International Section). The aim is to provide an experimental system to enable ICAO (the International Civil Aviation Organization) to decide how to use satellites for communications between aircraft and air traffic control on the ground. The system will include a ground station and computer, and the fitting of 20 aircraft with experimental equipment. The satellites will be in geostationary orbit, over the Atlantic and separated by 25° Long.

ECS Due for launch in early 1980 with 7-yr life, this will be the operational start of the European Communications Satellites programme, tested by OTS, with added experience provided by CTS, the Canadian communications satellite (L Jan 16 1976) which carried ESA test equipment.

Ariane Launcher

History Ariane is intended to provide Europe with a heavy launcher able to place satellites of 800 kg in geostationary orbit (1500 kg when the apogee boost motor is included), in the 1980s. It is a result of French insistence that Europe must have the independent ability to launch its own application satellites, and not have to rely upon either America or Russia to provide a booster and launch facilities; it was argued that there have been occasions in the

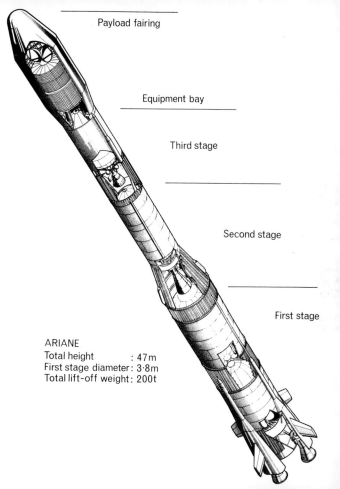

Payload fairing

Equipment bay

Third stage

Second stage

First stage

ARIANE
Total height   : 47m
First stage diameter : 3·8m
Total lift-off weight : 200t

past, and which may recur in future, when America was reluctant to launch commercial satellites likely to compete with her own systems. France launched the programme in 1972, following the collapse of ELDO and the failure of the Europa rocket. Development cost at Jul 1975 was estimated at FF3046M. Since France is paying 62·5%, the programme is being managed by CNES (Centre National d'Etudes Spatiales). Other countries contributing are Belgium (5%), Denmark (0·5%), W Germany (20·12%), Netherlands (2%), Italy 1·74%), Spain (2%), Sweden (1·1%), Switzerland (1·2%), UK (2·47%); other countries (1·37%). Launches will take place at the French Guiana Space Centre, Kourou; the first of 4 qualification launches (L0-1) is scheduled for Jul 15 1979; L02-4 are due in 1980, and experiments are being planned for synchronous orbit on the last 2 test flights.

General Description   Ariane is a 3-stage vehicle with total lift-off wt of 200 metric tonnes (202,000 kg), total ht of 47·388 m and body dia of 3·8 m. 1st stage, derived from the Europa III 1st stage, is powered by 4 Viking-2 engines each providing 60,000 kg static thrust. Protected by fairings with fins, they burn for 139 secs (propellant totals 140 tonnes of UDMH and N_2O_4). 2nd stage using 33 tonnes of same propellant, is powered by one Viking-4 engine giving 70,000 kg thrust for 131 secs. 3rd stage, using 8 tons of liquid hydrogen and oxygen, is powered by one engine giving 6600 kg thrust for up to 563 secs. The payload of up to 4 tons for low earth orbit, or 800 kg in stationary orbit is housed in a cylindrical/conical fairing.

Spacelab

History   Spacelab, short for Space Laboratory, is ESA's most exciting project. In mid 1980, on the 7th orbital flight of America's Space Shuttle, it should enable the first Europeans to follow the Russians and Americans into Earth orbit. They will get there as Mission Specialists—scientists responsible for operating European experiments. A co-operative venture with America's NASA, Spacelab provides for the design, development and production of a versatile space laboratory, which can be used either manned or unmanned. It will dramatically reduce costs because it can be prepared for its missions (50 are planned in a period of 5 years) quite separately from the Space Shuttle, which will be available for other flights during the preparation periods. When ready, it will be placed in the Shuttle payload bay, where it will remain, attached to and supported by the Shuttle and its 3-man crew throughout its missions; these will usually last 7 days, but may be extended up to 30 days. While these flights will be a major step towards 'internationalizing' America's manned spaceflight programme, NASA will control all its activities. 'European flight crew opportunities will be provided in conjunction with flight projects sponsored by ESA or by governments participating in the Spacelab programme' says NASA. On some flights there are expected to be up to 4 scientists and technicians aboard Spacelab, in addition to the Shuttle astronauts. On the first Spacelab flight, however, scheduled for the second half of 1980, there will be only 2 payload specialists: 1 European and 1 American, to look after 77 scientific and technological experiments—61 European, 15 American, and 1 Japanese. It seems almost certain that the

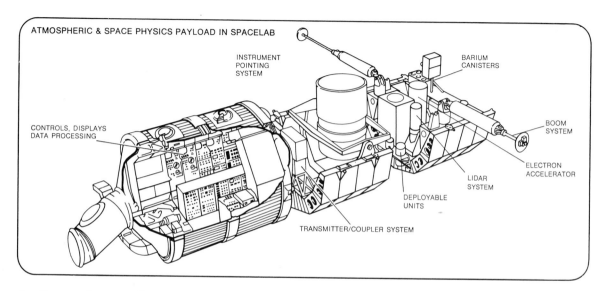

ATMOSPHERIC & SPACE PHYSICS PAYLOAD IN SPACELAB

INSTRUMENT POINTING SYSTEM

BARIUM CANISTERS

CONTROLS, DISPLAYS DATA PROCESSING

BOOM SYSTEM

ELECTRON ACCELERATOR

LIDAR SYSTEM

DEPLOYABLE UNITS

TRANSMITTER/COUPLER SYSTEM

first European in space will therefore be a German scientist. Germany, Italy and France all have prior claims over Britain, since the 9 Spacelab countries are contributing to its cost in the following percentages: W Germany 53·3; Italy 18·0; France 10·0; UK 6·3; Belgium 4·2; Spain 2·8; Netherlands 2·1; Denmark 1·5; Switzerland 1·0; Austria 0·8. Britain will thus be fortunate to get one of her own scientists aboard the second or third Spacelab mission.

General Description Spacelab consists of a module, pro-viding a pressurized laboratory in which up to 4 experimenters can work; and a pallet which will support telescopes, antennas, instruments and equipment requiring direct exposure to space. The pallet will normally be attached to, and operated from the module, but either can be used separately. Total Spacelab weight must not exceed 14,525 kg, since that is the landing payload for the Orbiter. For a nominal 7-day mission, module-only weight will be 5000 kg; pallet-only weight 9100 kg; module-plus-pallet 6000 kg. Max pallet length will be 15 m. The pressurized module sections are

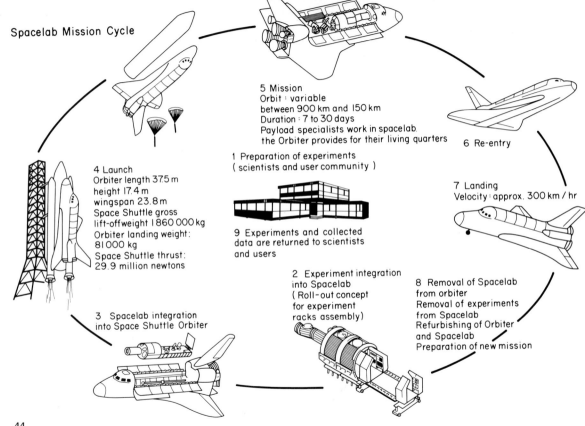

Spacelab Mission Cycle

5 Mission
Orbit : variable
between 900 km and 150 km
Duration : 7 to 30 days
Payload specialists work in spacelab.
the Orbiter provides for their living quarters

6 Re-entry

1 Preparation of experiments
(scientists and user community)

4 Launch
Orbiter length 37.5 m
height 17.4 m
wingspan 23.8 m
Space Shuttle gross
lift-off weight 1 860 000 kg
Orbiter landing weight:
81 000 kg
Space Shuttle thrust:
29.9 million newtons

7 Landing
Velocity : approx. 300 km / hr

9 Experiments and collected
data are returned to scientists
and users

3 Spacelab integration
into Space Shuttle Orbiter

2 Experiment integration
into Spacelab
(Roll-out concept
for experiment
racks assembly)

8 Removal of Spacelab
from orbiter
Removal of experiments
from Spacelab
Refurbishing of Orbiter
and Spacelab
Preparation of new mission

designed to accommodate a maximum of 22 and minimum of 5 cu m. To satisfy centre of gravity requirements, Spacelab must be placed to the rear of the Shuttle Orbiter, and will therefore be connected to the Orbiter cabin by a tunnel, being developed by NASA. Recent design changes led to decisions that the Spacelab crew will live below the Shuttle flight deck, and not inside Spacelab; and that to save weight and cost, the Shuttle will provide Spacelab's environment system. These decisions delayed, and possibly destroyed hopes that on some missions Spacelab, as a completely independent spacecraft, could be left in orbit by the Shuttle and collected at the end of a later flight.

Estimated total cost to ESA is AU386M ($515·7M).

Development An initial 6-yr contract worth £95M ($226 M) to design and develop Spacelab was awarded in Jun 1974 to ERNO of Bremen, whose major sub-contractors are Bell Telephone in Belgium, Matra in France, Aeritalia in Italy and Hawker Siddeley in England. Technical assistance to ERNO is being provided by the US company TRW. The group is to deliver to NASA one Spacelab flight unit, fully qualified and ready for installing experiments, by Apr 1979, in addition to 2 engineering models, 3 sets of group support equipment and initial spares. Design life will be for 50 missions of 1–4 weeks, or a nominal life of 10 years. NASA has undertaken not to develop its own Spacelab, but to buy any similar units needed for its own projects from ESA—which is hoping that a total of 3 will ultimately be ordered. Of the 725 Shuttle flights provisionally planned by NASA for 1980–94, 32%, or 160, will use Spacelab. Until 1984, Spacelab should provide a minimum of 1500 jobs in European industry as well as requiring 100 ESA staff.

European Crew Members By 1978 ESA must select up to 10 Europeans to take part in Spacelab flights. They will be scientists, technicians and engineers, who will join the US crew members on Shuttle/Spacelab missions. On the first Spacelab flight only one European is likely to be among the crew of between 4 and 7; later 4 Europeans may fly at the same time. Physical requirements will be much less exacting than those demanded for the early groups of astronauts, since G-forces during lift-off and re-entry should be easily tolerated by normal fit persons. NASA will be responsible for training the European team selected as crew members; since both Shuttle and Spacelab has been designed for women as well as men, the European team is likely to include both sexes.

FRANCE

History France is easily the world's third space nation. She consistently spends more on both satellites and launchers than any other country except Russia and America. By the end of 1975, she had been responsible for the successful launch of 17 satellites: 9 of them were entirely national; 4 more satellites had been launched for her by NASA, and 3 by Russia. France herself had launched one for W Germany. Her 9th successful national satellite however, was also her last, since the national Diamant launcher was abandoned in favour of the European Ariane launcher. With a remarkable facility for co-operating with both the major space powers simultaneously, French scientists regularly contribute to the major projects of both powers. Examples include the laser mirror on Russia's Lunokhod vehicles, experiments on the

1973 Mars probes, and the ultra-violet telescope to measure star brightness included in the US Skylab space station. France became the 3rd nation to achieve national orbital capability when the 18-tonne, 3-stage Diamant launcher placed the A-1 satellite in orbit on Nov 26 1965, 8 yr after Sputnik 1. The final collapse of ELDO in 1973, in which France was the driving force, coincided with a run of misfortunes in her national programme; the long-term future of her ambitious space centre at Kourou, in French Guiana (qv) was also threatened; but its prospects were restored with the re-emergence of the European Space Agency, with its headquarters in Paris and France once again the leading member.

Diamant Launcher   The 'A' version has an overall length of 18·75 m, 1st stage dia 1·40 m, and wt, with payload of 17,970 kg. It can place 80 kg in a 300km orbit. The 1st stage, Emeraude, was first launched, without complete success, on Jun 17 1964, but 21 subsequent firings were all successful; its Vexin liquid-propellant engine provides 28,000 kg thrust for 88 secs through a single gimballed chamber. 2nd stage, Topaze, has a polyurethane-type solid propellant motor in a steel casing, providing 14,500 kg thrust for 39 secs through 4 gimballed nozzles. 3rd stage, with similar solid propellant motor in glass-fibre wound casing, provides a variable thrust of 2500–5300 kg thrust for 44·5 secs through a single fixed nozzle. This rocket achieved the first 3 orbital launchings outside America and the Soviet Union.

Diamant B, 23·54 m long, and launch wt of 24,600 kg, has an L17 1st stage, with liquid bi-propellant motor, developing 35,000 kg thrust at lift-off for 112 secs. 2nd and 3rd stages are similar to Diamant A. First launching of Diamant B was at Kourou, on Mar 10 1970, when it placed France's first foreign payload in orbit.

In Dec 1971 and May 1973, however, there were 2 consecutive launch failures. Future versions (the BP-4) were planned with a 4th stage, giving Diamant the ability to place 120 kg into a 1600km circular orbit. This was successful in all 3 launches in 1975. Of the total of 12, 10 Diamant launches had been successful, and much useful technology was transferred to ESA's Ariane.

National Launches

A-1 (Asterix)   L Nov 26 1965 by Diamant from Hammaguir. Wt 41·7 kg. Orbit 528 × 1768 km. Incl 34°. First French satellite. Transmitted for 2 days. Orbital life 200 yr.

D-1A (Diapason)   L Feb 17 1966 by Diamant from Hammaguir. Wt 20 kg. Orbit 504 × 2753 km. Incl 34°. 3rd French satellite, and 2nd national launch. Carried out geodetic data research. Orbital life, 200 yr.

D-1C (Diademe 1)   L Feb 8 1967 by Diamant from Hammaguir. Wt 22·6 kg. Orbit 580 × 1340 km. Incl 40°. Geodetic satellite, operational in spite of low apogee. Orbital life 100 yr.

D-1D (Diademe 2)   L Feb 15 1967 by Diamant from Hammaguir. Wt 22·6 kg. Orbit 592 × 1886 km. Incl 39°. Provided laser and doppler data for 3 months; orbital life 200 yr.

WIKA/MIKA   L Mar 10 1970 by Diamant B from Kourou. Wt Wika 50 kg. Mika 64·8 kg. Orbits 307 × 1700 and 1746 km. Incl 5°. The W German Wika satellite was the first

ant B P-4 launcher

General view of the launch area Kourou

47

foreign payload to be orbited by France; it carried out 30 days of geocorona and upper atmosphere investigation, and had a 5-yr orbital life. Mika's task was to monitor the performance of the Diamant B launcher, which had 15,420 kg thrust, liquid 1st stage, solid propellant 2nd stage and 3rd-stage apogee motor to spin payload into orbit.

Peole   L Dec 12 1970 by Diamant B from Kourou. Wt 69·7 kg. Orbit 580×747 km. Incl 15°. Test flight to qualify the EOLE meteorology satellite, successfully launched by Scout B from Wallops, Va. on Aug 16 1971. This project, being conducted in collaboration with NASA, involves using a satellite to track hundreds of 4m pressurized meteorological balloons, carrying capsules on cables and measuring the temperatures surrounding them etc, as they drift at a constant height.

D-2A (Tournesol 1)   L Apr 15 1971 by Diamant B from Kourou. Wt 90 kg. Orbit 456×703 km. Incl 46°. Carried 5 experiments to study solar radiation and ultraviolet range. Orbital life 6 yr.

D-2A   L Dec 5 1971 by Diamant B from Kourou. Wt 97 kg. Intended for a higher orbit than the previous satellite, to continue solar radiation studies, but 2nd stage of launcher failed.

D-2A   L May 21, 1973 by Diamant B from Kourou. Twin satellites, Castor and Pollux, fell into the sea; apparently the launcher failed to produce sufficient thrust.

Starlette   L Feb 6 1975 by Diamant B from Kourou. Wt 68 kg. Orbit 804×1138 km. Incl 50°. A tiny 10″ dia spheroid, this was an early laser reflector for geodetic studies. Orbital life 2000 yr.

D-5A (Pollux) and D-5B (Castor)   L May 17 1975 by Diamant BP-4 from Kourou. Twin satellites, with wts of 35 and 77 kg, and placed in 270×1270km orbits with 30° incl. Pollux carried microrocket experiments and Castor an accelerometer. Orbital lives 80 days and 3 yr.

D-2B (Aura)   L Sep 27 1975 by Diamant BP-4 from Kourou. Wt 110 kg. Orbit 499×723 km. Incl 97°. France's last national launch carried experiments to study UV radiation from the Sun and hot new stars. Orbital life 15 yr.

Launches by US FR-1   L Dec 6 1965 by Scout from Vandenberg into circular 780km orbit to study ionosphere. **Eole**   L Aug 16 1971 by Scout from Wallops I. Orbit 678×903 km. Meteorological satellite interrogating 500 balloons drifting in atmosphere at 12,000 m.

Symphonie   A Franco-German project under which Symphonie-1 was launched Dec 19 1974 and Symphonie-2 on Aug 27 1975 by Thor-Deltas from Cape Canaveral into synchronous orbits at 11·5°W. With 237kg orbital wt, they each provided 2 TV channels or 300 two-way phone circuits covering Europe, the Middle East, Africa, S America and part of N America. The experiment was a merging of the French SAROS and German OLYMPIA projects, with ground stations in each country using a small 16m antenna to experiment with multiple access. When Symphonie-2 was launched the work of its predecessor was said to be '90% complete'. Originally Symphonies were to have been launched by Europa, the now-abandoned ELDO rocket.

Launches by USSR   **Aureole**   L Dec 21 1971 by C1 from Plesetsk. Orbit 410×2500 km. To study upper atmosphere and polar lights.   **SRET-1**   L Apr 4 1972 probably by C1 from Plesetsk. Orbit $460 \times 39,248$ km. To study radiation effects on solar cells.   **SRET-2**   L Jun 5 1975 piggyback with Molniya from Plesetsk to test Meteosat cooling system.

GERMANY
Federal Republic

History   It was Germany's pre-war rocket team, led by Dr Wernher von Braun and Dr Kurt Debus, after they had transferred their allegiance to the United States at the end of the war, which led to the building of the great Saturn 5 rocket and the Apollo Moonlandings. So inevitably German interest in space exploration has continued at a high level. 6 national satellites, 5 of them launched by NASA and 1 by France, culminated in the pair of Helios solar explorers, which are yielding much new knowledge of the Sun. Germany also joined France in the Symphonie project (*qv* under France). In Feb 1967, however, the German Government decided that there would be no more national, or even bilateral satellite projects such as Symphonie; two-thirds of the 1976–79 space budget of DM2200M (£425M) would be allocated to the European Space Agency, and only one-third to national space efforts. But Germany is playing much the major part (52·55%) in ESA's development of Spacelab, Europe's contribution to the American Space Shuttle project; in addition, DM40M (£7·8M) is being spent on the French-led Ariane launcher over an 8-yr period to 1984.

Azur   L Nov 8 1969 by Scout from Vandenberg. Wt 71 kg. Orbit 387 × 3150 km. Incl. 103°. The first in a projected series of German-designed and built research satellites. The solar-synchronous orbit enabled the 7 experiments to observe Earth's radiation belts, solar particle flows, etc, in a co-operative programme with the US to augment data from US Explorer and OGO satellites. Life 100 yr.

Dial Wika   L Mar 10 by Diamant B from Kourou. Wt 63 kg. Orbit 301 × 1631 km. Incl 5·4°. France's first foreign launch; it was intended to study hydrogen geocorona and the ionosphere, but excessive vibration during launch made data evaluation difficult. Orbital life 7 yr.

Aeros-1   L Dec 16 1972 by Scout from Vandenberg. Wt 127 kg. Orbit 223 × 867 km. Incl 97°. With 5 experiments (4 German, 1 US), Aeros-1 dipped into the Earth's upper atmosphere 3844 times, measuring short-wave UV radiation from the Sun which is dangerous to man, before re-entering on Aug 22 1973. On-board thrusters had been used at 2500 orbits to add an extra mission phase.   **Aeros-2**   (L Jul 16 1974 by Scout from Vandenberg; wt 127 kg, orbit 224 × 869 km; incl 97°). Intended to continue upper atmosphere research. Data was said to be 'acceptable', although the intended orbit to continue the work of Aeros-1 had been 230 × 900 km. Re-entry was on Sep 25 1975 after 436 days.

Helios-1   L Dec 10 1974 by Titan Centaur from C Canaveral. Wt 370 kg. Heliocentric orbit. First of 2 spool-shaped spacecraft, placed in solar orbits taking them nearer the Sun than any previous craft. On Mar 15 1975 Helios-1 passed within 48M km at 238,000 kph, survived temperatures hot enough to melt lead, and made the first close-in measurements of the Sun's surface and solar wind. The craft spins once every second to distribute the heat coming from the Sun evenly, 90% of which is deflected by optical surface mirrors. Heat from the central compartment is radiated to space in an axial direction from radiating areas via louvre systems, keeping the internal temperature below 30°C compared with an outside temperature of 370°C. Helios-1 detected 15 times more micrometeorites close to the Sun than there are near Earth, coming from many different

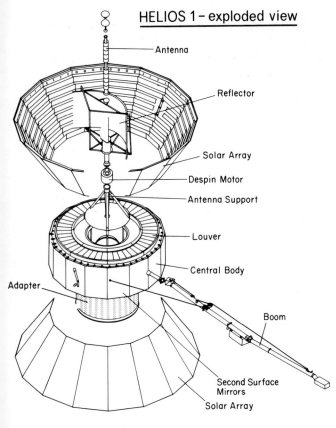

HELIOS 1 – exploded view

- Antenna
- Reflector
- Solar Array
- Despin Motor
- Antenna Support
- Louver
- Central Body
- Adapter
- Boom
- Second Surface Mirrors
- Solar Array

directions at different times. It is not known yet whether they are transported by the Sun's corona after being pulled in by the Sun's gravity from elsewhere in space. Helios 2 (L Jan 15 1976 by Titan-Centaur) was placed in a similar orbit taking it 3M km closer to the Sun and subjecting it to 10% more heat; it was hoped it would shed more light on the unexpected discoveries of Helios-1. Named after the Sun god of ancient Greece, this was the largest bilateral project in which NASA had participated, with Germany paying $180 M of the total $260M costs. Controlled from the German space centre near Munich, data will be correlated with the IMP Explorers 47 and 50 in Earth orbit, the Pioneer solar orbiters, and Pioneers 10 and 11 in the outer solar system.

GREAT BRITAIN

History Despite many years of political hesitation about the merits of a national space programme, by the end of 1976 Britain had orbited one national satellite and had 10 more launched for her by America; 7 of those were British-built. Combined with major contributions to the European Space Agency and to Intelsat, a reasonable national outlet had been found for the scientific talent which in earlier years had tended to drain away to America, where it played a major role in the manned Moonlanding projects. But Britain's first nationally launched satellite was also her last. Like the original US and Soviet programmes, it stretched back to wartime ballistic missile development. In Britain this was the Hawker Siddeley Blue Streak, finally cancelled as a missile in 1960, though continued until 1973 as a 1st-stage launcher for the European ELDO programme. A more

Ariel-3 in space chamber at RAE Farnborough

modest attempt at a national launch vehicle was announced in Sep 1964. It was decided to develop a small 3-stage vehicle from the successful Westland Black Knight research rocket. Construction and supply of 3 launchers, to be called Black Arrow, with the first 2 stages directly evolved from Black Knight, was ordered in Mar 1967. The purpose was to develop and test in space new components for communications satellites; and to develop a tool for space

178083

research. The successful accomplishment of these objectives in Oct 1971 was inevitably soured by the fact that in Jul the Government had announced that the programme would be cancelled after the final launch. But with this one satellite Britain became the 6th nation to achieve orbital capability—after Russia, America, France, Japan and China.

The British spacecraft programme now consists of the UK series (renamed 'Ariel' in orbit), which are scientific craft built for the Science Research Council; the Skynet geostationary series to provide military communications for the British forces; and the 'X' series for testing advances in technology.

A very successful programme of space research, plus Earth resources investigations for Germany, Argentina and other countries has been carried out with the British Aircraft Corporation's Skylark sounding rockets. By May 1976, 336 of these had been launched from Norway, Spain and Australia. The most recent at the time of writing was the much-heralded 250th Skylark launch from Woomera, intended to produce an energy-level map of X-radiation emitted from the exploded Puppis-A supernova. A cooperative project with NASA, the hope was to obtain information on the evolution of stars and formation of neutron stars. After weeks of delay, launch was made on May 12 1976; but the rocket's despin mechanism failed to stabilize the telescope platform.

At the end of its 2nd year, the British experimenters

BAC's 250th Skylark sounding rocket

Skylark launch

announced that a new sort of star, called a 'Rapid Burster' had been identified. Working with the Dutch ANS satellite, and America's SAS-1, about 12 Rapid Bursters had been identified on the opposite side of the Galaxy. They were a type of 'black hole' which emitted bursts of X-rays lasting around 10 sec at intervals of several hours. By then Ariel V had discovered 60 new sources of X-ray signals.

National Launches

Black Arrow/Prospero   Test flights and the final launch result were as follows:
X-1   L Jun 27 1969 from Woomera, Aus. Veered off course, and had to be destroyed 50 sec after lift-off. Intended to provide data on 1st- and 2nd-stage performance, with separation of dummy 3rd stage. **R-1**   L Mar 4 1970. Completely successful test of 1st- and 2nd-stage engines, and of 3rd-stage systems (propulsion test not included), and spin-up of dummy payload. 2nd-stage and payload impacted as planned, 15 min after launch, in Indian Ocean, 3050 km NW of Woomera. **R-2**   L Sep 1970. Intended to prove 3rd-stage propulsion system and inject into orbit X-2 development payload. First objective achieved but orbit not achieved, due to failure to maintain 2nd-stage (HTP) tank pressurization. **R-3 Prospero**   L Oct 28 1971, by Black Arrow from Woomera. Wt 66 kg. Orbit 547 × 1582 km. Incl 82°. Design aim of 1-yr life more than achieved; 812 passes were monitored from total of 4960 orbits in first year, and on-board tape-recorder replayed 697 times. It failed on May 24 1973 after 730 replays; but after 2 yr in orbit, real-time data was still being received at the Lasham, Hants, telemetry

Prospero, Britain's only nationally launched satellite

station, at the rate of 1 pass per week. Orbital life of satellite, 150 yr; rocket, 100 yr.

Launcher Description Overall ht 12·9 m. Dia 1st stage, 2 m. 1st stage: liquid-fuelled (HTP and kerosene) engine is Rolls Royce Bristol Gamma 8 with cluster of 8 thrust chambers, swivelled in 4 pairs for control in pitch, yaw and roll. Thrust 22,680 kg. 2nd stage: 2 RR Bristol Gamma 2 thrust chambers, also liquid fuelled, and fully gimballed for flight control. Thrust 6940 kg. 3rd stage: solid propellant motor developed from 2nd stage of Black Knight. This and payload are mounted within protective fairings which are jettisoned early in 2nd-stage firing. Guidance is fixed-programme, with gyro system maintaining course; by end of 2nd-stage thrust, the vehicle can coast up to orbital injection point, achieved by firing 3rd stage.

Prospero Description Wt 66 kg. Ht 71 cm. Dia 114 cm. Pumpkin shaped. Intended to prove new telemetry and power-supply system for satellites with electrical load of less than 30W; to test thermal surfacing, lightweight solar cells, etc. These are carried on 8 large and 8 small segments attached to main box section. 4 telemetry aerials mounted on base at 45° to spin axis. Internal equipment includes telemetry and data handling, micrometeoroid detector, electronics, etc.

US Launches

Ariel The first international satellite programme evolving from a 1959 NASA offer to launch foreign payloads when such experiments were of mutual scientific interest.

Ariel-1 L Apr 26 1962 by Delta from C Canaveral (orbit 389 × 1214 km, wt 60 kg, incl 54°), transmitted ionospheric and X-ray data until Nov 1964 and has a 15-yr life.
Ariel-2 L Mar 27 1964 by Scout from Wallops I (orbit 285 × 1362 km, wt 68 kg, incl 52°), also returned data until Nov 1964 and re-entered Dec 18 1967. **Ariel-3** L May 5 1967 by Scout from Vandenberg (orbit 498 × 606 km, wt 90 kg, incl 80°), was first all-British satellite and measured radio noise from lightning and from galactic sources at frequencies too low to penetrate the atmosphere etc. Spin-stabilized, it was very successful, operating for over 2-yr and re-entered in Dec 1970. **Ariel-4** L Dec 11 1971 by Scout from Vandenberg (orbit 485 × 599 km, wt 100 kg, incl 83°), continued electron density, radio noise and other experiments until Apr 1973, when it was decided enough data had been obtained. It was then switched on occasionally to obtain data in support of sounding rocket experiments. It has an 8-yr life. **Ariel-5** L Oct 15 1974, by Scout from San Marco (orbit 504 × 549 km, wt 129 kg, incl 3°), carried 6 experiments to study X-ray sources in space; it discovered a new type of star which could be a white dwarf or burnt-out star, and added considerably to knowledge of 'black holes'.

Skynet Military satellites designed to provide exclusive and highly secure voice, telegraph and facsimile links for both strategic and tactical communications between UK military headquarters and ships and bases in Middle and Far East. Drum-shaped cylinders, with 137 cm dia × 81 cm high, they are spin-stabilized but with a despun antenna. They had an unsatisfactory early history. **Skynet 1** L Nov 21 1969 from C Kennedy by Delta (wt 422 kg), was successfully placed in geosynchronous orbit over the Indian Ocean but

Thor-Delta launches Skynet 1B

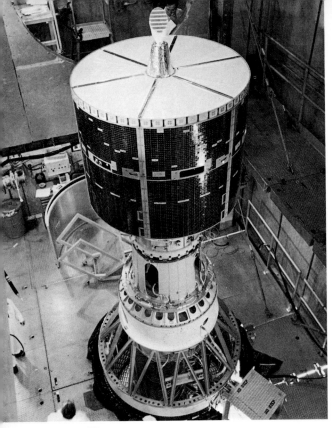

Skynet 2B

probably operated for less than a year. **Skynet 1B** L Aug 19 1970 by Thor-Delta from C Kennedy (wt 285 kg), had to be abandoned in its transfer orbit of 270 × 36,058 km, due to the failure of the apogee motor. Orbital life unknown. **Skynet 2A** L Jan 19 1974 by Thor-Delta from C Canaveral (wt 422 kg), was built by Marconi in England and was the first communications satellite built outside the US or USSR, and 3 times as powerful as Skynet 1s. A launch failure placed it in a tumbling 120 × 1857 km orbit and it re-entered after 6 days. **Skynet 2B** L Nov 23 1974 by Delta from C Canaveral, was successfully placed in synchronous orbit over the Indian Ocean, and with a life of at least 3 yr, it provides communications from the UK to W Australia. The Skynet system included 17 ground stations, with 12·8m master terminal at Oakhanger, air transportable terminals, for use on land as well as 1m shipborne terminals. By the time the system was operational, successive defence cuts and British withdrawals from the Far and Near East, had made it largely unnecessary. Its cost was never disclosed.

X Series The first and possibly only satellite in this projected series, X-4 (re-named Miranda in orbit, in the Shakespearian tradition of the UK series) was launched Mar 9 1974 by Scout from Vandenberg. Wt 93 kg; orbit 703 × 918 km; incl 97·81°, which made it near Sun-synchronous. British built by Hawker Siddeley for the Dept of Trade and Industry, its main purpose was to test a new type of 3-axis attitude control, as an alternative to the spin-stabilized systems used on previous UK spacecraft. Despite problems with 'the gas-reducing values—which affected control only minimally'—good data was said to have been obtained. Orbital life 150 yr. Total cost was over £5M, plus about £1·5M launch costs.

INDIA

The Indian Space Research Organization (ISRO) is developing the SLV-3 launcher, due to be test-fired from her own launch site at Sriharikota in 1980. It is intended to orbit educational broadcasting satellites, also starting in 1980. French Viking motors (60,000 kg thrust), developed for the ESA Ariane launcher, are being produced under licence. In Oct 1974 the Société Européenne de Propulsion (SEP), sold India the Viking production licence for £4·2M ($2·1M), of which ESA received 25%. An Earth resources satellite (SEO—satellite for Earth observation) is being prepared for launch in 1978.

India's first satellite, originally due to be launched in Dec 1974 by Russia, was twice delayed, but finally successfully orbited 4 months late:

Aryabhata L Apr 19 1975 by Skean + Restart stage, from Kapustin Yar. Wt ?300 kg. Orbit 563 × 619 km. Named after the 5th century astronomer and mathematician, this was intended for X-ray astronomy, study of neutron and gamma radiation from the Sun, and measurement of particle and radiation fluxes in the ionosphere. Indian scientists took part in preparation of the satellite 'under the scientific supervision of Soviet engineers'. Designed for 6 months, it had an orbital life of 10 yr. Tracking stations near Moscow, and at Hyderabad, S India, maintained communications; control was to have been handed over to Indian scientists on the 3rd day, but the Indian-designed transformer failed, and only a few days' useful life was obtained.

INDONESIA

History Satellite communications are providing 3000 inhabited islands in Indonesia with instantaneous telephone, radio, TV telex and data communications for the first time in history. Since it was not economically possible to cover thousands of kilometres of ocean with cables or microwave towers, the Government asked NASA to provide satellite communications for its 120M population on a reimbursable basis. The first satellite, Palapa-1, was operational in time for Indonesia's 31st independence celebrations on Aug 17 1976. It was named after a popular delicacy in the 14th century, which the then Prime Minister, Gajah Mada of the Kingdom of Majapahit, vowed not to eat until the whole of Indonesia was united.

Palapa-1 L Jul 8 1976 by Delta from Cape Canaveral. Wt 575 kg. Synchronous at 83°E. Identical to Canada's Anik and Western Union's Westar satellites, except for a modified 1·5 dia parabolic dish antenna to provide maximum illumination of the Indonesian landmass. The 12-transponder satellite provides 4000 voice circuits or 12 simultaneous TV channels; ht 3·7 m, dia 1·9 m. Each of Indonesia's 26 provinces has a transmit/receive Earth station. Its synchronous, 0·1 incl orbit places it over the equator.

Palapa-2 L Mar 10 1977 by Delta from Cape Canaveral. Wt 575 kg. Operational in synchronous orbit at 77°E, it completes the system.

Apollo/Soyuz Docking (Artists' Impression)

INTERNATIONAL

Apollo-Soyuz

Introduction   Both politically and technically the Apollo-Soyuz Test Project (ASTP), as the Americans dubbed the flight, was an enormous success. For America, however, the political success was marred in the last 10 min by crew errors on board Apollo which almost cost the crew their lives. The near-disaster also delayed Soviet-American plans for triumphal tours of both countries by the 5 crew members.

The flight brought short-term benefits to the space programmes of both Russia and America. But discussion about further joint manned flights quickly ran into difficulties. Neither country is anxious to embark on joint ventures likely to be of more benefit to one side than the other; such flights are costly and time-consuming and likely to divert resources and technicians from more urgent—and more practical—national projects. However, General Tom Stafford, at the end of the astronaut's visit to Moscow in Oct 1975, said talks had begun on joint projects to study the Moon and planets, meteorology and space medicine; General Leonov hoped it would be 'only the beginning of large-scale international co-operation in space research'. But the prospect of a manned Soviet-American mission to Mars still seems a long way off.

History   The plan for a joint US/Soviet spaceflight, later named the Apollo/Soyuz Test Project, was included in an agreement on the peaceful exploration of outer space signed by President Nixon and Premier Kosygin on May 24 1972.

Apollo/Soyuz Official Emblem

Target launch date chosen was Jul 15 1975. Earlier discussions had been on the basis that Apollo and Soyuz

ASTP: Saturn 1 B Lift-off from Pad B

spacecraft would dock with a Salyut space station, but this was dropped in favour of the simpler, direct link-up of the 2 spacecraft with the aid of a docking module 3·15 m long, 1·4 m max dia, and 2012 kg wt. For a year or more, with Soviet scientists clearly having difficulty in shedding the deep-rooted secrecy surrounding their space activities, there was much scepticism in the West about the likelihood of the project being carried through. But as Russia's own space programme ran increasingly into a period of setbacks and failures, and America's was so successful that its future funds were heavily cut, ASTP became more attractive to both sides. Russia badly needed to improve managerial areas such as quality control, about which she could learn much from America. For America, ASTP had immediate political benefits, gave her scientists their first opportunity to learn something about the Soviet space programme, and provided some much-needed manned spaceflight activity during the long gap between completion of Skylab and the start of Space Shuttle flights. It also enabled good use to be made of left-over Apollo spacecraft and other hardware which would otherwise have been wasted.

Early in 1974 NASA reported that the joint project had reached the half-way point, was providing more than 4000 jobs, and was on schedule. US costs were within the original estimate of £104M ($250M).

Mission Summary It was finally planned and executed, with remarkable precision, as a 9-day flight. Because of its superior manoeuvring ability, and much larger supply of propellant, it was agreed that Apollo would play the major, active role, with Soyuz remaining largely passive throughout the mission. With its total CSM supplies of 1290 kg of manoeuvring propellant (compared with about 227 kg on

Soyuz), Apollo was able to carry out both dockings and many other manoeuvres. There was much public confusion because Soyuz was in the 'active mode' for the 2nd docking; but the actual manoeuvres, as described in detail later, were performed by Slayton. Launched $7\frac{1}{2}$ hr before Apollo, Soyuz waited in orbit for the Americans to locate and dock with her 52 hr later. Having been launched, like the 3 Skylab crews before them, by a Saturn 1B rocket, the Apollo crew, 1 hr later, had to make the familiar Transposition and Docking manoeuvre; this time, not to extract the Lunar Module carried on Moonflights in the S4B upper stage, but the Docking Module which replaced it. Soyuz and Apollo duly docked, and the crews worked together, for nearly 48 hr. After that they continued their 'unilateral' programmes; Soyuz landed on the 6th day; Apollo on the 9th. Summary flight plan follows:

ASTP MISSION EVENTS

Apollo Ground Elapsed Time (AGET) was used only for Day 1. The remainder of the mission was referenced to Soyuz Ground Elapsed Time (SGET; in my text GET). The table given below is the official US pre-flight schedule. Variations in actual times were astonishingly small, and are given in brackets after the main events.

Major Events	Time from Soyuz Lift-off, S.G.E.T., Hr:Min
—————————TUE/JUL 15—————————	
Soyuz Lift-off (63°E, 46°N)—TV	00 : 00
Soyuz Insertion	00 : 09
Apollo Lift-off—TV	07 : 30
	(00 : 00)
Apollo Insertion	07 : 40
	(00 : 10)
Begin Apollo Transposition, Docking, and Extraction of Docking Module—TV	08 : 44
	(01 : 14)
Apollo Evasive Manoeuvre (AEM)—TV	10 : 04
	(02 : 34)
Apollo Circularization Manoeuvre (ACM)	11 : 15
	(03 : 45)
First Apollo Phasing Manoeuvre (NC1)	13 : 11
	(05 : 41)
—————————WED/JUL 16—————————	
Soyuz Circularization Manoeuvre (SCM)	24 : 26
Apollo Conducts Independent Experiments	
Apollo Phasing Correction Manoeuvre (PCM)	32 : 22
—————————THUR/JUL 17—————————	
Second Apollo Phasing Manoeuvre (NC2)	48 : 34
Apollo Corrective Manoeuvre (NCC)	49 : 18
Apollo Coelliptic Manoeuvre (NSR)	49 : 55
Apollo Terminal Phase Initiation (TPI)	50 : 54
Begin Apollo Braking (TPF)—TV	51 : 23
Apollo/Soyuz Stationkeeping	51 : 32
Apollo/Soyuz Docking on Apollo Rev 29—TV	51 : 55
	(51 : 49)
U.S. Crew (Stafford and Slayton) Transfer to Soyuz for Initial Greeting—TV	54 : 50
—————————FRI/JUL 18—————————	
Joint Docked Activities	
U.S. Crew in Soyuz for Final Farewells—TV	80 : 10
—————————SAT/JUL 19—————————	
Apollo/Soyuz First Undocking—TV	95 : 42
	(95 : 45)
Begin Joint Solar Eclipse Experiment—TV	95 : 43
Apollo/Soyuz Redock—TV	96 : 08
Apollo/Soyuz Final Undocking—TV	99 : 06
Begin Joint UVA Experiment	99 : 21
Apollo Separation Manoeuvre (SEP)—TV	102 : 16

Apollo and Soyuz Conduct Independent Activities

————————————MON/JUL 21————————————

Soyuz Deorbit	141 : 46
Soyuz Landing in Kazakhstan	142 : 31
	(142 : 32)

Apollo Conducts Independent Experiments

————————————TUE/JUL 22————————————
Apollo Conducts Experiments

————————————WED/JUL 23————————————

Apollo Press Conference—TV	192 : 03
Apollo Jettisons Docking Module	199 : 21
CSM Separation Manoeuvre for Doppler Experiment (DM1)	199 : 56
CSM Stable Orbit Manoeuvre for Doppler Experiment (DM2)	204 : 15

————————————THUR/JUL 24————————————

CSM Deorbit Manoeuvre (ADM) Apollo Rev 138	224 : 18
CM/SM Separation	224 : 24
CM Landing near Hawaii	224 : 58

MISSION DURATION	Days Hr Min
Soyuz Lift-Off to Landing	5.22.31
Apollo Lift-off to Landing	9.01.28
Docking to 1st Undocking	1.19.47

For America, the big disappointment was that, throughout the 52 hr during which Apollo and Soyuz were within visual range, TV viewers throughout the world saw continuous pictures of Soyuz transmitted from Apollo, and not one single view of Apollo taken from Soyuz. This was

Orbital view of Apollo and Docking Tunnel taken by Soyuz

basically due to technical troubles with the Soyuz TV system which began during launch, and which made the outside colour TV camera unusable. During Apollo's final manoeuvres around Soyuz, Houston asked Moscow to get one of the cosmonauts to show shots of Apollo by hand-holding an interior TV camera at one of the windows. Moscow turned it down because it would have meant opening the hatch between the orbital and descent modules, and also because the crew was too busy with other activities. While the Soviet decision, in retrospect, seems to have been perfectly justified, at the time NASA officials found it difficult to conceal their dismay and annoyance that the biggest TV audience in history had had splendid pictures of Soviet activities, provided by American technology, with no reciprocal pictures being provided of America's much more sophisticated and reliable Apollo.

Crews Both countries selected and announced their ASTP crews 2 yr before the flight. It was the first time names of cosmonauts were known in the West before they had actually flown. The average age of the 2 prime crews was 45—an indication that both sides felt that the occasion called for men of diplomacy as well as experience.

US Crews Prime Commander, Thomas P. Stafford, who flew on Geminis 6 and 9 and Apollo 10. CMP, Vance D. Brand, back-up commander on Skylab 3 and 4, making his first flight. Docking Module Pilot, Donald K. Slayton, making his first flight at 51, and 16 years after being chosen as one of the orginal 7 Mercury astronauts. Back-up Crew Commander, Alan Bean; CMP, Ronald Evans; DMP, Jack Lousma. Support Crew Richard Truly; Robert Overmyer; Robert Crippen; Karol Bobko. The US leader of the Joint Working Group was Dr Glynn Lunney of NASA.

ASTP Onboard photo: Brand in Command Module (or 'in CM')

USSR Crews   1st Prime Crew Commander, Alexei Leonov, first man to walk in space on Voskhod 2; with Valeri Kubasov, who was flight engineer on Soyuz 6. 2nd Prime Crew Commander, Anatoli Filipchenko, who was commander on Soyuz 7; with Nikolai Rubavishnikov, who was test engineer on Soyuz 10. 1st Back-up Crew Vladimir Dzhanibekov and Boris Andreyev. 2nd Back-up Crew Yuri Romanenko and Alexander Ivanchenko. In addition, General Vladimir Shatalov was named as Director of Cosmonauts' Training, Alexei Yeliseyev as Soviet Flight Director, and Valeri Bykovsky as ASTP Training Officer. The USSR leader of the Joint Working Group was Prof K. D. Bushuyev of the Soviet Academy of Sciences.

Preparations   Joint working groups of engineers and scientists were established, and between Jul 1972 and Oct 1974, 16 meetings were held—8 in each country. In Apr 1974, Russia for the first time issued advance notice of the exact time and date planned for a manned launch: 15.37 from Baikonur cosmodrome on Jul 15 1975. To ensure that one was ready, 3 Soyuz spacecraft would be prepared for flight. Gradually, early restraints and tensions evaporated. But an undercurrent of resentment at certain aspects of Soviet policy continued right through the flight among some US team members and astronauts. This was based on Soviet insistence that, while America had the right to follow its traditional 'open' policy, with full news coverage, the Russians were equally entitled to continue their own well-established policy of very limited access to launch and control centres, with strictly controlled news coverage. This came to a head when Tom Stafford, the Apollo commander, at one of the final meetings in Moscow, declared that unless the NASA astronauts and support team were allowed all the facilities they needed to visit Baikonur and inspect the spacecraft and launch equipment, the whole project was off. There was considerable NASA management concern about the possible political impact when Stafford's stand brought the meeting to an abrupt end; followed by relief next day when the Russians expressed surprise that the Americans had got the impression that the facilities they wanted would not be allowed. However, it should be recorded that when the day did come for the American astronauts to visit Baikonur, their aircraft arrived and departed in the dark. The US team consisted of 11 astronauts and 2 support staff, and one of them later described it as a 'stark and sterile' tour. After being accommodated overnight in a comfortable 3-storey hotel at Leninsk, a modern city for 50,000 space workers and their dependants, the group was told to leave their cameras behind before starting the 1-hr bus drive to the launch site. Although Tyuratam contains 80 operational launch pads, and it is believed, a new Saturn 5-type pad 122 m tall, the group saw 'nothing but railroad tracks that disappeared over the horizon', a tracking site and a liquid oxygen production plant. At the launchpad they found both Soyuz spacecraft in test stands with one SL-4 launcher horizontal on its railroad transporter. Only vague answers were given about the whereabouts of the backup booster and launchpad. However, the primary and backup US crews each spent an hour in the 2 Soyuz spacecraft. Checks were made that US equipment fitted correctly in the primary Soyuz, and arrangements completed for a group of US technicians to make similar checks on the backup Soyuz a few weeks later. Soviet cosmonauts had been shown the Launch Control room when they had visited Pad 39 at the Cape; but the astronauts were not taken to Tyuratam's launch control room; and questions about it were again met with vague answers.

An insight into the atmosphere during the pre-flight discussions was provided during a Moscow press conference immediately after the first docking had been achieved. Discussing it, Viktor Blagov, launch director at that time, said (according to the interpreter): 'We were somewhat excited . . . knowing that Thomas Stafford is a man of high spirits. During one of the visits to Houston, Texas, by our team, Stafford showed them a film of his manoeuvring next to an Agena rocket stage. I attribute this to the usual American characteristic of enterprise. Our docking operations are carried out somewhat more smoothly, and I am happy to say that during the final stage of today's docking Tom Stafford was converted to the Russian faith.'

Another major hurdle during Soviet-American preparation discussions was US insistence on being given a detailed report on the Soyuz 11 failure. This was finally provided, showing both causes and modifications; Soviet scientists said modifications were verified on Cosmos flights 496 and 573, both unmanned Soyuz flights, and again on the Soyuz 12 mission. As further reassurance, the Russians disclosed their intentions (without giving details) of making 2 or 3 Soyuz flights to test their ASTP hardware. Soyuz 13 was the first of those flights. (*See* Soyuz 15–18.)

Russia also built special housing for US flight crews at Zvezdny Gorodok (Star City), 45 min drive NE of Moscow, for use during joint training of manned spacecraft pilots. An American problem, at all 3 manned spaceflight centres (Kennedy, Johnson and Marshall), was that the 17-month gap between the last Skylab mission and ASTP made it difficult to maintain 'a cadre of manned flight experience and knowledge' among both the civil and contractor personnel. Arrangements were made to recall experienced personnel from other jobs when the time came, and in the event this worked well.

Hardware Each country independently designed its docking system to meet the requirements of a jointly defined 'common interface', which included structural rings, seals, latches and guide petals. The system did away with the docking probe which had caused problems throughout the Apollo programme (and finally on ASTP when used for the CSM/Docking Module linkup). This will probably emerge as the major technical advance achieved by the joint flight. Each country's design was reviewed and accepted by the other by autumn 1973. By then, NASA had decided that the new docking system would be used on all future US manned missions, whether they were routine or rescue flights. It would also be used for any future international flights, when the Space Shuttle came into service. In the meantime, the Skylab rescue spacecraft was modified as a backup ASTP craft.

Apollo CSM Modified Skylab version, with steerable high-gain antenna, used on Moonflights but removed for Skylab, brought back again. This provided communications via ATS-6 and Mission Control for 55% of each orbit. Control for Docking Module also added to CM. Standard probe-and-drogue carried on the CM was the only one ever used twice, since it was brought back by Apollo 14 after giving trouble in lunar orbit. Refurbished and re-used on ASTP as an economy measure, it nevertheless gave trouble again during the CSM/DM link-up.

Docking Module Length 3·15 m; Max Dia 1·4 m. Wt 2012 kg. Basically an airlock with docking facilities at each end to provide transfer facilities between the Apollo and Soyuz spacecraft. It contained control and display panels, VHF/FM transceiver, life-support systems and storage compartments. Equipment included oxygen masks, fire extinguisher, floodlights and handholds, junction ('J-Box') for linking Soyuz communications circuits to Apollo's, TV equipment, etc. Also housed in the DM was the multipurpose Electric Furnace which enabled the crews to conduct 7 high-temperature experiments using weightlessness to investigate crystallization and material processing. 4 spherical tanks attached outside the DM contained 18·9 kg of nitrogen and 21·7 kg of oxygen for use during crew transfers. Because Apollo's normal atmosphere in orbit is 100% oxygen at 5 psi, and the Soyuz atmosphere an oxygen/nitrogen mix at 14·7 psi, crew transfers provided a major problem. To avoid developing what divers call 'the bends', the cosmonauts needed to spend a considerable time, possibly 2 hr, in the DM airlock, pre-breathing oxygen to purge suspended nitrogen from their blood-streams before passing into Apollo. The problem was overcome, however, when the Russians agreed to lower the Soyuz pressure to 10 psi during the joint exercise. Transfers were then possible in 'shirtsleeves'. The DM system was always operated by an Apollo crewman, with 2 men in the DM during a transfer. The rules were that there must always be one cosmonaut in Soyuz and one astronaut in Apollo; hatches at both ends of the DM, with pressure equalization valves, made it possible to transfer without disturbing the atmospheres in either spacecraft.

Docking System While the DM was entirely designed and constructed in the US, the compatible docking assembly was designed and tested jointly by NASA and Soviet

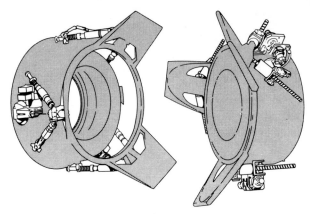

American and Soviet Docking Systems

engineers. Both spacecraft can use it in either an active or passive mode for docking. Each half consists of an extendable guide ring with 3 petal-like plates, with a capture latch inside each plate. The spacecraft in the active mode advances with its guide ring extended, or thrust forward, on 6 hydraulic attenuators, or retractable arms. As the spacecraft touch, the 6 capture latches (3 on each craft) engage on the guide ring. The active craft then retracts its guide ring; this pulls the 2 spacecraft together to squeeze pressure seals together between Apollo and Soyuz. 8 structural latches then engage to hold them firmly and safely together.

Communications No chances were taken of misunderstandings occurring for lack of communications. Houston and Kalinin were linked for the mission with 14 voice transmission lines; 2 TV lines; 2 Telex lines; 2 teletypewriter lines; and 2 Datafax lines. In addition, Apollo and Soyuz carried radios compatible with the tracking stations of both countries. Each country sent a 9-member specialist team to the other country's Mission Control, to explain any Apollo problems to the Russians, and any Soyuz problems to the Americans.

The Countdown The most complicated countdown in history involved the co-ordination of 2 Soyuz and 1 Apollo spacecraft at launch sites 16,000 km apart. The occasion was marked by the assembly, at Cape Canaveral and Houston, of the largest gathering of news reporters since Apollo 11. Russia established a news centre at the Hotel Intourist, a Moscow hotel, where over 600 Soviet and foreign newsmen were accredited. At Mission Control at Houston there was concern, as launch day approached, that the Soyuz 18 crew had reached their 50th day in Salyut 4 with no indication that they were preparing to return. Dr Glynn Lunney, Apollo Director, sought and obtained assurances from Prof Konstantin Bushuyev, USSR Technical Director, that there were separate control centres for Salyut 4 and ASTP; Salyut 4 would not be using any communications facilities also required for ASTP; and in any case, absolute priority would be given to the joint flight.

Newsmen were delighted to receive their first-ever Soviet Press Kit—204 pages long, and in some ways better than NASA's 113-page Kit. But it should be remembered that full copies of the official NASA Flight Plan, plus endless briefings (with subsequent transcripts) by astronauts and mission directors, were also given by the Americans. There were some notable omissions in the Soviet Press Kit; it listed

15 Soyuz manned flights between 1967 and 1975 which 'accomplished' their flight programmes with no mention that in the case of Soyuz 1 and Soyuz 11 the cosmonauts were killed during re-entry. It was not made clear that both dockings would be performed by the Apollo crew; in the case of the 2nd docking, for instance, it was stated 'Soyuz docking assembly is set in the active position . . . Then crews perform redocking'.

The 2 commanders, Tom Stafford and Alexei Leonov, had regular reassuring chats over the 'dedicated lines' and 'key circuits' linking the Cape (via Houston) and Baikonur. The agreement under which each spoke the other's language worked well; the Apollo crew were said to have spent over 1000 hr learning Russian. NASA used 14 tracking stations (including 1 ship), employing 2800 personnel; the Soviet Union employed 7 ground stations and 2 ships. In addition NASA was using the ATS-6 satellite to relay communications from Apollo to Houston while it was over Europe. This was to meet Russia's requirement for the dockings to take place over or near her territory, to provide direct communications. NASA at first insisted that since they were doing the dockings, they should take place over US territory, thus providing them with direct communications; but compromised when ATS-6, after providing experimental educational TV programmes to Alaska was being moved to provide a similar experiment for India, and was made available for ASTP.

At the Cape the period chosen for this launch was the worst of the year; thunderstorms built up almost daily, breaking in the afternoon, just about the time set for lift-off. In the week before Jul 15, launch would have been possible on only 2 days. Decisions were made to modify the rules laid down after the Apollo 12 launch; and experiments were made by USAF Phantoms, trying to 'seed' the clouds so that they could be made to discharge their electricity just before the launch. In the event the weather cleared and none of this proved to be necessary. Other worries were about morale among the Cape's technicians, since 1800 were due to be laid off once the launch had been made, followed by another 500 after the Viking launches a month later. Slight embarrassment was caused by Deke Slayton, Chief Astronaut for many years until he gave up the post to make this first flight, when he said frankly at a news conference that he thought Russia had 'a lousy political system', and he wanted no part of it. Newsmen who had gone to Moscow tried to insist that they should be allowed to travel to Baikonur for the launch, on the ground that Soviet newsmen would be at the Cape. NASA was surprised when a Soviet spokesman flatly denied that there would be any Russians at the Cape, since 9 had asked for, and been given, accreditation. A check showed that they had not arrived. Apparently they were redirected, at the last moment, to Houston, to cover the flight there with 21 other Soviet radio, TV and newspapermen.

The background to all this was that the Russian authorities, having reluctantly permitted US astronauts and technicians to visit Tyuratam the previous April, continued to refuse permission for any US programme personnel to attend the launch. So NASA was surprised when, a day or two before it, 4 Americans in Moscow were invited to Tyuratam. The 4 were the US Ambassador in Moscow, Walter J. Stoessel, and his wife; William Shapley, a NASA Administrator; and Egon Koebner, US Science counsellor in Moscow. Unlike the astronaut group, they were flown to Tyuratam in daylight, and their subsequent accounts provided a few more details about Russia's prime launch centre. It also showed that for those who are there, Soviet launch practice is much more relaxed than at the Cape. As they flew in on launch day they could see the town and landing strip, but no launch facilities of any sort—'nothing but miles and miles of desert' said Mr Shapley. About 3 hr before lift-off they were taken to the Soyuz checkout area, where they talked with Leonov and Kubasov—something which would be strictly prohibited in the case of the US astronauts, for fear of infection. Then, only 1 hr before lift-off, they toured the launchpad—which again would be strictly against NASA safety rules. While at the launchpad, within 60 m of the Soyuz/SL4, already fuelled with liquid oxygen and kerosene, they were joined by Maj Gen Vladimir Shatalov, Chief Soviet Cosmonaut, and other officials. They finally viewed the launch from a stand only 850 m away. They saw no sign of Soyuz 2, the backup spacecraft and launcher, but were assured it was available.

The US astronauts, and no doubt the Soviet cosmonauts too, took great precautions against the embarrassing possibility of developing space sickness during their joint activities. In the case of the astronauts this took the form of aerobatic flying in their T58 jet trainers. The author tried in vain to establish whether cosmonauts make similar flights. Finally Soviet space clocks were moved on by one thousandth of a second, so that they were exactly synchronized with NASA's clocks. The 2 countries, with the world watching, were ready for their historic flight.

THE FLIGHT

Soyuz Pre-docking 4 years of intense preparations were rewarded with on-time lift-offs from both launchpads, although they were 16,000 km apart. Russia's Soyuz 19 left the pad within 10 sec of the planned time, 12.20 GMT on Jul 15 1975. 7½ hr later, Apollo lifted off within 3·8 sec. Total

Soyuz lift-off wt was approx 317,520 kg including approx 6690 kg for the spacecraft. The Apollo crew, not due to wake until 2 hr after the Soyuz launch, slept through it; the Russians confounded the sceptics by giving us live TV (a Soviet 'first') of their launch, preceded by a countdown commentary and pictures of Leonov and Kubasov going aboard. Launch Control at the Cape had to provide a 'Go' so far as they were concerned 20 sec before Soyuz lift-off. The failure of a planned live TV transmission from Soyuz showing the cosmonauts during the launch was the first of a number of failures in this area. This was because of a faulty switching device common to all 4 Soyuz TV cameras (2 colour, 2 black-and-white). Ground-based simulations in Russia produced a solution which was only partially successful: Kubasov used adhesive bandage tape from the first-aid kit to insulate the wire joints—an interesting example of the way Soviet space technicians appeared to have learned from American 'trouble-shooting' techniques.

The SL-4 launcher's 3rd, or upper stage, was shut down 8 min 50 sec after lift-off; the antennas and solar panels were extended; and control of Soyuz passed from the Launch Centre to Soviet Mission Control Centre at Kalinin, near Moscow.

The initial orbit 221×186 km, with $51 \cdot 78°$ incl compared with the planned 228×186 km and $51 \cdot 48°$ incl. Soyuz's orbital life would have been only 30 orbits, and the Kalinin Information Officer commented: 'There is no immediate necessity for emergency raising of the orbit.' In fact the orbital error was corrected on the 4th orbit, with a 7 sec firing; this produced a near-nominal 191×232 km. On the 17th orbit a schedule burn raised this to a satisfactory 225×222 km, compared with the planned 225km circular orbit for docking.

Before those corrections were made, however, Leonov and Kubasov, on the 2nd orbit, had to open up the hatch to the orbital module and check the pressure seals. They took off their suits 110 min into the flight, though they had been wearing them for 5 hr. They then had to pump out condensation from the suits before drying and packing them. On the 3rd orbit, when Soyuz 19 passed over US territory, the crew talked to Mission Control at Houston, and Moscow talked to them through US ground stations, thus opening up the Soviet-American communication system. Until the 4th revolution correction Soyuz was placed in the drift mode so that no reaction control system propellant was used; after that it was spin-stabilized towards the Sun, charging the batteries, and continuing to save RCS propellant. On this orbit—a busy one for Leonov and Kubasov—they began the pressure reduction needed for the linkup; it took 2 hr 34 min to bring it down from 867 to 539 mm. They also started the 3 biological experiments on different forms of primary life; containers with microorganisms, zone-forming fungi, spawn of small fish and caviar eggs were placed in nutritive liquid so that their progress and development in Zero-G could be observed.

At the end of their first day in space, the crew had to start repairing the TV system, with the result that they were $1\frac{1}{2}$ hr late getting to bed. The first colour TV transmission was made next day on the 20th orbit; shortly after the crew relaxed by talking to Pyotr Klimuk and Vitaly Sevastyanov as they passed within radio range of Salyut 4. 'If you need something repaired you can turn to us for help', Leonov told them.

Docking day began badly for the Soyuz crew, just as it did for the Apollo men. They were up earlier than planned, grappling again with more TV problems. (The failure of

Soyuz spacecraft in Earth orbit seen from Apollo

Soyuz to provide pictures of Apollo was to prove the major failure, especially from a political point of view, of the whole mission.) Their 35th orbit—the last before docking—was a strenuous one. Unlike the Apollo crew, Leonov and Kubasov had to don spacesuits, then turn on the docking target and flashing beacons for the benefit of Apollo's docking system. At that point 28 km separated the spacecraft. Then the Soyuz crew moved into their Descent Module, and closed and checked No. 5 hatch, separating them from the Orbital Module.

Apollo Launch   The Apollo crew watched a VT recording of the Soyuz launch over breakfast, passed their medical checks, and boarded their spacecraft the usual $2\frac{1}{2}$ hr before lift-off. The 32nd and last Saturn rocket, although 7 years old, provided the 'cleanest' launch from a technical point of view. It was the last occasion, too, when NASA required a launch team of 500. Because of threatened thunderstorms, a Weather Avoidance Hold of 5 min 24 sec was built into the countdown at $T-4$ min, so that if necessary, launch could be made early to beat an approaching storm. In the event skies cleared and the 10 aircraft and 8 helicopters patrolling over the Atlantic had only good news to report. Total Apollo/Saturn lift-off wt was approx 567,000 kg. In orbit at the end of the first day total Apollo/Docking Module wt was 14,679 kg. As in the case of Soyuz, the initial orbit was less than perfect—170×152 km, compared with the planned 167×149 km. But Deke Slayton, an original Mercury 7 astronaut, in space at last and savouring the view, called back at $T+9$ min: 'Man, I tell you, this is worth waiting 16 years for.'

A series of minor problems that kept the Apollo crew working at full stretch, trying to regain lost time before the docking, actually started just before launch. During the checkout procedures Brand incorrectly operated some switches which it was feared had caused an air bubble to form in the Service Module's RCS system; Stafford had to clear this when in orbit by simultaneously dumping propellant through all 16 thrusters. (A bubble in the propellant system could have caused damage to the thruster during a manoeuvre.) Transposition and docking went well; less than 2 hr into the flight Stafford reported that he had had 'a good hard docking' with the Docking Module. Problems after that, however, included overheating of the cabin—it went up to 77°F before Mission Control found a remedy—and difficulty with the urine dump. There was dismay when the crew finally had to go to bed, $1\frac{1}{2}$ hr late, with the docking probe between the CM and DM still not dismantled. There was concern in Moscow and Houston, since it seemed possible that while Apollo and Soyuz would still be able to dock, the crews would be unable to exchange visits. While the Apollo crew slept trouble-shooting teams got to work at Houston, and newsmen unearthed the astonishing fact that the docking probe was from Apollo 14, the only one ever used twice, since it was the only one ever brought back to Earth. That was because it had given so much trouble in lunar orbit. However the probe itself was vindicated this time when it was found that an improperly installed wire was blocking a hole into which a tool had to be inserted to collapse the probe latches. When the crew woke for their second day in space Houston was able to tell Brand how to clear the wire and remove the probe. The crew spent their second day trying to make up lost time, and were only 30 min behind at the end of it. Good humour prevailed despite their difficulties. A large Florida mosquito which had hitched a ride into space was the butt of many jokes; Slayton

threatened to bring it home and give it a pair of astronaut wings.

Experiment problems   There were minor problems when the Apollo crew settled down to start their series of 27 scientific experiments. Stafford had some trouble extracting a frozen sample for a W German electrophoresis experiment designed to study the possibility of producing vaccines and serum in space for medical use on Earth. And Slayton had difficulty with the door of the Multipurpose Electric Furnace in the Docking Module, in which 7 experiments (one jointly with the Soviet Cosmonauts) were to be carried out on growing crystals in space, weightless processing of magnets etc.

Docking   Docking day, Jul 17, started badly for the Apollo crew. Though already short of sleep, they were woken 1 hr 19 min early by a series of master alarms indicating trouble with the guidance system. Mission Control trying to sort them out, urged the crew to get some more rest; Brand commented: 'That sort of alarm sure wakes up a crew . . . and makes them very alert.' Stafford worked good-temperedly on the alarms as they recurred, and told Houston not to worry about the crew. Brand had already shaved, and Slayton was shaving while he (Stafford) prepared breakfast, before they had been due to wake.

At that time, on Apollo's 23rd revolution, the 2 craft were 1120 km apart, and closing at a rate of 222 km per revolution.

With the knowledge that even during Skylab, Apollo astronauts had sometimes had to make 6 attempts before achieving a successful docking, NASA had agreed with Soviet scientists that if necessary, attempts to achieve the first international link-up would continue for 3 orbits—$4\frac{1}{2}$

hr. Then, if it still had not been achieved, Apollo and Soyuz would pull away for safety reasons, while the crews rested. Then there would be another attempt on Jul 18. In the event none of this contingency planning was necessary.

Apollo, catching up with Soyuz at a rate of 255 km per orbit, was still over 480 km away when the first direct exchanges took place. The crew confined them to establishing 2 channels of reliable communications.

At 13.01 GMT Brand told Astronaut Richard Truly, the Capcom: 'OK Dick. We've got Soyuz in the sextant.' At first, said Brand, Soyuz was 'just a speck—hard to distinguish from the stars.' The 2 spacecraft were then 407 km apart; Soyuz in a 225×226 km orbit, Apollo in a 164×226 km orbit, still to complete its final circularization burn.

The series of circularization and phasing manoeuvres, made mainly by Apollo since the launches, culminated at 48.31 GET with an SPS burn of less than 1 sec, yet increasing Apollo's speed by 14 kmh. Instead of trailing 28 km below Soyuz, Apollo at last moved ahead. Stafford, using his small thrusters, and a technique first developed during Gemini, began a braking burn to put Soyuz in a station-keeping position 20 m–50 m from Soyuz.

Capcom Dick Truly called Apollo, 'I've got 2 messages for you. Moscow is GO for docking; Houston is GO for docking. It's up to you guys. Have fun.' The politicians, anticipating success, could not wait; we were informed that President Ford would be talking to the spacemen as soon as they had met.

By means of Apollo's TV cameras, viewers—hundreds of millions now—at this point could see the Docking Module on Apollo's nose, with its 3 guide petals extended. Ahead, Soyuz, bright green, its solar panels giving it the appearance of a hovering humming bird, was steadily growing larger. A medley of American and Soviet voices, speaking each other's languages and being translated, filled the loudspeakers.

The first exchanges between Apollo and Soyuz consisted of Stafford twice asking Leonov, in Russian, to tell him when he began his final manoeuvre; Leonov replied crisply in English 'Read fine,' and then: 'Soyuz initiating orientation manoeuvre.' A few moments later Leonov announced: 'Soyuz docking system is ready.' Stafford replied in Russian: 'We are also ready; Apollo is ready,' and found time to add that his crew could see Soyuz and 'it's very beautiful.' Stafford, concentrating hard on both his manoeuvres and Russian, said, 'I'm approaching Soyuz'—and then was blotted out by laughter when a Russian voice, probably an interpreter's, said: 'Oh, please, don't forget about your engine.' While Stafford counted down the distance, from 5, to 3 and then 1 metre, Mission Control, worrying about its world TV audience, was vainly calling Slayton to close down the 'f' stops because the picture was too bright. Stafford, still in Russian, announced 'Contact;' then Leonov in English: 'Capture.' Stafford: 'We also have capture,' and added 'We have succeeded; everything is excellent.' Leonov: 'Soyuz and Apollo are shaking hands now.' Stafford: 'We agree.'

The new docking system had worked so smoothly that soft-docking was achieved 6 min early, at 16.09 GMT. In Houston the time was 11.09; in Moscow it was 19.09. GET was 51.49. The 2 craft had linked over the Atlantic, 1000 km west of Portugal, at the start of Soyuz's 36th orbit and Apollo's 29th revolution.

At last Mission Control gained Slayton's attention and got the TV picture turned down. Stafford asked Mission Control to tell Prof Bushuyev that it was a soft-docking. 3 min later Apollo's docking mechanism was retracted. Stafford, having given the orders, in Russian: 'Close active hooks,' to a cry in English from Leonov of 'Perfect; beautiful,' reported first in Russian, then in English: 'Docking is completed.' Leonov: 'Well done Tom; it was a good show. We're looking forward now to shaking hands with you on board Soyuz.' Stafford: 'Thank you Alexei. Thank you very much to you and Valeri.'

Kubasov, reporting to Moscow, gave a vivid description of watching the docking. As Apollo 'gets closer and closer, a couple of metres, and at last there comes a slight push. The CAPTURE light goes on fast and then, in a couple of seconds the USA astronaut Vance Brand informed us that he had activated the retraction mode. The spacecraft retraction began which went on for several seconds ... Our INTERFACE light went on, and in just a couple of seconds the Apollo active hooks were activated and the rigid retraction of the spacecraft occurred with a force of about 20 tonnes. Thus the seal was compressed, which was indicated by the SEAL COMPRESSED light. Now everything is fine.'

Elation turned to alarm when Slayton and Stafford began opening No 2 Hatch between Apollo and the DM. 'Deke smells something pretty bad up in the docking module,' Stafford reported. 'We're going to put on oxygen masks right now; and we're going to close that hatch.' Stafford said it smelt like burnt glue—something like acetate.' After a quick check of their consoles, Houston said they had no indications to suggest an emergency. The crew got their oxygen masks ready, but in the end did not wear them, nor close the hatch. The smell gradually cleared, and it was decided that, the DM having been closed for $4\frac{1}{2}$ hr, it could have been connected with an earlier experiment in the furnace and the subsequent stowing of the heated samples next to some Velcro.

While this was going on, followed by Stafford and Slayton sealing themselves in the DM, and changing the atmosphere to a 10 psi mix of oxygen and nitrogen to match that of Soyuz, Leonov and Kubasov were opening up their hatch into the orbital module, and removing, drying and stowing their spacesuits. The transcript of exchanges between Kalinin and Soyuz shows they too were having their troubles; the crew hunted in vain for a TV extension cable, and were finally told to stop looking for it, since it was not on board. At the same time Soviet Mission Control was persisting with instructions about procedures when the 2 crews met, since a message from Mr Brezhnev, Soviet Communist Party Secretary, was being inserted into the schedule. Some tension occasionally built up between the cosmonauts and Soviet Mission Control, and an example of this was when Moscow, after 6 calls before getting a reply (sometimes there were 10 calls), wanted to know what stage the crew had reached. 'We are at Panel 18, final pressurization of our tunnel,' Leonov replied. 'Valeri is carrying it out. We are preparing to receive our guests, but it is necessary either to work or to complain about what you can't do.' Mission Control: 'We prefer the first alternative. Work better.'

The long-anticipated space handshake was a slightly confused affair; it came nearly 3 hr after the docking, and required much strenuous work by both crews. In Apollo things were made much more difficult because of high squeal on the ATS-6 circuit, and problems over establishing direct communications with Soyuz. Stafford and Slayton remained good-humoured despite enormous difficulties with TV cables and power umbilicals, and calls from Houston to move either themselves or the cameras to ensure that the world had a good view when Hatch 3 finally

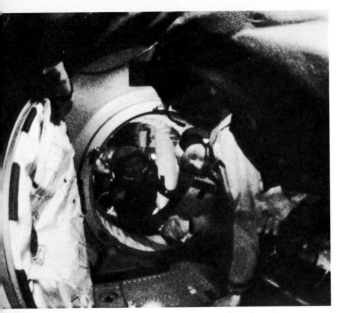

Stafford (foreground) and Leonov shake hands in space

opened. When it did, Leonov could be seen, slightly uncertain. 'Come in here and shake hands,' said Stafford in English. Then, peering into Soyuz and seeing a similar tangle of cables there, he commented to Slayton: 'Looks like they got a few snakes in there, too; they're almost as bad off as we are.' Remembering that he must speak Russian, Stafford then said: 'Alexei, our viewers are here,' and finally got Leonov to put his head and shoulders through the hatch into the Docking Module. At last came the handshake, with both commanders saying: 'Glad to see you.' Then Leonov greeted Slayton; the tension broke at last into laughter. 'Very, very happy to see you,' said Leonov. 'This is Soyuz and the United States,' said Slayton. (The docked spacecraft were over Amsterdam at the time—not over Bognor Regis, England. Some observers had been so confident that would be the location that Moscow Radio's London correspondent was sent there to describe the town's excitement.) Then Stafford went into Soyuz. More confused movement, during which Kubasov became visible, as Stafford entered Soyuz's orbital module. All 4 carried documents outlining their procedures, and, in the case of the cosmonauts, the text of what they were to say. Congratulations from Mr Brezhnev to all 5 spacemen were read by a Soviet TV newscaster after Leonov had complied with an urgent instruction from Moscow to 'get Tom out of the camera's way'. Then, with Slayton also in the Orbital Module, and all 4 squeezed around the table, Houston took over. Consternation, and 'Oh Jesus, I'm sorry' from Stafford, because the picture was upside down, until he reversed the camera. Slayton was asked to give his headset to Leonov so that the Soviet Commander could hear, and President Ford came on the line. He talked personally to all 5 men in turn (Brand, of course, being sealed alone in Apollo), and asked Slayton, whether as

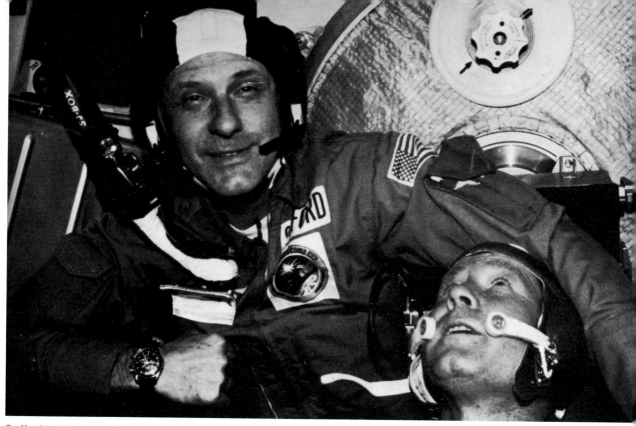
Stafford and Leonov in Soyuz Orbital Module

the world's 'oldest space rookie' he had any advice for young people who hoped to fly on future space missions. 'Decide what you really want to do, and never give up till you've done it,' replied Slayton.

With only 10 min of TV time left via ATS-6, Stafford announced that the planned ceremony of flag exchanges would be done 'real fast'. Each spacecraft carried 10 national flags to hand over; in addition a UN flag launched in Soyuz was passed to the Apollo crew to take home. Commemorative medals and plaques were exchanged and joint flight certificates signed. The Apollo crew passed over a packet of fast-growing pine tree seeds—enough for a $\frac{1}{2}$-hectare plantation.

The 4 astronauts then settled down to dinner—green borscht soup and turkey for Stafford and Slayton, lamb soup and chicken for Leonov and Kubasov. As they finished Moscow told them they had all been made honorary citizens of Samarkand. Slayton then moved back into the DM to start the 5 joint scientific experiments by placing samples of aluminium and germanium in the Multi-purpose Furnace. These samples were taken up by Soyuz and returned to the Russians after 20 hr in the furnace, being heated up to 1050°C and being melted and then allowed to resolidify— the object being to discover whether the processing of materials in weightlessness gives better results than on Earth. The furnace was also intended for 6 similar US experiments related to metal-forming, mixing of materials and crystal-growing.

After nearly 3 hr together, Stafford too returned to the DM, hatches 3 and 4 were closed, and the crews prepared for a well earned rest. But the Soyuz crew was clearly much concerned to discover a pressure leak into the tunnel between hatches 3 and 4; they re-opened and closed their

Kubasov signing joint ASTP rendezvous certificate

hatch, but still recorded a pressure drop of 1 mm of mercury per minute. In Apollo, where such a pressure change was too low to register, there was less concern, though Moscow and Houston had some anxious discussions. Finally it was agreed that there was probably no leak at all, and that the pressure drop was caused by changing temperatures in the docking tunnel. Both crews settled down to sleep about 2 hr late, but agreed to get up on time, even though that meant only about $5\frac{1}{2}$ hr rest.

The 3rd day in space began in a more relaxed fashion in both craft. After breakfast, Leonov and Kubasov both described, in TV reports to the Soviet public, the previous day's events. Then, accompanied by much noise and squeal on the intercom, and somewhat confused attempts between the spacecraft and Mission Controls to locate the source of interference, Vance Brand transferred to Soyuz, and Leonov to Apollo. At Houston it was possible to watch TV transmission coming back from both ends of the docked spacecraft at the same time. Tom Stafford took Soviet viewers on a tour of Apollo, not forgetting to point out to them that America had used this spacecraft to go to the Moon and also for their record visits to Skylab. Brand and Kubasov obviously had much in common as the 2 working engineers; as Kubasov gave US viewers a run-down in English, Brand told Soviet viewers that Soyuz was a good spacecraft and very comfortable. Then, after more complicated transfers, the 2 commanders sat together in Soyuz, with Slayton, Brand and Kubasov in Apollo, to hold a TV news conference. They answered 2 lists of questions, one prepared by newsmen covering the flight in Moscow, the other prepared by newsmen in Houston. Kubasov, who did the first space welding on Soyuz 6, and had taken part in the joint metallurgy experiments with the Multipurpose Furnace,

forecast that space factories would provide man with many new alloys and metals which it would not be possible to produce on Earth. Leonov, the artist, made the most memorable contribution, displaying what he called a 'cosmic portrait gallery'—sketches of Stafford (both with and without hair), Slayton and Brand. So far as exterior views were concerned, Leonov said he would like to fly longer and higher (than the low-Earth orbit used for the link-up) so that his artist's eye could have a better look at Earth.

More TV tours—by Brand for Soviet viewers of Cape Canaveral and other parts of America; by Kubasov of

Leonov in Soyuz Orbital Module showing drawing of Stafford

Stalingrad and the Volga (though Kubasov was 3200 km behind with his travelogue)— and then came the 4th and final transfer. Kubasov and Slayton moved into Soyuz. The 5 spacemen, battling good-humouredly against a flow of instructions from Moscow and Houston, a continuous tangle of TV cables, and with interruptions from Air Traffic Controllers over Canada, Russia and elsewhere, somehow got through an overcrowded schedule, which included joining up the halves of metals sent up from the 2 countries. At last, at 80 hr 29 min GET, came the last handshakes and goodbyes, and the crews returned to their own spacecraft. Stafford had spent a total of 7 hr 10 min in Soyuz; Brand 6 hr 30 min, and Slayton 1 hr 35 min. Houston's hopes of giving the Apollo crew their first good night's rest were dashed when they had to wake them up 2 hr after they had settled down to close a waste-management valve left open; 1 hr later they were woken again by a break-through of ATC transmissions from Atlanta Airport. On board Soyuz too, TV problems continued, with warnings from Moscow that some of the planned transmissions would have to be cut. The Apollo crew got less than 5 hr unbroken sleep before starting preparations for 6 hr of delicate space manoeuvres for which Slayton had prime responsibility. With the Soyuz crew in pressure suits, which were not considered necessary in Apollo, at 95.45 GET—right on the original timeline— Slayton released the capture latches and allowed Soyuz and Apollo to drift apart; then, with the RCS thrusters, he moved rearwards to about 50 m and kept station while Leonov placed the Soyuz docking mechanism in the active mode. The Soyuz crew carried out the Solar Eclipse Experiments, taking photographs of the solar corona as Slayton placed Apollo between the Sun and Soyuz, thus 'occulting' the Sun. Slayton reported anxiously to Houston that Soyuz appeared to be out of the proper docking attitude in pitch and yaw, and that the Sun, very bright behind Apollo, was reflecting on the Soyuz docking target, blurring his vision. Stafford, in Russian, told Soyuz, 'Am approaching,' Leonov replied in English, 'Keep coming.' Only a slight bump was noticeable on TV as the 2 craft redocked 32 min after undocking, and Houston commented: 'It was a beautiful docking.' Later, however, it was learned that Slayton contacted the Soyuz petals off-centre, and the docked vehicles yawed and slewed as he drove in. He had used 145 kg of RCS propellant, compared with the estimated 54 kg, for the undock/redock manoeuvres. Later, Moscow appeared somewhat critical of the docking, and as always worried about Soyuz pressurization, asked the crew to check whether any damage had been done. At the time, Leonov seemed happy about the docking, but was angry about interference during the crucial docking and undocking operations from a Soviet airport (believed in the West to be a new, long-range bomber base, using powerful transmitters on the same wavelength as Soyuz). During the 3-hr period of the 2nd Apollo/Soyuz docking, the crews had lunch and made final preparations for the ultra-violet absorption experiment. As the 2 craft separated the plan was for Apollo to bounce light beams off a Soyuz retroflector; the beams returned to an optical absorption spectrometer on Apollo which measured the atomic oxygen and atomic nitrogen particles in Earth's upper atmosphere. Soyuz had to be at right-angles to Apollo, and Slayton had to manoeuvre Apollo so that its thrusters did not contaminate the atmosphere between the 2 craft. The complexity of it all was reflected by everybody shouting at once as Slayton moved gently away. Astronauts, cosmonauts, mission controls and interpreters provided a baffling, almost frightening medley of voices; Leonov came through saying, 'Tom be careful,' as Stafford called out the distances—a mere 20 m at first. While Slayton was having piloting problems, Stafford called Soyuz 8 times asking them to turn off their orientation and beacon lights, which were interfering with the experiment, before he could gain their attention. 'You were calling our spacecraft?' Leonov replied. 'Yes, of course,' said Stafford. At 500m, however, communications improved, with Leonov turning the lights on and off promptly in response to Stafford, and the experiment went better. Slayton moved in again to only 50 m for still photos of Soyuz; then carried out final separation with a 5·8 sec, 2·3 fps RCS burn. This placed Apollo 1000 m above Soyuz, so that, with a lower orbit, Soyuz drifted out ahead of Apollo at an increasing rate. This occurred at 102.16 GET, on Apollo's 61st revolution. As the 2 craft moved away, Leonov called to Brand: 'Thank you very much for your big job.' A few minutes later Vance, describing the scene, said: 'He's pulling ahead of us and you can see the dark Earth in the background; it's just as if an airliner's going underneath us.' During the manoeuvres, Slayton had used a record total of 290 kg of RCS propellant; but since the CSM had been launched with a total of 1290 kg, there were still comfortable reserves left for Apollo's remaining 189 hr of solo flight. Soyuz continued for another 36 hr.

Soyuz post docking As they moved steadily away from Apollo Leonov and Kubasov spent 4 hr photographing the US-Soviet fungi specimens and tidying up the spacecraft, then settled down to sleep early, to adjust their schedule to Moscow time. Next day's work included photography of both Earth and Sun, and preparations for a test-firing of Soyuz's main engine prior to re-entry. Once again Moscow became concerned about a small pressure drop in Soyuz,

which it was feared might have been the result of the second heavy docking. This was referred to at a Moscow news conference as 'a hitch', and 'heavy'. The Kalinin flight control shift stayed on 2 hr late until it was established that the pressure drop had been caused, not by a leak, but by a temperature change. The re-entry rehearsal also included an elaborate pressure check. The crew donned spacesuits once more, sealed themselves in the descent module, then lowered the pressure in the orbital module to see if any leaks developed between the 2 vehicles. At 118 hr 19 min the main Soyuz engine was test-fired for 4 sec, changing the orbit from a circular 225 to 222 × 217 km. Moscow said this was standard procedure on all Soyuz missions to check the engine condition before actual retrofire. Despite protestations from the cosmonauts that they felt perfectly well, Moscow thought they were tired, and urged both to take a mild sleeping pill. They were also concerned that Leonov's heartbeat had become slightly irregular, and asked him to take a medicine increasing calcium in the blood. Leonov was asked to sit quietly for a few moments while biomedical data was transmitted; he did so, protesting: 'I don't feel any fatigue; I don't have a headache. Everything is normal.' 4 hr later he was to be ordered to take the medicine. After a final sleep period, the cosmonauts began preparations for descent, donning spacesuits for the 5th time since launch and closing the hatch to the orbital module. At 140.40 GET Leonov reported that the spacecraft's orientation was within 2–3° of the desired attitude (which NASA engineers noted was not nearly so precise as an Apollo re-entry). With Soyuz about 12,900 km from final touchdown point, on its 96th revolution, the fully automatic re-entry procedure began with a 105 sec burn, Leonov reporting they could both hear and see the engine firing; 10 min later the orbital and service

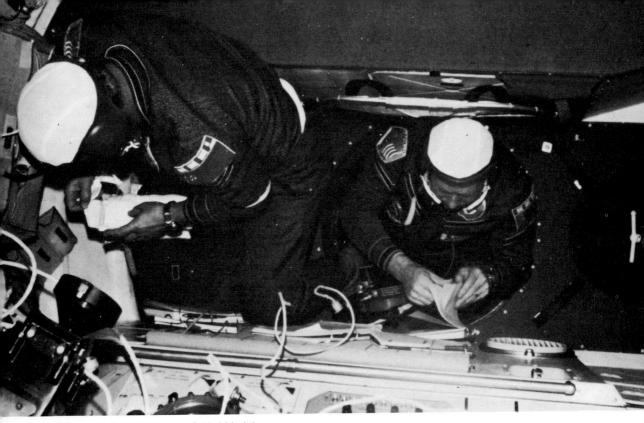

Kubasov (*left*) and Leonov in Orbital Module

modules were jettisoned by explosive charges from each side of the descent module; at 18 min entry into the atmosphere began; the cosmonauts reported that they had expected the G-loading to be greater than it turned out to be. Radio blackout lasted 7 min; the main drogue, followed by the single, red-and-white striped main parachute, was deployed at about 9750 m. A Soviet Air Force Mi-8 helicopter sighted the descending vehicle shortly after. At 3050 m the graphite-fibre heatshield was jettisoned, to expose the final braking rockets. And more TV history was made with live coverage throughout the world during the final 8 min of the descent. This had been improved by a Soviet decision a few hours earlier to move the target point 160 km E, apparently to avoid cloud cover. 2 helicopters with TV cameras—there were 9 helicopters altogether—flew around, and even provided shots from above. American viewers gasped, and thought the spacecraft had crashed when a huge dustcloud enveloped Soyuz 19, obscuring the final Earth contact; it was caused, of course, by the retrorockets, fired at 1·9 m to slow the descent from 7·9 to 1 m per sec at actual touchdown. This was at 142.32 GET, 87 km NE of Arkalyk in Kazakhstan, and 9·6 km from the target point. Trucks arrived from all directions, and a large crowd assembled, as Leonov and Kubasov were seen to climb out, waving aside offers of assistance, and then sign their names on the outside with chalk provided by cameramen. Leonov added the word *Spasibo* (Thanks). They arrived by IL-18 back at Baikonur Cosmodrome within 3 hr, where Leonov told a news conference: 'It was not an easy job but very gratifying.' Kubasov said the artificial solar eclipse was 'strikingly beautiful', although 'flare-ups' of Apollo's engines somewhat distorted the picture. When astronauts and cosmonauts were reunited in Moscow at the end of Sep,

Leonov had been promoted General, and Stafford Major-General. In Oct Leonov and Kubasov made a triumphant return visit to the US, and duly delivered to the United Nations headquarters in New York the UN flag which had been flown into space aboard Apollo, handed over to Soyuz in orbit, and landed in Russia.

Apollo post docking   When joint activities finally ended, at about 103.30 GET, the Apollo crew, just half-way through their 10-day flight, settled down to a packed programme involving 16–17 hr per day on scientific experiments. They did succeed however, in sleeping better, averaging 8 hr per night; and twice had to be woken by Mission Control. While the Soyuz crew settled down to rest as the craft drifted apart at 7·2 km per revolution, the Apollo crew had another 6 busy hr of work. Slayton had difficulty lining up the telescope for the extreme ultraviolet survey (searching for radiations from red giants, pulsating white dwarfs, etc) and the others were observing and photographing the Australian desert, looking for coral sea eddies, the poisonous 'red tide' off Florida, and operating the furnace. Stafford complained that their timeline had been 'success oriented', with too much work crammed in on the assumption that everything would go well. Mostly, however, Stafford, Brand and Slayton worked enthusiastically during the remainder of the flight, covering more than 60 sites in their study of the Earth's resources. These included iceberg areas; as natural floating reservoirs of fresh water, it might be possible to insulate them and tow them to areas with a water shortage. Wave patterns around Hawaii, the optical effects of sunrise, oil slicks and water pollution were all included. The biggest scientific achievement—'badly needed', according to those who thought the ASTP money would have

Apollo view of Mediterranean Coast N of Beirut

been better spent on another Skylab mission—came on the last day. On the last of 30 extreme ultraviolet targets, after much argument among scientists as to whether it had been missed or not, the crew fitted the sighting in on the last possible pass, and in the words of the Principal Investigator, the signals 'blew us off the console panel'. The Apollo crew had confirmed earlier SAS-C observations, and opened up the last closed area of the spectrum for observation. Some astronomers in the past had even doubted the existence of extreme ultraviolet sources. Just as they had slept through the launch of the Soyuz crew, so Stafford, Slayton and Brand slept through the safe landing of Soyuz on Apollo's 7th day in space. On their 8th day they obviously enjoyed the first genuine TV news conference ever to take place. Newsmen, sitting in the auditorium at the NASA News Centre, watching the crew on a huge screen, and being televised themselves at the same time, were able to put their own questions direct to the crew. They admitted to some trying moments, with the spacecraft jammed with scientific equipment, trying to make it look like a TV studio as well. 'We were always bumping into each other,' said Brand; 'we needed a traffic cop.' And Stafford added that 16-hr working days in that confined space had been 'really a bear'. Slayton, asked when a man would be too old to fly, replied that his 91-year-old aunt would manage very well. It became clear that Stafford, Slayton and Brand were the healthiest and best-tempered crew ever sent into space. Another 8th-day event which filled Mission Control was a complicated experiment connected with jettisoning the Docking Module—which, since it had been bedroom, exercise room and storeroom, the crew hated to lose. The aim of the experiment was to check recent theories that the Earth's upper mantle consists of moving 'plates', which cause continental drift, by measuring mass anomalies (similar to the Moon's mascons, or mass concentrations) in the Earth's crust. These cause perturbations of a spacecraft's orbit. On a single spacecraft these perturbations show only large mass anomalies; by doppler tracking of 2 spacecraft travelling close together in similar orbits, it was hoped their relative motions would show gravitational anomalies in the 100 to 1000 km range, caused by smaller mass concentrations. Many of these are along the Pacific rim. First the crew packed the DM with waste material and the docking probe, then closed the hatch and donned spacesuits for the first time since lift-off—a precaution against possible loss of pressurization when the explosive bolts were fired to separate the CSM and DM. Then the whole vehicle was rotated 360° to spin the DM for doppler tracking. Jettison was performed at 199.23 GET. Then came 2 manoeuvres by Apollo—DM1 and DM2, at 200 and 204.12 GET. These short SPS burns placed the 2 vehicles 300 km apart in an equidistance orbit of 209 × 222 km. After these highlights had provided the scientists with the data they needed, the crew reverted to the more earthy task of taking one another's leg measurements. After a final night's rest, preparations began for re-entry and splashdown on their 9th and last day—although the crew continued Earth resources observations of ocean currents off Japan, in the Gulf of Mexico, Red Tide of New England, and oil slicks in the N Atlantic, until $2\frac{1}{2}$ hr before the retrograde burn. They also found time, as they passed over London, to send greetings to 'the BBC and all our good friends in England', and to carry out the first 'lost property' search in space. Moscow asked if 4 film cassettes had been left behind by the Soyuz crew in Apollo. The crew found them, and promised to include them with US film in the 'quick release bags', which are the first things to be handed out when the hatch is

opened. The burn came at the start of rev 138 with a 7-sec retrograde firing of the SPS engine which slowed the spacecraft by 190 fps, or 130 mph. The SM was jettisoned 7 min later. Then, during final re-entry, and unknown at the time to Mission Control, the Recovery Forces and the world radio and TV audience, 9 days of both political and technical success turned into near disaster. When Stafford was reading off the checklist during the descent there was an unusually high noise level in the spacecraft. For some reason—and at a news conference he publicly accepted responsibility—Brand failed to operate the 2 ELS (Earth Landing System) switches which at 9000 m start the automatic sequence to bring out the parachutes. At 7000 m Stafford asked Slayton if he could see the drogues; 3 sec later, when he could not, Slayton told Brand to hit the drogue deploy button, to bring them out manually. The ELS system would have turned off the RCS (reaction control system) thrusters automatically as the drogues came out; but since the ELS had not been activated, the thrusters began firing, automatically trying to correct the swinging of the spacecraft on the drogues. One of the thrusters, the plus-roll motor, fired nearly continuously for 6 sec. Stafford, noting this, cut off the propellant flow—and inadvertently made the problem even more serious. At 7000 m a pressure-relief valve, only 90 cm from the plus-roll motor began to open, to allow outside air to enter the spacecraft and equalize the inside pressure of 5 psi pure oxygen, with the normal outside atmosphere. Although the RCS system was now cut off, the nitrogen tetroxide oxidizer (one of the most poisonous gases known to man) continued to boil out, and was drawn into the spacecraft through the now-open pressure-relief valve. The sound of all 3 crewmen coughing violently could be heard on the on-board flight recorder, which was later recovered and made public. Slayton put the cabin oxygen flow rate into high, which helped to disperse the gas. But things were not improved when Apollo, in Stafford's words, 'hit the water like a ton of bricks', and turned upside down. Splashdown occurred at 9.18 GMT, 7.16 km from the US recovery ship New Orleans. Stafford, anxious to get out the oxygen masks, undid his straps, and immediately fell into the docking tunnel. He crawled back out of it, and succeeded in getting the masks out of a locker: by then, Brand had lapsed into unconsciousness. He came round in about 40 sec, after Stafford and Slayton had got his mask on as well as their own. When the spacecraft, with the crew still inside, was finally winched aboard New Orleans, 45 min later, all 3 stepped out to a brass band welcome, and long telephone conversations with President Ford. Mission Control, and the US Navy recovery team, had still not noticed, behind the high background noise and their own activities, that the astronauts had been in trouble, and that Stafford had at one point said: 'Get this . . . hatch open!' They were noticeably rubbing their eyes when talking to the President, but conducted an animated exchange with him for about 10 min before reporting the gas inhalation to the doctors at the post-recovery medical. It was unfortunate that Dr James Fletcher, head of NASA, opened the post-splashdown news conference while this was going on with the words: 'Needless to say, we have just witnessed another flawless Apollo splashdown.' The traditional splashdown parties at Houston were in full progress, when officials and newsmen were recalled from them with the news that the astronauts were ill. The slow-acting gas had turned their lungs white, and, in Slayton's case, blistered them. Doctors insisted that they remain in Honolulu, mostly at an Army Hospital (and at first in the intensive care unit there), for a

fortnight. Their wives and some of the children were flown out to them. Finally they gave their delayed news conference in Washington, on Aug 9, when Brand accepted responsibility, and Stafford intervened to say the re-entry sequence was the responsibility of the whole crew. Doctors were reassuring about the possibility of long-term ill-effects. But worse was to come for Deke Slayton, whose appointment as Chief of Space Shuttle Test Flights had been announced during the mission, with an indication that he would be able to do some of the flight-testing himself. X-rays showed a small shadow on his left lung. Doctors later said that examination of pre-flight X-rays showed it was already there, 'but hardly discernible' before the mission—though without explanation of whether it had been noticed then. An operation established that it was a benign lesion, and that there was no cancer; and Slayton was able to accompany Stafford and Brand on a visit to Russia, for a reunion with Leonov and Kubasov, in early Oct.

BIOSATELLITE

Cosmos 782   L Nov 25 1975 by A2 from Plesetsk. Wt 4000 kg. Orbit 227 × 405 km with 62·8° incl. A direct result of ASTP collaboration, 7 nations, including Russia and America, took part in this international biological experiment. It carried an 'orbital centrifuge' so that the effects of creating artificial gravity in orbit (an important step forward for long-term manned flight) could be tried out on rats and other forms of life. NASA scientists, whose work in this field had been hampered ever since Bios 3 (see page 139) caused so much public resentment that the programme had to be cancelled, provided 2 44kg containers. Animals, plants and cells (including white rats provided by Czechoslovakia) were placed both in the centrifuge and outside it, so that comparisons could be made of the effects on their living conditions of providing artificial gravity during 22 days in space. Biologists from both the US and Soviet Union studied the ageing effects of the flight on the liver of flies. Plant tissues with cancerous growth grafted on them were provided by Moscow and Colorado Universities to study indications that cancerous growths slow down if the force of gravity is increased. France also provided biological experiments, and specialists from Hungary, Poland and Romania took part. Cosmos 782 was successfully recovered on Dec 15, after $19\frac{1}{2}$ days. The flight was cut short because of snowstorm conditions in the Siberian recovery area.

INMARSAT Maritime Satellites

History   The need for a global service of satellites serving the world's maritime ships, on an international basis somewhat similar to INTELSAT's service between countries, led to agreement to establish the International Maritime Satellite Organization (INMARSAT) in 1979. It will probably be based in London, because of its links with IMCO (the Inter-Governmental Maritime Consultative Organization). Conferences in 1975 and 1976, attended by 45 countries, and representatives from international bodies like the UN and UNESCO, agreed on broad principles. Many countries, however, were concerned to ensure that neither of the super powers had such a large stake that they could dominate the organization. Currently the larger stakes are

USA 17%, USSR 12%, Britain 11%, Norway 9·5% and Japan 8·4%. By mid 1977 America was the only country with operational maritime satellites (see Marisat, under Comsat General). It is likely to be 1984 before Inmarsat can provide its own service. The European Space Agency is well advanced with its Marots satellite, which is purpose-built for INMARSAT; no doubt Comsat General will also seek to provide and manage the overall service on INTELSAT lines; but Russia has recently developed her first geostationary satellites, and will no doubt also be a competitor for providing the hardware.

INTELSAT Telecommunications System

History   The International Telecommunications Satellite Consortium by 1976 was providing a global commercial communications system for 107 countries and territories. The 23 satellites launched by Intelsat over a period of 12 yr had brought clear and reliable TV, telephone and other communications to countries and areas which had little hope of enjoying good reception before the age of the satellite. An example of the benefits was that in Britain the cost of a transatlantic telephone call had dropped from £3 to 75p per min, with the added advantage of being able to dial it oneself. Intelsat itself is the prime example of the way in which a new technology both creates and satisfies new demands. It now carries more than two-thirds of all long-distance international communications. It was established in Aug 1964 by 14 countries on the basis that each country invested in the satellite system in proportion to its expected use, and shared accordingly in the revenues. Since Intelsat 1 (the famous Early Bird) was launched the following Apr, 4 successive generations of more advanced satellites have been deployed, with the first launch of a fifth series due in late 1979.

Early Bird provided 240 high-quality voice circuits, and made transatlantic TV possible for the first time when placed in synchronous orbit over the Atlantic. Only 7 yr later, the 4th generation Intelsats (4s were in position by the end of 1972) were each able to carry an average of 5000 telephone calls, plus TV. From 5 TV transmissions a month in 1965 via Early Bird, the average was 100 a month by 1970. In 1972, with 225 satellite pathways available, there were 6790 transmission hours of TV, and 3725 satellite circuits were being leased on a full-time basis. During the 2 weeks of the 1972 Olympic Games at Munich, more than 1000 hr of TV were transmitted to 25 countries. The system played a major part in all 10 manned Apollo flights, both by providing support communications and in making world-wide TV coverage possible. The manned flights created an enormous demand for live TV coverage; and as quality improved and colour was added, there was a rapid growth in demand for satellite transmissions of news and sports events, both 'live' and in the form of edited and recorded packages, sent shortly after the event. Another landmark was in Nov 1975, when the Intelsat 4 system was used for the first time to relay facsimile pages of the Wall Street Journal from the paper's Massachusetts headquarters to Florida. They were used as the basis of the 50,000-copy south-eastern edition. The parent company, Dow Jones, said that satellite transmission was 70% cheaper than conventional telephone landlines and microwave links, as well as being much quicker.

The speed of the technical advance, spurred on by the commercial demand, is best illustrated by the fact that just over 10 yr after Early Bird had provided the first 240 circuits, Intelsat 4A was able to provide 6250 2-way voice circuits, plus colour TV channels. The fact that communications satellites were a commercial proposition had been established as early as 1962. The setting up of Intelsat followed in 1964, with Comsat (the Communications Satellite Corporation) representing the US interest with 63% ownership, and managing the system on behalf of the member nations. As Intelsat grew, its members became concerned that the US might develop an international monopoly of such systems; and as ownership was shared with new members, Comsat's ownership dwindled to 30%.

In Feb 1973 it was agreed that Intelsat should have an independent Secretary General and a Board of 25 Governors, with Comsat's control being reduced to technical and operational management. Any country belonging to the International Telecommunications Union can join Intelsat, and only 4 are not ITU members—China, N Korea and the German Democratic Republic. The Soviet Union sends representatives to Intelsat meetings, but has not so far joined. As this book goes to press, Russia's recently announced plans for a rival Statsionar (qv) system are causing much concern to Intelsat, because of the possibility that the 2 systems will overlap and cause frequency interference. With the 4-gc/6-gc bands allocated for satellite communications becoming overcrowded, preparations were started to shift the Intelsat 5 series to the 11-gc/14-gc bands. A request from the United Nations that Intelsat should provide free satellite channels for UN communications and international emergencies is being considered.

More than 70 Earth stations, usually owned by the 60 countries in which they are located, were in operation by

1976, transmitting and receiving the flow of telephone calls, TV and other signals being passed through the satellites, and the number of stations was increasing all the time. Satellites are launched by NASA from Cape Canaveral, with Intelsat paying NASA for both the rocket and launch service. For the first 3 series Delta rockets were used; for subsequent launches, Atlas-Centaur rockets.

Intelsat 1 (Early Bird) Dia 70 cm. Ht 60 cm. Wt 68 kg at launch, 38·5 kg after apogee motor fire. Capacity 240 circuits or 1 TV channel. Early Bird, the only one in this series, was launched Apr 6 1965; became operational over Atlantic at 325°E Jun 28 1965. The world's first commercial communications satellite, it made transocean TV possible for the first time. Its antenna was focused N of the Equator to service N America and Europe. Design life was 18 months, but operated satisfactorily for $3\frac{1}{2}$ yr before being placed in orbital reserve.

Intelsat 2 The series extended satellite coverage to two-thirds of the world, though only the last 2 achieved the planned 3-yr life. Wt 162 kg at launch, 86 kg after apogee motor fire. F-1 L Oct 26 1966, failed; F-2 L Jan 11 1967, operational for 2 yr over Pacific. F-3 L Mar 22 1967, operational $3\frac{1}{2}$ yr over Atlantic. F-4 L Sep 27 1967, operational $3\frac{1}{2}$ yr over Pacific.

Intelsat 3 Capacity increased to 1200 circuits or 4 TV

Above   Intelsat 1

Below   Intelsat 2

87

channels, with 5 yr design life (which was not achieved). Wt 293 kg, 151 kg after apogee motor fire. F-1 L Sep 18 1969, failed. F-2 L Dec 18 1968, operational $1\frac{1}{2}$ yr over Atlantic. F-3 L Feb 5 1969, operational for unstated period over Indian Ocean. F-4 L May 21 1969, operational 3 yr over Pacific. F-5 L Jul 25 1969, failed. F-6 L Jan 14 1970, operational 2 yr over Atlantic. F-7 L Apr 22 1970, operational $1\frac{3}{4}$ yr over Atlantic, F-8, failed.

Intelsat 4 Capacity increased to 5000 circuits or 12 TV channels, with 7-yr design life. Wt 1414 kg, 730 kg after apogee motor fire. F-2 L Jan 25 1971, operational 2 yr over Atlantic; then in reserve. F-3 L Dec 19 1971, primary satellite over Atlantic for 3 yr, then in reserve. F-4 L Jan 22 1972, operational 2 yr over Pacific, then in reserve. (Assembled by British Aircraft Corporation at Bristol, F-4 was first commercial satellite assembled outside US, and provided TV link between US and China for President Nixon's visit to Peking.) F-5 L Jun 13 1972, operational 3 yr over Indian Ocean, then in reserve. F-7 L Aug 23 1973, major path satellite for 2 yr over Atlantic, then in reserve. F-8 L Nov 21 1974, operational at 174°E over Pacific from Dec 1974. F-6 L Feb 20 1975, failed to achieve transfer orbit. F-1 L May 22 1975, after being held in storage for 4 yr, operational over Indian Ocean at 63° from Jul 1975.

Intelsat 4A Capacity increased to 6250 circuits plus 2 TV channels, with 7-yr life. Wt 1515 kg, 825 kg after apogee motor fire. F-1 L Sep 25 1975. Primary path satellite over Atlantic at 24°W from Feb 1976. F-2 L Jan 29 1976; spare for Primary over Atlantic at 29°W from Apr 1976. Total of 6 launches planned, last 3 in 1977.

Above   Intelsat 3

Below   Intelsats 1, 2, 3, 4, 4a: comparison of size and antennas

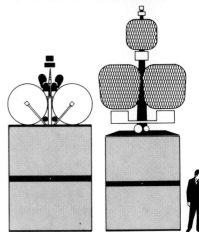

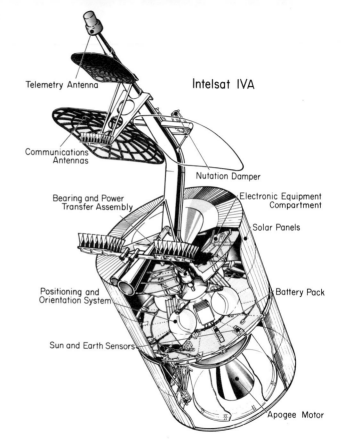

Telemetry Antenna

Intelsat IVA

Communications
Antennas

Nutation Damper

Bearing and Power
Transfer Assembly

Electronic Equipment
Compartment

Solar Panels

Positioning and
Orientation System

Battery Pack

Sun and Earth Sensors

Apogee Motor

Intelsat 5 7 satellites, cube-shaped instead of spinning cylinders, will double the Intelsat 4A capacity. Costing $470M, with first 2 launches in 1979. There will be competition between NASA and ESA to provide launchers for at least the first 4; last 3 are due to be orbited by the Shuttle.

Summary There are usually 4 satellites in operation, with 2 over the heavily used Atlantic, and 1 each over the Pacific and Indian Ocean, with spares available in orbit. Interruptions to service do occur occasionally: in 1975 static discharges caused by solar storms affected the directional control systems of the antenna twice on a Pacific Intelsat 4, and once on an Atlantic Intelsat 4 for 'relatively short' periods. There was also a temporary transfer from the Pacific Intelsat 4 to a standby satellite traced to 'a satellite design deficiency in a moon discriminator in an earth sensor'. This was corrected by modified operating procedures. Overall cost of Intelsat 4 was $235M, with each satellite averaging $13·5M plus launch costs. Overall cost of Intelsat 4A was $280M for 6 satellites. Each mission cost $23·5M for the spacecraft and $23M for the Atlas-Centaur launch. By 1975, Comsat's annual revenue was over $46M.

Synchronous Satellites It was in 1945 that Arthur C. Clarke, the space writer, suggested that a satellite placed 35,680 km from Earth, would orbit at the precise speed required to appear to be hanging in space above one point on the planet's surface. Such a 'synchronous', or 'stationary' satellite would always be in position to relay radio, TV and telephone signals, and no elaborate tracking devices would be needed. Compared with conventional underwater cables, satellite communications offered flexibility and limitless capacity. The first synchronous satellite, **Syncom 1**, was successfully placed in orbit in Feb 1963, but its radio equipment failed to work. **Syncom 2**, launched in Jul 1963, was then placed in a 35,567km orbit, and during the following 3 weeks manoeuvred into position over Brazil by the firing of jets from small, onboard hydrogen peroxide rockets. On Sep 13, **Syncom 2** and **Relay 1** were used to link Rio de Janeiro, New Jersey and Lagos, Nigeria, in a 3-continent conversation. But Syncom 2, not quite in the plane of the equator, appeared to describe a figure 8 as the Earth turned beneath it. **Syncom 3**, weighing 38·5 kg, was placed in true equatorial orbit, with no north-south swing, on Aug 19 1964: perigee 35,670 km; apogee 35,700 km; period 1436·2 min; incl 0·1°. Drifting over the Pacific Ocean near the International Dateline on Oct 10, it telecast the opening-day ceremonies of the Olympic Games in Tokyo.

Launch Technique This can be illustrated by a description of the launch procedure for the Extended Capacity Intelsat 4A. The 1506kg satellite is mounted on the Atlas-Centaur launch vehicle, protected by a 3m fibreglass nose fairing. Once clear of Earth's atmosphere, this is jettisoned during the first burn of the Centaur upper stage. Between them, the Atlas booster and the Centaur place the satellite in a 185 × 644km parking orbit. After a 15-min coast period, Centaur is again ignited. This provides the satellite with sufficient velocity so that, on separation from Centaur, it will coast to an altitude of 35,680 km above Earth's surface. Final synchronous equatorial orbit is achieved by firing the solid-fuelled apogee motor, which provides 34 sec of thrust at 5670 kg. The altitude and speed of the spacecraft then matches the rotational speed of Earth, so that it remains over the same point. Small gas jets are then used so that the spacecraft drifts in orbit and is finally positioned accurately at any desired point on the equator. This process can take up to 2 months, and the spacecraft is not brought into operational use until this has been achieved. The jets are also used for 2 other vital manoeuvres. One is to orient the spacecraft so that it is correctly lined up with the Earth, and its solar cells are in position to gather heat from the Sun to charge its batteries. The other is to set the spacecraft spinning at the rate of 60 revolutions per minute; this stabilizes its position, on the same principle as a gyroscope; 40% of the spacecraft is 'despun' or contra-rotated, by means of a ball-bearing and power-transfer assembly, so that the highly directional antennas can be kept pointed towards the Earth. The satellite carries 136 kg of hydrazine propellant for these manoeuvres; once this is exhausted, the spacecraft ceases to be operational, despite its orbital life of 1M yr. (When the Space Shuttle becomes fully operational, it should be possible to recover or refuel such satellites.)

INTERCOSMOS

History A series of international research satellites, which have successfully provided the 9 member countries with opportunities to expand their research activities, and to apply space techniques to their national economies. The Intercosmos programme was adopted at a Moscow conference in Apr 1967 by Bulgaria, Hungary, Cuba, GDR, Mongolia, Poland, Romania, the Soviet Union and Czechoslovakia. Subjects to be covered were problems of space physics, meteorology, communications, medicine and biology; observations were to be made of cosmic rays, radiation belts, Earth's ionosphere and magnetosphere, and the way the Sun influences Earth. First launch was Cosmos 261, on Dec 20 1968, from Plesetsk, which studied air density and polar auroras. Subsequent launches averaging 2 per yr have been from both Baikonur (Tyuratam) and Kapustin Yar, and by the end of 1976 Intercosmos 16 had been launched. The programme overlaps the work of both the Proton and Prognoz series; defending this when the series started, Academician Boris Petrov said that while the layman might think the flight programmes repeated themselves, the later flights were a continuation of earlier research with improved accuracy of measurements and greater capacity to handle the information. At the end of 1975 Petrov was able to add that the series 'had helped to make quite a few significant discoveries' about the Sun, world climate and radio waves. With maximum solar activity expected at the end of the '70s, scientists needed to be well armed with knowledge about physical conditions around the Earth. Intercosmos 14 was aimed at providing answers to unresolved problems in the 'Sun-Earth' system. He disclosed that, in addition to the observations being made in the member-countries' own observatories, 'collectives of experts' would be working at the Intercosmos flight headquarters sited in the environs of Norilsk, beyond the Arctic Circle, where parallel ground observations would be conducted. By that time 2 Vertikal rockets and several dozen meteorological rockets had been launched during the programme to provide complementary data.

Spacecraft description Few details have been given. With the exception of the big Intercosmos 6, which was recovered, Western observations suggest they usually weigh 400–550 kg and are cylindrical in shape, with hemisphere-shaped compartments at each end. Length estimated at 1·8 m; dia 1·5 m. Electrical power is provided by batteries rather than solar arrays, which no doubt accounts for their short life. Intercosmos 6 was believed to be an improved follow-on to mid-60s Proton type.

Intercosmos 1   L Oct 14 1969 by B1 from Plesetsk. Wt ?400 kg. Orbit 640 × 260 km. Incl 48°. E Germany and Czechoslovakia contributed instruments, and launch was attended by representatives of 7 participating countries. Studied Sun's ultraviolet and X-ray radiation, and its effects on Earth's upper atmosphere. Dec. Jan 2 1970.

Intercosmos 2   L Dec 25 1969 by B1 from Kapustin Yar. Wt ?400 kg. Orbit 1102–206 km. Incl 48°. Investigated Earth's ionosphere concentrations of electrons and positive ions. Dec. Jun 7 1970.

Intercosmos 3   L Aug 7 1970 by B1 from Kapustin Yar. Wt 409 kg. Orbit 1320 × 207 km. Incl 49°. Instrument and operational team jointly provided by USSR and Czechoslovakia. Studied composition and temporary variations of charged particles (protons, electrons, alpha-particles); low-frequency electro-magnetic waves; and Earth's magnetic field. Dec. Dec 6 1970.

Intercosmos 4   L Oct 14 1970 by B1 from Kapustin Yar. Wt ?400 kg. Orbit 628 × 250 km. Incl 43°. Carried more advanced equipment, made in GDR, Czechoslovakia and USSR with greater data storage capacity and improved telemetry. Continued studies of Sun's UV and X-rays and their effects on Earth's atmosphere. Found that amount of oxygen at 96 km altitude was less than expected, and varied with time of day. Dec. Jan 17 1971.

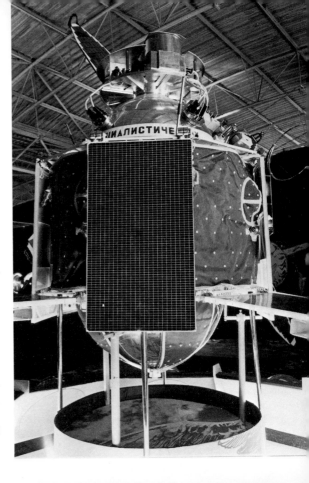

Intercosmos 1

Intercosmos 5 L Dec 2 1971 by B1 from Kapustin Yar. Wt 1500 kg. Orbit 1200 × 205 km. Incl 48°. Equipped with Czech and Soviet instruments, and operated by a joint Soviet-Czech team, it continued the study of charged particles started by Intercosmos 3. Dec. Apr 7 1972.

Intercosmos 6 L Apr 7 1972 by A1 from Tyuratam. Wt 4082 kg. Orbit 203 × 256 km. Incl 51°. New, recoverable type; scientific payload, wt 1070 kg, brought back after 4 days. A nuclear photo-emulsion unit recorded several thousand collisions of cosmic ray particles, among them a 'space wanderer' with a pulse of 1000 million million electronvolts. The photo-emulsion sheets, pasted on to glass, were divided for study among Polish, Hungarian, Czech, Romanian and Soviet scientists.

Intercosmos 7 L Jun 30 1972 by B1 from Kapustin Yar. Wt ?400 kg. Orbit 267 × 568 km. Incl 48°. With Czech instruments, and controlled by a Soviet-GDR-Czech team, this continued upper-atmosphere studies of Intercosmos 1 and 4. Ellipsoid, with 6 panels, and an estimated 10-week life, it studied short-wave solar irradiations which are absorbed by oxygen molecules and do not reach Earth's surface.

Intercosmos 8 L Dec 1 1972 by B1 from Plesetsk. Wt ?400 kg. Orbit 214 × 679 km. Incl 71°. Instruments and operating team provided by Bulgaria, GDR, Czechoslovakia and USSR, to continue study of Earth's ionosphere: concentration and temperature of electrons between the satellite and Earth's surface. This time an orbit more inclined to the equator was chosen so that the spacecraft would pass though the ionosphere at higher latitudes, and the polar light zones in the areas, or Earth's geomagnetic poles; in these regions Earth's magnetic field has 'a less barring effect' on particles.

Intercosmos 9 L Apr 19 1973, by B1 from Kapustin Yar. Wt ?400 kg. Orbit 202 × 1551 km. Incl 48°. Soviet-Polish satellite, named Intercosmos Copernicus 500, to celebrate 500th anniversary of birth of Polish scientist. Elliptical orbit 'stretched towards the Sun' enabled the instruments to record radio-frequencies of solar radiations which fade out in the ionosphere before reaching Earth's surface. It also recorded the concentration of charged particles (electrons and ions) at the Earth's 'space threshold'. Dec. Oct 15 1973.

Intercosmos 10 L Oct 30 1973 by C1 from Plesetsk. Wt 550 kg. Orbit 1477 × 265 km. Incl 74°. With an orbital life of 3 yr, it ceased transmitting after 2 yr and was later said to have provided new data about the origin of powerful electric currents. It had been sent up to study electromagnetic connections between the magnetosphere and ionosphere, and to measure magnetic field variations in some specific frequency ranges. During its flight, meteorological rockets, made in both the USSR and GDR, were launched to provide comparative results.

Intercosmos 11 L May 17 1974 by C1 from Kapustin Yar. Wt 550 kg. Orbit 526 × 484 km. Incl 50°. Timed to mark the 250th anniversary of the USSR Academy of Sciences, and carrying instruments designed by the GDR, USSR and Czechoslovakia, it was intended to study solar UV and X-ray radiations, and Earth's upper atmosphere. Orbital life 6 yr.

Intercosmos 12 L Oct 31 1974 by C1 from Plesetsk. Wt 550 kg. Orbit 243 × 707 km. Incl 74°. This launch marked 5 yr of space co-operation. Similar objectives with Romanian instruments included for the first time. Dec. after $8\frac{1}{2}$ months.

Intercosmos 13 L Mar 27 1975 by C1 from Plesetsk. Wt 550 kg. Orbit 296 × 1714 km. Incl 83°. This continued studies by Intercosmos 3, 5 & 10 of cosmic radiation, spectral composition of protons in Earth's radiation belts, and low-energy electrons. It contained Soviet and Czech equipment, and had a 4-yr orbital life.

Intercosmos 14   L Dec 11 1975 by C1 from Plesetsk. Wt 700 kg. Orbit 345 × 1707 km. Incl 74°. Intended to study low-frequency electro-magnetic fluctuations in Earth's magnetosphere, and structure of the ionosphere. Orbital life 5 yr.

Intercosmos 15   L Jun 19 1976 by C1 from Plesetsk. Wt ?700 kg. Orbit 487 × 521 km. Incl 74°. Announced as a new generation, with more room for international instruments; even more important, its information could be received and processed outside the Soviet Union for the first time. Ground stations had already been built in Hungary, the GDR and Czechoslovakia; later stations in Bulgaria and Cuba would be added. With an orbital life of 6 yr, this was apparently mainly intended to test the new systems.

Intercosmos 16   L Jul 27 1976 by C1 from Plesetsk. Wt ?700 kg. Orbit 465 × 523 km. Incl 50°. Mainly intended to study solar radiation and its effect on Earth's upper atmosphere, it carried an ultraviolet spectrometer polarimeter jointly developed by Russia's Crimean Observatory and Sweden's Lund Observatory. It had a small telescope, with 100mm focal length and a mirror device which measures the direction of light-beam oscillations in the ultraviolet spectral region.

ITALY

History   Italy has built a small but effective space programme around her international launch platforms in the Indian Ocean off San Marco, Kenya. In addition to the 4 San Marco satellites used to explore atmospheric density between 1964–74, Italy has arranged to pay NASA to launch the geostationary Sirio satellite from C Canaveral about 1977. This will be 2 yr later than planned, but will experiment with telephone and TV transmissions on little-used and unfamiliar frequencies. The Rome, Florence, Milan and Bologna Universities have contributed experiments to the Heos, TD-1A, OSO-6, COS-B and GEOS satellites. An average of 6 sounding rockets per year are launched from a range on Sardinia, as well as sounding rockets and research balloons as part of her contribution to the European Space Agency (qv).

San Marco 1   L Dec 15 1964 by Scout from Wallops I. Wt 24 kg. Orbit 194 × 697 km. Incl 38°. Measured atmospheric density until its decay on Sep 13 1965.   **San Marco 2**   L Apr 26 1967 by Scout from San Marco. Wt 129 kg. Orbit 185 × 211 km. Incl 3°. The first launch from the San Marco platform, it was the first of 3 ingeniously designed to measure atmospheric density; they consisted of 2 concentric spheres of different mass joined by 3 flexible arms, so that the drag on the outer sphere could be measured as it dipped into the atmosphere. Re-entry was on Oct 14 1967.   **San Marco-3**   L Apr 24 1971 by Scout from San Marco. Wt 164 kg. Orbit 222 × 718 km; incl 3°. Dec. Nov 29 1971.   **San Marco-4**   L Feb 18 1974 by Scout from San Marco. Wt 164 kg. Orbit 231 × 910 km. Incl 3°. With Explorer 51 taking measurements in the auroral zone, and San Marco-4 on the equator, it was possible to compare the different responses to energy coming from the Sun, and to study the theory that most solar energy enters the atmosphere at the poles. It transmitted on command for 806 days until decay on May 4 1976.

JAPAN

History Japan was the 4th country, after the Soviet Union, United States and France, to achieve national satellite capability. Despite many setbacks and disappointments, her first test satellite 'Osumi' was finally launched on Feb 11 1970; by Feb 1977 there were 10 Japanese satellites in orbit. Over a period of 16 yr Japanese scientists developed a series of 'Kappa' sounding rockets, followed by the bigger 'Lambda-3' with satellites as their long-term aim. Between Sep 1966 and Sep 1969 there were 4 failures to orbit a test satellite with versions of the Lambda-S; the 5th attempt resulted in the brief life of Osumi. This led to hopes that technologies for the much-delayed and more advanced Mu-4S rocket had been proved. Its first attempt on Sep 25 1970 failed when the 4th-stage motor did not ignite, but was followed by a success 5 months later.

Despite the fact that Osumi was $2\frac{1}{2}$ yr later than originally planned, Japan's space budget over the 16-yr period that led to final success totalled only £78·8M ($189M). The first operational satellite 'Shinsei' came in Sep 1971. The ambitions of Japanese scientists to become the world's 3rd space nation, almost achieved, had been foiled partly because there were 2 rival space programmes—with Tokyo University working on solid-fuel rockets, and the Government's Science and Technology Agency on liquid-fuel rockets. This was resolved by the setting up on Oct 1 1969 of NASDA (National Space Development Agency) to promote development of satellites, launchers, tracking systems, etc. By 1973 a 15-yr programme had been outlined, including 4 interplanetary craft (one of which would land on Venus), a big launch vehicle able to put 10 tonnes into Earth orbit, and a 'Japlab' or Japanese Space Laboratory, somewhat similar to America's Spacelab. The more immediate aims concentrated on a series of scientific satellites, culminating in a synchronous test in 1976, and an experimental geostationary communications satellite in 1978; in the meantime, however, Japan finally adopted the practice of other nations by arranging for 3 US-built meteorological communications and broadcasting satellites to be launched for her in 1977 by American Thor-Delta rockets. By 1974, Japan's annual space budget had soared to $186·6M, and it was an open question whether Japan or China would become the world's third nation to send men into space.

Launches

Osumi L Feb 11 1970 by Lambda 4S-5 from Kagoshima. Wt 24 kg. Orbit 340×5140 km. Incl $31°$. Named after the district of Kyushu in which the space centre is located, Osumi was the 4th stage of the rocket, containing a 12kg instrument package as test satellite. It transmitted for 7 orbits and 17 hr. Orbital life 80 yr.

Tansei-1 (MS-T1) L Feb 16 1971 by M-4S-2 from Kagoshima. Wt 63 kg. Orbit 1110×990 km. Incl $30°$. Named Tansei ('Light Blue') for the colours of Tokyo University, this spherical test payload mainly to demonstrate the launcher's capability, carried thermometer, accelerometer

Kiku (ETS-1)

and transmitter, and operated for its full lifetime of 1 week. Orbital life 1000 yr.

Shinsei (SS-1)   L Sep 28 1971 by M-4S-3 from Kagoshima. Wt 65 kg. Orbit 1870×870 km. Incl $32°$. Japan's first scientific satellite, named Shinsei or 'New Star' to mark the occasion. With 1·2m ht, and solar arrays mounted on an octahedral structure, it sent back ionosphere, solar and cosmic ray measurements; the on-board data recorder failed after 4 months but real-time data continued. Orbital life 5000 yr.

Denpa (SS-2)   L Aug 19 1972 by M-4S-4 from Kagoshima. Wt 75 kg. Orbit 6570×240 km. Incl $30°$. Scientific satellite 2, it carried out radio exploration, observing plasma rays and density, electron particle rays and geomagnetism.

Tansei-2 (MS-T2)   L Feb 16 1974 by M-3C-1 from Kagoshima. Wt 56 kg. Orbit 3240×290 km. Incl $31°$. Test flight for improved launcher and satellite system. Orbital life 9 Yr.

Taiyo (STRATS)   L Feb 24 1975 by M-3C-2 from Kagoshima. Wt 86 kg. Orbit 3135×255 km. Incl $31°$. Called 'The Sun', this Solar Radiation and Thermostatic Structure satellite was intended for a 250×2000 km orbit, but successfully observed solar soft X-rays, UV radiation etc. Orbital life 5 yr.

Kiku (ETS-1)   L Sep 9 1975 by N-1 from Tanegashima. Wt 83 kg. Orbit 1098×968 km. Incl $47°$. Called 'Chrysanthemum', an Engineering Test Satellite, this was both NASDA's first launch and the first launch from the new Space Centre with the new rocket. Orbital life 800 yr.

Ume (ISS-1)   L Feb 29 1976 by N-2 from Tanegashima. Wt 135 kg. Orbit 1013×994 km. Incl $70°$. Ionospheric Sounding

Satellite to examine world-wide critical frequencies etc, but within a month became almost inoperable due to overheating of solar cells when orbit was raised. Orbital life 1200 yr.

Tansei-3 L Feb 19 1977 by MU-3H from Kagoshima. Wt 134 kg. Orbit 796 × 3821 km. Incl 65°. Tokyo University scientific satellite. Orbital life 2000 yr.

Kiku (ETS-2) L Feb 23 1977 by N-1 from Tanegashima. Wt 130 kg. Japan's first successful geostationary satellite, manoeuvred to 130°E Long. For communications experiments.

Launchers

Lambda Developed from Japan's famous family of Kappa sounding rockets, the 4-stage, solid-fuel Lambda 4S, used for her first satellite, consisted of 1st-stage core with 36,970 kg thrust, augmented by 2 13,150kg strap-on boosters; 2nd stage providing 11,800 kg thrust; 3rd stage with 6580 kg thrust; and 4th stage with 816 kg thrust. Ht 16 m; wt 9390 kg. First 3 stages were unguided; 4th stage was provided with attitude control only.

Mu M-4S, with overall ht of 23·56 m, wt 43·6 tonnes, used for Japan's 2nd, 3rd and 4th satellites, was a combination of 4 different solid propellant rockets, and had no sophisticated guidance system. Controlled only by radio commands, it was capable of putting a 120kg payload into a 500km circular orbit. 1st stage, M-10, had an average thrust of 85 tonnes; 2nd stage, M-20 33·1 tonnes; 3rd stage, M-30, 13·2 tonnes; 4th stage, a spherical motor, 2·7 tonnes. An improved Mu, the M-3C, using only 3 solid stages, with 20·2 m ht and 41·6 tonnes wt, was used for the 5th and 6th satellites. M-3C uses liquid-injection thrust vector control on the 2nd stage.

N Mitsubishi Heavy Industries is producing this 3-stage,

Ume (ISS-1) mated with 3rd stage motor of N Launch Vehicle

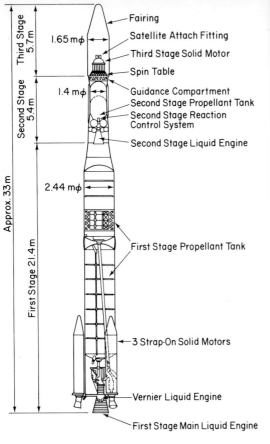

Third Stage 5.7m
Second Stage 5.4m
Approx. 33m
First Stage 21.4m

1.65 mφ
1.4 mφ
2.44 mφ

Fairing
Satellite Attach Fitting
Third Stage Solid Motor
Spin Table
Guidance Compartment
Second Stage Propellant Tank
Second Stage Reaction Control System
Second Stage Liquid Engine

First Stage Propellant Tank

3 Strap-On Solid Motors

Vernier Liquid Engine
First Stage Main Liquid Engine

Above ETS-1 Launch at Tanegashima
Left Diagram of N Launch Vehicle

Thor-based launcher under licence from McDonnell Douglas. After delays and problems with earlier national launchers, it was decided to utilize American technology. With total length of 33 m, 90 tonnes wt, and max dia 2·4 m, the first 2 stages are liquid-propelled. 1st stage for ETS-1 launch used a Rocketdyne MB-3 engine with 3 strap-on solid motors to provide lift-off thrust of 150 tonnes. 2nd stage, designed by Japan's NASDA with Rocketdyne assistance, is powered by a storable liquid rocket; 3rd stage by a solid rocket. The aim is to put 130kg satellites into geostationary orbits. An uprated N booster, Nkai-1, with longer 1st stage fuel tanks and 9 solid strap-ons, is planned for 1981; Nkai-2, with improved 2nd stage, is planned by 1985.

NETHERLANDS

History Up to 1970 the Netherlands' space activities were carried out mainly within the European organizations. In 1971, however, the Government decided to fund 2 ANS (Astronomical Netherlands Satellites) at a cost of 76M Guilders, with NASA help. Dutch laboratories have also carried out scientific experiments on NASA and Japanese sounding rockets, and contributed X-ray, UV and radiation experiments to 4 European satellites and to America's OGO-5.

ANS-1 L Aug 30 1974 by Scout from Vandenberg. Wt 129 kg. Orbit 258×1173 km. Incl 98°. Designed and built by a Dutch industrial consortium, the Netherlands' first satellite was intended for a near-polar, Sun-synchronous circular orbit of 500 km. Its studies of the UV spectrum of young, hot stars and both soft and hard X-rays from cosmic sources, were controlled from the European Space Agency's ESOC centre at Darmstadt, Germany; orbital life $2\frac{1}{2}$ yr.

SWITZERLAND

History Switzerland's space projects have so far all been on a collaborative basis with Europe and NASA, though she has built the Zenit sounding rocket which has been launched at least a dozen times since 1970. Her most famous contribution has been the instrument packages provided by Berne Physics Institute for the Apollo and Skylab programmes. These included the simple but effective sheet of aluminium set up by the Apollo astronauts as soon as they stepped on the Moon, and collected before they left, to register the composition of the solar wind.

Apollo 8
Lift Off—
double
exposure
showing
Moon

US SPACE PROJECTS

APOLLO Manned Spacecraft

History A 3-man spacecraft was first proposed by NASA in Jul 1960, for Earth orbital and circumlunar flights, to be launched by a Saturn 1-type rocket with 680,400 kg thrust. On May 25 1961 President John F. Kennedy proposed that the US should establish as a national goal a manned landing on the Moon by the end of the decade, and this in turn led to the development of Saturn 5, with 3·40 million kg of thrust. North American Rockwell was selected as prime contractor for the Apollo Command and Service Modules, and Grumman Aircraft for the Lunar Module. On May 28 1964 a Saturn 1 placed the first Apollo Command Module in orbit from Cape Kennedy; parallel development of the spacecraft and rockets was making remarkable progress until, on Jan 27 1967, an electrical arc from wiring in a spacecraft being ground-tested at Cape Kennedy started a fire which became catastrophic in the 100% oxygen-atmosphere. Astronauts Grissom, White and Chaffee were burned to death, and the first manned flight was delayed for 18 months. On Nov 9 1967 the first unmanned test of the combined Apollo/Saturn 5 vehicles—designated Apollo 4—was successfully accomplished. Apollo 5, in Jan 1968, successfully tested the Lunar Module systems, including firings in Earth orbit of both the ascent and descent propulsion systems; in Apr 1968 Apollo 6, the 2nd unmanned test of the combined Apollo/Saturn 5, was only partially successful since vertical oscillations, or 'Pogo' effects, were encountered with the first stage. Nevertheless, the first manned flight, Apollo 7, went ahead in

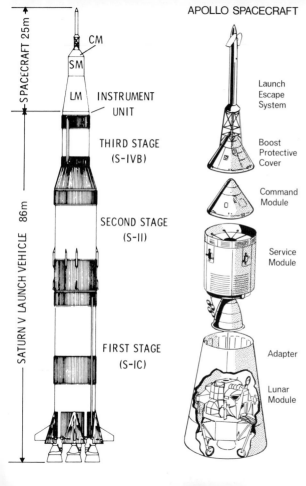

APOLLO SPACECRAFT

SPACECRAFT 25m

SATURN V LAUNCH VEHICLE 86m

CM

SM

LM INSTRUMENT UNIT

THIRD STAGE (S-IVB)

SECOND STAGE (S-II)

FIRST STAGE (S-IC)

Launch Escape System

Boost Protective Cover

Command Module

Service Module

Adapter

Lunar Module

Oct 1968, and the first lunar landing by Apollo 11 was astonishingly achieved only 9 months later. A mission-by-mission account follows on pages 107–32.

Spacecraft Description
Apollo 11 Launch Weights

CM:	5558 kg
SM:	23,204 kg
LM Adapter:	1837 kg
LM:	15,094 kg
Escape System:	4041 kg
	49,734 kg

Spacecraft Weight into Earth Orbit: 45,697 kg
CM Weight at Splashdown: 4976 kg

The complete Apollo spacecraft is 25 m tall, and consists of: 1) the Command Module; 2) the Service Module; 3) the Lunar Module; 4) the Launch Escape System; 5) the Spacecraft Lunar Module Adapter.

The Launch Escape System consists of a 10m tower weighing 3629 kg and a solid-rocket motor 4·72 m long, providing 66,675 kg of thrust. If a fire or abort situation occurs during the countdown or in the first 100 sec of lift-off, the Commander presses the abort button and the Escape Tower lifts the spacecraft about 1·6 km clear of the pad and rocket; the main parachute system is then used for descent. The Spacecraft Lunar Module Adapter serves as a smooth aerodynamic enclosure, linking spacecraft and launch vehicle during lift-off, and protecting the Lunar Module, which is extracted from it during transposition and docking, shortly after the spacecraft leaves Earth orbit on its journey to the Moon.

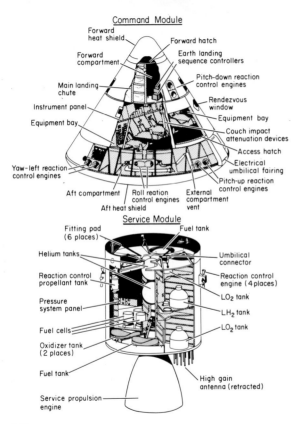

Command Module

Forward heat shield

Forward hatch

Earth landing sequence controllers

Forward compartment

Main landing chute

Pitch-down reaction control engines

Instrument panel

Rendezvous window

Equipment bay

Equipment bay

Couch impact attenuation devices

Access hatch

Electrical umbilical fairing

Yaw-left reaction control engines

Pitch-up reaction control engines

Aft compartment

Roll reation control engines

External compartment vent

Aft heat shield

Service Module

Fitting pad (6 places)

Fuel tank

Helium tanks

Umbilical connector

Reaction control propellant tank

Reaction control engine (4 places)

Pressure system panel

LO_2 tank

LH_2 tank

Fuel cells

LO_2 tank

Oxidizer tank (2 places)

Fuel tank

High gain antenna (retracted)

Service propulsion engine

Command Module The control centre and living quarters for the 3-man crew; 3·48 m from nose to heatshield. One man spends the entire mission in it; the others leave it only for the lunar landing; it is the only part finally to return to Earth. Lift-off weight rose to about 5806 kg including crew; splashdown weight about 5307 kg. As on an aircraft, the Commander, who normally operates the flight controls, is on the left; the CM pilot, responsible for guidance and navigation is on the centre couch; the LM pilot, responsible for management of subsystems, is on the right. Their couches face the main display console, nearly 2·13 m long and 0·91 m wide, containing the switches, dials, meters etc, for the inter-related systems of guidance and navigation, stabilization and control. During flight the cabin's 5·95 m³ has a 100% oxygen atmosphere at 0·35 kg/cm²; but as a result of the 1967 fire, during ground tests and countdown a less flammable 60/40 oxygen/nitrogen atmosphere of 1·05 kg/cm² is used, and gradually changed after lift-off by the environmental control system, which maintains a comfortable 'shirtsleeve' temperature of 21–24°C. The CM's outer shell consists of stainless steel honeycomb between stainless steel sheets covered on the outside with ablative, or heat-dissipating material. The heat-shield on the base of the cone is 6·98 cm thick, and of a type of reinforced plastic (phenolic epoxy resin) which turns white hot, chars, and melts away so that the 1649°C re-entry temperatures do not penetrate the surface of the spacecraft. The CM's inner shell is aluminium alloy sheets separated by a layer of insulation. Food, water, clothing, waste management and other systems are packed into bays which line the walls. There are 2 hatches, 1 at the side for entry, and 1 at the top, or nose, for use when docked with the LM. There are 5 windows: 2 side windows, 83·9 cm², for observation and photography; 2

triangular, 20·5 by 33 cm, used for rendezvous and docking; and a hatch window.

2 reaction control engines and the Earth landing system are housed in the forward, or nose, compartment; 10 reaction control engines, propellant tankage, helium tanks, water tanks, and the CSM umbilical cable, are housed in the aft compartment.

Service Module A relatively simple cylindrical structure 7·4 m long, including the 2·8m SPS (Service Propulsion System) nozzle extension. The SM contains the main space-craft propulsion system and supplies most of the consumables. Total weight rose to 25,033 kg for the last 3 missions, designated 'J', of which 18,413 kg was propellant for the 9300 kg thrust SP engine. SPS fuel is 50/50 hydrazine and unsymmetrical dimethyl-hydrazine; oxidizer is nitrogen tetroxide. Since the SM is attached to the CM's heatshield, it cannot be entered by the astronauts, and is jettisoned and burnt up during re-entry. Around the centre section containing the SPS engine and helium tanks, are 6 pie-shaped sectors: Sector I—oxygen tank 3 and hydrogen tank 3, J-mission SIM bay; Sector II—space radiator, +Y RCS (Reaction Control System) package, SPS oxidizer storage tank; Sector III—space radiator, +Z RCS package, SPS oxidizer storage tank; Sector IV—3 fuel cells, 2 oxygen tanks, 2 hydrogen tanks, auxiliary battery; Sector V—space radiator, SPS fuel sump tank, −Y RCS package; Sector VI—space radiator, SPS fuel storage tank, −Z RCS package.

The Scientific Instrument Module (SIM) was carried for the first time on Apollo 15. Wedge-shaped, 2·87 m tall, and 1·52 m across the front, it carried 8 experiments, consisting of 4 spectrometers, panoramic and mapping cameras, laser altimeter, and a 35·4-kg subsatellite for ejection into lunar orbit.

Docking Mechanism This subsystem enables the CM and LM to be connected twice during a normal lunar mission—at the beginning of translunar flight, and when the ascent stage returns from the lunar surface. The probe, with a head housing 3 capture latches, is mounted in the CM docking tunnel and guided into a socket at the bottom of the drogue in the LM docking tunnel. The capture latches hold the 2 modules together until the probe retraction device (a nitrogen pressure system located in the probe) is operated. Then 12 latches on the CM docking ring automatically operate at contact to form a pressure-tight seal between the modules. When the CM hatch is opened the probe and drogue are dismantled to allow the astronauts to pass through the docking tunnel into the LM.

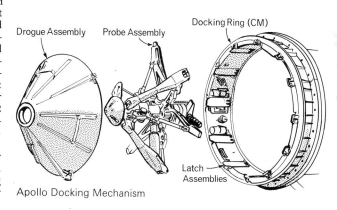

Apollo Docking Mechanism

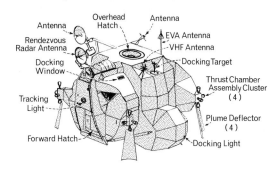

LUNAR MODULE - ASCENT STAGE

Antenna
Overhead Hatch
Antenna
EVA Antenna
VHF Antenna
Docking Target
Rendezvous Radar Antenna
Antenna
Docking Window
Thrust Chamber Assembly Cluster (4)
Tracking Light
Plume Deflector (4)
Forward Hatch
Docking Light

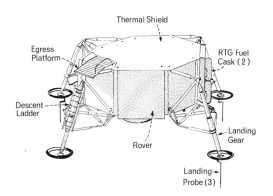

LUNAR MODULE - DESCENT STAGE

Thermal Shield
Egress Platform
RTG Fuel Cask (2)
Descent Ladder
Landing Gear
Rover
Landing Probe (3)

Lunar Module A 2-stage vehicle 6·985 m high and 9·45 m wide diagonally across the landing gear. Its job is to convey 2 astronauts to and from the CSM in lunar orbit and to provide living quarters and a base of operations on the Moon.

Total weight, with crew, rose from approximately 14,696 kg on the early missions, to approx. 16,647 kg on the last 3 'J' flights, when the LM was carrying the Lunar Rover as well as extra scientific equipment, batteries and consumables needed for a 3-day staytime on the Moon. Because the vehicle is operated only in the vacuum of space, aerodynamic symmetry was unnecessary; design was dictated only by the need to provide a descent stage able to serve as a launching platform for an ascent stage containing a pressurized crew cabin. Built by Grumman Aerospace Corp, the LM is designed to operate for 73 hr while separated from the CSM, with a nominal 67 hr spent on the Moon.

Ascent Stage The LM's control centre. Constructed of welded aluminium alloy, surrounded by a 7·6cm thick layer of insulating material; a thin outer skin of aluminium covers the insulation. The cabin provides 6·65 m³ of space, and is manned by the Commander at the left flight station, and the LM Pilot at right. Even in lunar gravity couches are unnecessary, so armrests and harness-like restraints are used to support the astronauts, giving them sufficient freedom of movement to operate all the controls while in an upright position. There are 2 inward-opening hatches: the 81cm diameter docking tunnel hatch at the top; and the 81cm square forward hatch, located between the astronauts, through which they must back, on hands and knees, to the platform and ladder when descending to the Moon. Each crewman has a triangular window, canted down for sideways and forward visibility; a 3rd docking window, 12·7 by 30·5 cm, is directly over the commander's position. The LM has essentially the same subsystems as the CSM, including propulsion, environmental control, communications, reaction control and guidance control.

The ascent propulsion engine develops 1587 kg thrust, can be fired for a total of 460 sec, and can be shut down and restarted up to 35 times.

Descent Stage   The unmanned portion of the LM, representing two-thirds of the total LM weight at Earth-launch. Since it has to support the ascent stage for the landing, and later act as its launching platform, it needs a larger engine and more propellant. The gimballed engine provides 4490 kg max thrust, and a minimum of 580 kg thrust for the complex 720-sec descent manoeuvres from lunar orbit. Its 8890 kg of propellant provides 950 sec of life, with 227 sec of surface hovering time (compared with 120 sec on early missions). In an emergency occurring *before* lunar touchdown, as in Apollo 13, it can be used to replace the main SM engine. The Descent Stage's egress platform, or 'porch', is mounted above the forward landing gear strut, which is provided with a ladder of 9 steps for descent to the lunar surface. The 4 landing legs are released explosively; the main struts are filled with crushable aluminium honeycomb to absorb the landing impact; there are also footpads 94 cm in diameter to provide surface support. All the pads except the forward one are fitted with a 173cm long lunar surface sensing probe which, upon contact with the surface, signals the crew to shut down the descent engine. The landing radar, which controls the descent from a height of 12,190 m and speed of 3840 kph, is housed in the Descent Stage. Around the Descent Engine are 4 equal-sized bays, or quadrants, which, in addition to propellant tanks, contain the MESA, or Modularized Equipment Stowage Assembly, consisting of

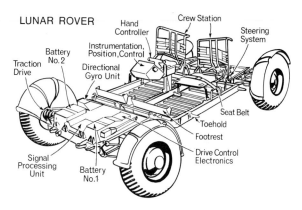

LUNAR ROVER

TV equipment, lunar sample containers; the Lunar Rover Vehicle; and the ALSEP or Apollo Lunar Surface Experiment Package.

Lunar Rover

Earth Weight:	209 kg
Payload:	490 kg
Length:	310 cm
Max Speed:	16·9 kph
Range:	92 km

The Lunar Roving Vehicle (LRV), built by the Boeing Company, was first used with great success on Apollo 15. This manned spacecraft on wheels had to be built to all the exacting specifications of Apollo hardware, in order to operate in temperatures reaching 121°C, in a vacuum which

Apollo 11: splashdown jubilation in Mission Control

ruled out air-cooling. Built and delivered in only 17 months, it carries more than twice its own weight 181·5 kg for each astronaut and his equipment, plus 127 kg of tools, equipment, TV and communications gear, and lunar samples. (The average family car carries only half its own weight.)

2 complete 36-volt battery systems, each of which can power the vehicle on its own, provide a total of 1 hp (1·01 cv), each wheel being individually driven by an electric motor of $\frac{1}{4}$ HP (0·253 cv). It has 78 hr operational lifetime during the lunar day, and provides the astronauts with a 9·65km exploration radius from their touchdown point—the limitation being the walk-back distance in the event of a breakdown. Design difficulties were increased by the need to take it to the Moon folded into a tight, pie-shaped quadrant of the Lunar Module's descent stage, yet able to unfold itself on the Moon, with the astronauts merely pulling sequentially on 2 nylon operating tapes and then removing a series of release pins. Special 1-G trainers had to be built for Earth tests, since the lightweight lunar vehicles could not rest on their own wheels on Earth.

A T-shaped hand-controller located on the control and display console between the 2 seats enables the LRV to be driven by either astronaut. Pushing forward sets the vehicle in motion; pushing sideways turns it; and pulling backwards applies the brakes. It can climb and descend slopes of 25° and negotiate obstacles and small crevasses—though the seat belts are essential safeguards against the pitching and bouncing experienced when travelling in one-sixth gravity. A dead-reckoning navigation system, set before they drive off and using the sun-angle as a bearing, enables them, no matter how much they twist and turn, to know exactly the direction and distance back to the Lunar Module.

By enabling the astronauts to expend less energy, and

thus use their oxygen and cooling water at a slower rate, the LRV doubles the astronauts' staytime on the Moon. Its remotely controlled TV unit also enables Mission Control and the public to watch whatever is being done on the Moon, the one drawback being that the umbrella-like antenna can only be unfurled for direct transmissions to Earth when the vehicle is stationary so that views of the changing scene during the drives are not possible, though radio contact does continue.

Future Development Many possibilities are being explored. Remote control to enable use of vehicles to continue after the astronauts return is the first and most obvious; another possibility is to send the vehicle, by remote control, to the next landing site, so that a subsequent mission would have to take only fresh batteries instead of a completely new vehicle. Remotely operated versions are inevitable for initial Martian exploration. For future lunar exploration it would be advantageous to convert the Lunar Module itself into a vehicle, so that the astronauts could travel in pressurized conditions.

APOLLO MISSIONS

Apollo 7   Oct 1968   The first manned flight of the Command and Service Modules conducted in Earth orbit, lasting nearly 11 days (260 hr and 163 orbits). Walter Schirra, Donn Eisele and Walter Cunningham successfully tested the operational qualities of the space vehicle, measured the performance of the Service Module's main propulsion engine by firing it both automatically and manually on eight occasions; simulated the all-important manoeuvre of extracting the Lunar Module (which was not in fact carried) from the S4B 3rd-stage rocket; and checked out the perfor-

mance of the heat-shield during re-entry. The astronauts cross-infected each other with severe colds, and there was some irritation in exchanges between them and Mission Control over the requirement for TV transmissions on top of many experiments. Schirra resigned from NASA soon after the flight. But it set the pattern of success which carried the Apollo Project right through to its successful Moonlandings.

Apollo 8   Dec 1968   Man's first flight round the Moon. Frank Borman, James Lovell and William Anders were the first men to be launched by the 3000-tonne Saturn 5 rocket. NASA's immense courage in electing to make this a lunar flight, with 10 orbits round the Moon if all was going well, was justified by the complete success of the mission. The critical lunar orbit insertion (LOI) manoeuvre was achieved by firing the SPS engine behind the Moon on Dec 24; after that, the astronauts spent 20 hr circling the Moon, filming and photographing the farside, never before seen by man, as well as obtaining pictures of craters, rills and potential

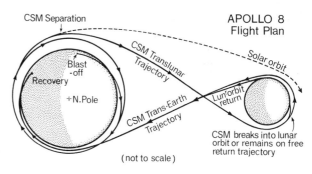

APOLLO 8
Flight Plan

CSM Separation

Blast-off

Recovery

+N.Pole

CSM Translunar Trajectory

Solar orbit

CSM Trans-Earth Trajectory

Lunar orbit return

CSM breaks into lunar orbit or remains on free return trajectory

(not to scale)

Apollo 9: LM 'Spider' ready for extraction from 3rd stage

landing sites on the nearside. It was at this time that Frank Borman took the famous 'Earthrise' picture, showing the Earth rising above the Moon's horizon; and, on Christmas

morning, stirred the whole world with his reading of the 'In the beginning' passage from Genesis. Splashdown in the Pacific, at the end of the 147-hr mission, was just 11 sec earlier than the time, computed months before, given in the Flight Plan. The performance of the spacecraft and its rocket engines, and the ability to make precise mid-course corrections when required, provided the final evidence that NASA had the technical equipment and knowledge to land men on the Moon. All that was left, was proof that the so-far untried Lunar Module would perform equally well.

Apollo 9 Mar 1969 The first manned flight with the Lunar Module. James McDivitt, David Scott and Russell Schweickart were launched after a 2-day postponement because of sore throats and nasal congestion. Although the 241-hr flight was confined to Earth orbit, separation of the Lunar Module from the Command Module, followed by rendezvous and docking, were carried out in conditions simulating those to be used on the later lunar missions. After the spacecraft was placed in Earth orbit, transposition and docking was achieved for the first time—that is, the Command Module separated from the S4B 3rd-stage rocket, turned around, and then nosed in to dock with the Lunar Module and withdraw from it from its housing inside the S4B. McDivitt and Schweikart then opened up the docking tunnel, and went through into the LM. With the 2 vehicles still docked, they test-fired the LM's 4535kg thrust rocket engine. A 2-hr spacewalk by Schweickart was cancelled because he had suffered nausea and vomiting, but, with both spacecraft depressurized, he climbed out of the LM, stood on the porch, and tested out the ladder on the LM's landing leg. Scott, standing up with head and shoulders through the open CM hatch, took some memorable pictures

Scott taking pictures of Schweickart on LM Porch

of Schweickart's activities which also provided a satisfactory test of the Extravehicular Activity (EVA) suit in space conditions. On the 4th day of the flight, with McDivitt and Schweickart again in the LM, and Scott in the CM, the 2 craft

were separated. When 182 km apart the LM's Descent Stage was jettisoned and the Ascent Stage simulating a take-off from the Moon, successfully fired its 1590kg thrust rocket engine to rendezvous and dock with the CM.

Apollo 9: LM's first free flight

Apollo 10: CM seen from LM on Moon's farside

Apollo 10 May 1969 This 8-day (192-hr) flight by Thomas Stafford, Eugene Cernan and John Young was the successful dress rehearsal for the actual Moonlanding 2 months later. The first time the complete Apollo spacecraft had orbited the Moon, it took man to within 14·5 km of the lunar surface. The mission closely followed the Apollo 11 flight plan. The CM remained in a 111km lunar orbit for nearly 32 revolutions, with Young in control, while Stafford and Cernan undocked and made a simulated landing in the LM by twice descending to within 14·5 km of the surface. A moment of great hazard occurred when the LM's Descent Stage was jettisoned after the 2nd close-approach, prior to firing the Ascent Stage engine to rejoin the CM. The Ascent Stage began pitching violently, and it took Stafford about 1 min to stabilize it; it was afterwards found that a control switch, omitted from the detailed check list, had been left in the wrong position. The spacecraft successfully docked after 8 hr of separation. Apart from its technical success, this flight was notable because, for the first time, all 3 astronauts remained in excellent health, both during the flight and afterwards—the first time this had happened on any US manned flight. Another first was colour TV, including a memorable sequence showing the LM firing its attitude control jets as it moved away from the CM prior to its descent towards the lunar surface.

Apollo 11 Jul 1969 The historic flight by Neil Armstrong, Edwin Aldrin and Michael Collins which culminated in Armstrong and Aldrin becoming the first men to step on the Moon. Launch from Cape Kennedy was achieved without any technical problems on Wed, Jul 16. Lunar orbit insertion was successfully achieved on Sat, Jul 19, and the spacecraft placed in a 100 by 121km orbit; Apollo 8 and 10 experience had shown that by the time the Lunar Module docked with the Command Module after the landing, the gravitational effect of the Moon's 'mascons' (mass con-

centrations) would have gradually changed the orbit into a circular 111 km. On Sun, Jul 20, while Collins remained in control of the CM (code-named Columbia), Armstrong and Aldrin entered the LM (code-named Eagle). The 2 spacecraft separated on the 13th lunar orbit, and Eagle's descent engine was fired behind the Moon. As he neared the surface Armstrong decided to take over manual control because the spacecraft was approaching an area in the Sea of Tranquillity strewn with boulders. At 21.17 BST (16.17 EDT) on Sun, Jul 20, with less than 2% of descent propellant remaining, more than 500M people heard Armstrong tell Mission Control:

Apollo 11: deploying scientific instruments

At 03.56 BST on Mon, Jul 21 (22.56 EDT on Sun, Jul 20) he became the first man to step on the Moon. As he placed his left foot on the dusty surface, a breathless world could just distinguish his words: 'That's one small step for a man; one giant leap for mankind.' Aldrin joined him on the surface 18 min later, and the 2 astronauts practised the best way of moving about in one-sixth gravity by walking, running and trying out a 'kangaroo hop'. They erected a 1·52m US flag, extended on a wire frame since there is no wind on the Moon, saluted it, and received a congratulatory telephone call from President Nixon. A plaque fixed to the spacecraft's ladder was unveiled, which read: 'Here men from the planet Earth first set foot upon the Moon, July 1969 A.D. We came in peace for all mankind.' It bore the signatures of the 3 Apollo 11 astronauts and of President Nixon. Armstrong and Aldrin collected 21·75 kg of rock and soil samples, and placed on the surface of the Moon a microdot disc containing messages from most of the world's leaders, as well as a scientific package. This included a laser reflector to enable scientists to measure Earth-Moon distances by reflecting light beams from it; and a seismometer to measure meteorite impacts and moonquakes. (The seismometer was so sensitive that one of the first signals it transmitted to Earth was the impact of the astronauts' heavy boots when they threw them out of the LM to help lighten it before take-off.)

'Contact light. O.K., engine stop . . . Houston, Tranquillity Base here. The Eagle has landed.' Mission Control replied: 'Roger, Tranquillity. We copy you on the ground. You've got a bunch of guys about to turn blue. We're breathing again. Thanks a lot.' After checking the LM's systems in case an emergency take-off was required, Armstrong and Aldrin elected not to take the 4-hr rest-period scheduled in the flight plan, but to go straight ahead with preparations for their 2¼ hr of activity on the Moon's surface. The arduous process of donning spacesuits and backpacks, and depressurizing Eagle, took a good deal longer than expected. But finally, Armstrong, on hands and knees, backed carefully out of the spacecraft on to the small platform or porch outside, and descended the 3·05m ladder attached to the landing leg.

The launching of the Ascent Stage went as planned, and docking was achieved 3½ hr later; there was some vibration at that point, and Collins was heard to say 'all hell broke loose'. After vacuum-cleaning their clothing and equipment, and taking many other precautions to avoid bringing back to Earth any germs or contamination from the Moon, Armstrong and Aldrin transferred to the CM; the LM was jettisoned, the Service Propulsion System was fired behind

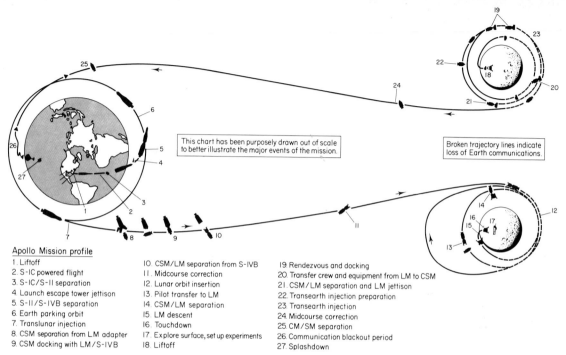

This chart has been purposely drawn out of scale to better illustrate the major events of the mission.

Broken trajectory lines indicate loss of Earth communications.

Apollo Mission profile

1. Liftoff
2. S-IC powered flight
3. S-IC/S-II separation
4. Launch escape tower jettison
5. S-II/S-IVB separation
6. Earth parking orbit
7. Translunar injection
8. CSM separation from LM adapter
9. CSM docking with LM/S-IVB
10. CSM/LM separation from S-IVB
11. Midcourse correction
12. Lunar orbit insertion
13. Pilot transfer to LM
14. CSM/LM separation
15. LM descent
16. Touchdown
17. Explore surface, set up experiments
18. Liftoff
19. Rendezvous and docking
20. Transfer crew and equipment from LM to CSM
21. CSM/LM separation and LM jettison
22. Transearth injection preparation
23. Transearth injection
24. Midcourse correction
25. CM/SM separation
26. Communication blackout period
27. Splashdown

the Moon, and an uneventful return journey achieved. Splashdown was at 17.50 BST on Thur, Jul 24 (12.50 EDT), only 30 sec later than predicted before the start of the 195-hr mission. After donning Biological Isolation Garments, the astronauts were taken by helicopter to the recovery ship, where they immediately entered the Mobile Quarantine Facility (a modified trailer), in which they travelled to the Lunar Receiving Laboratory at Houston, Texas. Thus they were kept in isolation—a precaution which proved to be unnecessary—for 21 days after lift-off from the lunar surface, to ensure that no lunar contamination was brought back to Earth. During their 8-day mission the Apollo 11 astronauts had travelled 1,533,225 km.

Apollo 12 Nov 1969 This 2nd lunar landing mission, by Charles Conrad, Richard Gordon and Alan Bean, was in some ways an even more remarkable flight than Apollo 11. It started sensationally, for as the launch was being made through a rain squall, the Saturn 5 rocket was struck by lightning. The spacecraft's electrical system was put out of action for a short time, and Mission Control lost contact briefly. For the first time during a manned launch they were very close to an 'Abort' situation. It was feared that the lightning strike might have damaged the Apollo computer's 'memory' containing the data required for the flight trajectories and the Moonlandings. But after being checked during the first 2 hr in Earth-parking orbit, all systems were found to be in good order, and astronauts and Mission Control agreed that the flight should continue. Because it was an all-Navy crew, on this occasion when they separated in lunar orbit, the Command and Lunar Modules were code-named Yankee Clipper and Intrepid, with Conrad and Bean making the landing. Because Apollo 11 had touched down 6·5 km beyond the target point, the planners had decided that a series of very small manoeuvres, such as making a water dump, having the LM make a turn so that the CM could inspect it before they moved apart etc, had had a cumulative effect upon the orbit. Extra course corrections were made, and a 'soft' undocking carried out.

The final result, at the end of the 579,365km journey, was the incredible achievement that Intrepid touched down, as planned, only 183 m from the Surveyor 3 spacecraft which had been soft-landed in the Ocean of Storms 31 months

earlier. Conrad expressed concern later about his landing difficulties, which were increased by a much bigger dust-storm blown up by the descent engine than in the case of Apollo 11; but he did have a safety-margin of 58 sec of hovering time left compared with Apollo 11's 20 sec. Conrad and Bean were on the Moon for $31\frac{1}{2}$ hr, during which they made 2 excursions, totalling $7\frac{1}{2}$ hr on the surface. In addition to collecting over 34 kg of rocks and soil samples, they deployed the first ALSEP package—6 scientific experiments, with a nuclear-powered battery giving the equipment an operational life of at least a year—and brought back the TV camera and other parts of the Surveyor spacecraft, so that scientists could discover the effects on them of long-term exposure to the solar wind, and of the extreme variations of temperature in vacuum conditions. Conrad became the first man to fall on the Moon, but was quickly helped by Bean, and remarked that, despite his cumbersome spacesuit, it was 'no big deal'.

When they had rejoined Yankee Clipper, Intrepid was deliberately crashed on to the lunar surface. The impact 72·5 km from the newly installed seismometer, set the Moon 'ringing like a bell', as one geologist put it; the vibrations continued for 51 min, an effect completely unanticipated. One theory advanced was that it indicated that the Moon was an unstable structure, and that the impact had set off a series of collapses and avalanches. Although the astronauts were again kept in quarantine for 3 weeks after leaving the lunar surface as a precaution against bringing back any possible contamination to Earth, it was again found to be unnecessary. Anxiety on the subject was so far relaxed that on this occasion, nearly 13·6 kg of Moonrock samples were distributed to scientists around the world by normal mail services and by diplomatic bags. The only important thing that went wrong on this 10-day mission—2 days longer than Apollo 11's—was that Intrepid's colour TV camera was burnt out when Bean accidentally pointed it at the Sun, so that viewers on Earth were unable to watch lunar activities.

Apollo 13 Apr 1970 This 3rd Moonlanding attempt was intended to be the first of 3 landings devoted to geological research on the lunar surface. Instead, an explosion on board when the spacecraft was 329,915 km from Earth all but cost the lives of the crew, and turned the mission into a 3½-day rescue drama surpassing any space-fiction story as tens of thousands of technicians in the US aerospace industry worked to bring the crippled spacecraft safely home. The flight was on the verge of postponement for a week before the launch date because the Commander, Jim Lovell, about to make his 4th spaceflight, Command Pilot Thomas Mattingly and Lunar Pilot Fred Haise, neither of whom had flown before, had been in contact with German measles. Finally it was decided that only Mattingly had no immunity to the disease, and he was replaced by the back-up command pilot, Jack Swigert. Apollo 13 was then launched on time, Mrs Marilyn Lovell declaring that she was not in the least worried because it started at 13.13 hr Houston time. Once again it was a less-than-perfect launch; the centre engine amid the cluster of 5 on the 2nd stage shut down 2 min early. The other 4 burned to depletion to make up the lost thrust, and the 3rd stage, Saturn 4B engine, was automatically fired an extra 10 sec to put the spacecraft into the required orbit. However, it was decided that the 3rd stage had plenty of fuel left for the Translunar Injection manoeuvre, and the flight continued. Its progress towards the Moon then became so uneventful and routine that public interest in the expedition rapidly waned. But on Apr 13 at 55

Apollo 13: the home-made air conditioner

hr 54 min into the flight came the explosion. It was 9 p.m. at Mission Control at Houston, 4 a.m. in Britain; the astronauts had just been congratulated on a routine, but successful TV transmission, and were being advised on how to locate Comet Bennett, when Swigert interrupted sharply: 'Hey, we've got a problem here.' There had been 'a pretty large bang', followed by a 'Main B Bus undervolt'. (The

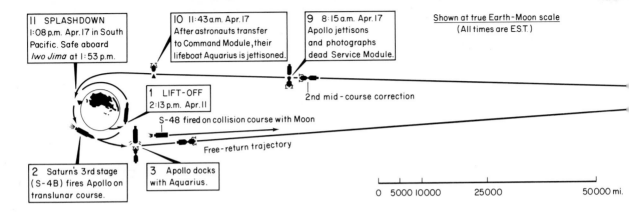

11 SPLASHDOWN
1:08 p.m. Apr.17 in South Pacific. Safe aboard *Iwo Jima* at 1:53 p.m.

10 11:43 a.m. Apr.17
After astronauts transfer to Command Module, their lifeboat Aquarius is jettisoned.

9 8:15 a.m. Apr.17
Apollo jettisons and photographs dead Service Module.

Shown at true Earth-Moon scale
(All times are E.S.T.)

1 LIFT-OFF
2:13 p.m. Apr.11

2nd mid-course correction

S-48 fired on collision course with Moon

Free-return trajectory

2 Saturn's 3rd stage (S-4B) fires Apollo on translunar course.

3 Apollo docks with Aquarius.

0   5000 10000   25000   50000 mi.

spacecraft's 3 major electrical harnesses are known as A and B, and called 'buses' by engineers). Only later was it established that the module had been ripped open by an explosion amid its oxygen tanks; but 10 min after, amid mounting tension as oxygen readings in the spacecraft and at Mission Control fell to zero, the crew reported they could see 'a gas of some sort' venting into space.

Only 80 min after the crisis began, journalists were told that the Moonlanding was abandoned, and that the aim must be to swing the spacecraft round the Moon and use the Lunar Module's power systems and supplies to bring it back to Earth. 10 min after that Mission Control told the crew: 'We're starting to think about the LM lifeboat,' and Swigert replied: 'That's something we're thinking about too'—the only spoken acknowledgement at any time of the acute

danger they faced. With only 15 min of power left in the CM, Mission Control instructed the crew to make their way into the LM. At first it was thought that, even by using the LM as a lifeboat to tow the crippled spacecraft home, only 38 hr of power, water and oxygen were available—about half as much time as would be needed to get the craft home. But as technicians brought their computers and simulators into use, techniques were devised for powering down the systems so that, although it meant increasingly cold and uncomfortable conditions, there would be an ample margin for the return. A major problem arose because the LM's air-conditioning equipment could not extract the poisonous carbon dioxide from the docking tunnel and CM as well; so a do-it-yourself air conditioner was devised on the ground from canisters, spacesuit hoses etc., known to be available in

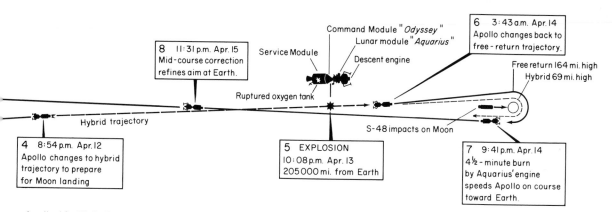

Apollo 13: Flight Sequence showing explosion

the spacecraft, and instructions on how to make it were read up, hour after hour, to the crew. 4 firings of the LM's descent engine were carried out in an elaborate series of manoeuvres—accompanied by heart-stopping moments on 2 occasions when the astronauts, tired and chilled, made mistakes.

The CSM, code-named Odyssey, jettisoned the SM 3½ hr before splashdown, and as it moved away there were exclamations from the astronauts as they saw—and photographed—for the first time the extent of the damage. Finally, the LM, code-named Aquarius, was jettisoned and burned up, 1 hr before re-entry began. At that time, no one knew for certain whether the explosion had damaged the heat-shield. Evidence that it had not was dramatically provided when the spacecraft, descending on its 3 para-

chutes, suddenly filled the TV screens of the world's hundreds of millions of viewers. Apollo 13 ended as a triumphant failure; Lovell, in 4 flights had completed 715 hr, or nearly 30 days, in space; NASA pointed out that US manned spacecraft had flown 106·215M km without losing an astronaut in space. But it had been a very near thing, and hastened dramatic changes in the whole pattern of future Apollo flights.

It was 5 years later, in NASA's official Apollo history, before the full details of what went wrong were revealed by Jim Lovell himself. The CMS's Oxygen Tank No 2 overheated and blew up (putting No 1 out of action as well) because its heater switches welded shut during excessive pre-launch electric currents. Originally Tank No 2 had been installed in Apollo 10; there was trouble with it then, and it

Apollo 13's damaged SM, seen when jettisoned, shows damage from oxygen tank explosion

went unheeded, and the tank, damaged from 8 hr overheating, was a potential bomb the next time it was filled with oxygen. That bomb exploded on Apr 13 1970—200,000 miles from Earth.'

Apollo 14 Feb 1971 The 3rd successful Moonlanding expedition, commanded by Alan Shepard, America's first man in space, and the only one of the original 7 Mercury

Apollo 13's LM, seen when jettisoned, served as lifeboat to bring crew safely home

was removed for modification and damaged during the removal. 'I have to congratulate Tom Stafford, John Young and Gene Cernan, the lucky dogs, for getting rid of it,' wrote Lovell. By the time the tank was fitted in Apollo 13, the permissible voltage of the heaters had been raised from 28 to 65v DC, but unfortunately—'while contractors and NASA test teams nodded'—the thermostatic switches on the heaters were not modified to suit the change. During pre-launch tests Tank No 2 refused to empty, and the test director decided to 'boil off' its oxygen by using the electrical heater within the tank. For 8 hr 65v DC power was applied from ground support equipment to switches designed for 28v. That was why the switches ultimately welded shut, and this is how Lovell concluded his description of this famous drama: 'Furthermore, other warning signs during testing

118

astronauts finally to set foot on the Moon. Shepard's leadership, together with the immense physical endurance he displayed, proved that criticisms of his selection as Commander at the age of 47, and after recovery from serious ear trouble, were completely unjustified. He had only flown one 15-min space 'lob', and his companions, CM Pilot Stuart Roosa, and LM Pilot Edgar Mitchell, had had no previous spaceflight experience at all, so this was easily the least experienced of the Apollo crews. For the first time an Apollo launch was late; the countdown was held for 40 min 2 sec after reaching $T-8$ min. This was in conformity with new rules following the lightning strike when Apollo 12 was launched. There were again rainclouds and squalls in the Cape area, and the launch was finally made through a hole in the weather, and against a background of distant lightning flashes.

The flight to the Moon was accompanied by a series of technical problems, the first and most serious occurring soon after TLI—3 hr 14 min after lift-off, and 11,590 km from Earth. At the end of the Transposition and Docking Manoeuvre, Roosa had to make 6 attempts before he finally succeeded in docking with the Lunar Module. He reported that the probe had been correctly aligned with the hole in the drogue, but instead of the capture latches locking the 2 craft together, the CM bounced out again. The reasons for this became the subject of anxious discussion between Apollo 14 and Mission Control throughout the flight to the Moon, with consideration as to whether it was safe to continue in case the docking manoeuvre proved equally difficult and fuel-consuming when Antares, the LM, returned from the lunar surface. One possibility was considered to be that Roosa carried out his dockings too gently; another that there was some form of 'contamination', such as ice, on the capture latches, since the probe, when dismantled and brought inside the spacecraft, was found to be in perfect order. During the remainder of the flight to the Moon, it was possible to cancel the 3rd midcourse correction, and make other fuel economies which finally compensated for the extra quantity used during the docking problem. Perhaps the most important experiment made during the outward journey was a study of the optical flashes thought to be due to cosmic rays penetrating the optic nerves, and frequently noticed by astronauts.

For the first time on this mission the CSM was placed in a very low lunar orbit, with an apocynthion of 112 km, but low point of only 15,250 m. This meant that the LM would have more fuel available for hovering and selecting a safe landing point. After separation on the 12th lunar orbit Roosa took Kitty Hawk, the CSM, up to a 112km circular orbit. Antares had problems with the abort system and landing radar, but Shepard finally brought the spacecraft to a safe touchdown at 10.18 GMT on Feb 5. The landing point was only 26·5 m from the target. Although it was 75 sec late, there were 60 sec of fuel left, compared with Armstrong's 20 sec on Apollo 11. When Shepard stepped on the Moon 50 min late because of trouble with his portable communications system, he commented: 'It's been a long way but we're here.' The first EVA, from depress to repress, lasted 4 hr 44 min; it took longer to deploy ALSEP than expected, and Mitchell's 'thumper' device, which should have fired small explosive charges on the surface of the Moon to be recorded on the seismometer, was not completely successful; only 14 of the 21 charges went off. Next day Shepard and Mitchell loaded up the Modularized Equipment Transporter (MET), and towing it, set off for the rim of Cone Crater, nearly 1·6 km away and 122 m above the

TYPICAL SPACE SUIT FOR LUNAR EXTERIOR OPERATIONS   Later Apollo Missions

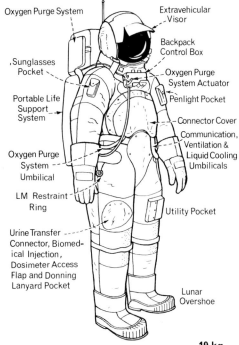

Oxygen Purge System

Extravehicular Visor

Backpack Control Box

,Sunglasses Pocket

Oxygen Purge System Actuator

Penlight Pocket

Portable Life Support System

Connector Cover

Communication, Ventilation & Liquid Cooling Umbilicals

Oxygen Purge System Umbilical

LM Restraint Ring

Utility Pocket

Urine Transfer Connector, Biomedical Injection, Dosimeter Access Flap and Donning Lanyard Pocket

Lunar Overshoe

19 kg
Approx. Weight of Garment 42 lbs

touchdown point. After 2 hr 10 min they were 50 min behind schedule, and tiring. Shepard's heart rate reached 150, Mitchell's 128; they finally turned back short of the rim, and one of the highlights of the mission, rolling stones down the inside of 38m deep Cone Crater, had to be abandoned. Shepard made up for it when TV viewers saw him become the first lunar golfer. He produced 2 golf balls and used the handle of the contingency sample equipment to tee off. Their second EVA lasted 4 hr 41 min.

There were no docking or other problems on the return flight, the only unusual aspect being that the docking probe was dismantled, stowed and brought back for examination, despite the extra discomfort, and the risk associated with the possibility that it might break away. A record 44 kg of lunar rocks and soil was brought back to Earth. The crew were the last lunar landers to have to endure the 3-week quarantine procedure after their return.

Apollo 15   Jul 1971   The first of 3 'J Series' missions, intended to exploit the scientific potential of the Apollo hardware. Lasting 12 days 7 hr, it was 2 days longer than any previous Apollo flight; and such an unqualified success that one estimate was that David Scott, the Commander, Alfred Worden, the CMP, and James Irwin, LMP, had brought back as much scientific information as Sir Charles Darwin acquired during his 5-yr voyage in *Beagle* in 1831–6. The all-US Air Force crew had in fact named the LM 'Falcon' after their Air Force mascot, and the CSM 'Endeavour' after the ship that carried James Cook on his 18th-century scientific voyages. Despite pre-launch problems—the launchpad was struck by lightning 11 times—the countdown itself was flawless, and lift-off took place only 0·008 sec late. The Saturn 5, developing 3,556,160 kg thrust, was the most

Apollo 15's open SIMBAY studying Sea of Fertility

the damaged instrument, even though that was for use during the hazardous touchdown, but because of the danger of glass splinters getting into the hatch seals and causing a pressure leak when the LM was finally jettisoned. Alarm over a water leak and signs of flooding were quickly allayed by instructions from Mission Control on how to tighten up a valve. The crew donned their spacesuits, when, a few hours before entering lunar orbit, they blew a 2·90m by 1·52m panel off the SM, to expose 8 scientific experiments in the SIMBAY—the bay containing the Scientific Instrument Module. The experiments included mapping and panoramic cameras, and spectrometers, 2 of them on extendable booms more than 6·1 m long, for measuring the composition of the lunar surface, solar X-ray interaction, and particle emissions while the CSM spent 6 days orbiting the Moon. The SIMBAY also carried the 35·4kg subsatellite, which was successfully ejected following the conclusion of the lunar landing to spend a year studying the Moon's mascons and other phenomena.

Falcon made a successful touchdown at 22.16 GMT on Jul 30, within a few hundred metres of the target, despite much bigger perturbations than expected in Apollo 15's lunar orbit, because it was passing over more mascons than ever before; and despite the fact that separation of Falcon from Endeavour failed at the first attempt because an umbilical, or power line, was not tight. Separation was completed 35 min late on the 12th revolution in time for the planned descent on the 14th. Falcon had about 3050m clearance over the top of the 3960m Apennine Mountains as it dropped down to land in a basin near Hadley Rille, a 366m chasm named after an 18th-century English mathematician. Falcon came to rest with a 10° tilt because one footpad was in a 1·52m crater—a tilt which caused Scott and Irwin much

powerful to date; but total lift-off weight, at 2,946,088 kg was not a record. At 47·7 tonnes, the spacecraft was 2 tonnes heavier than previous vehicles sent to the Moon—made possible by such measures as reducing fuel reserves and the number of retro-rocket motors on Saturn's first 2 stages.

Translunar injection, transposition and docking were completed without any of the problems encountered on Apollo 14. A midcourse correction was brought forward 2½ hr after a warning light had indicated a short-circuit in one of the 2 systems for firing the SPS. For a time the lunar landing was in doubt; but Scott, by manually operating circuit breakers, established that the faulty system could be used for back-up purposes; the mission could therefore continue by using the B, instead of the A system, for SPS firings. Broken glass on an altimeter in the LM, when Scott and Irwin carried out their first inspection, caused some concern throughout the mission—not so much because of

Hadley Delta: Irwin loads first Lunar Rover

inconvenience with the swinging hatch door when they were crawling in and out in their cumbersome space suits for 3 EVAs, which gave them a total of 18 hr 36 min on the surface.

They had some difficulty deploying the Lunar Rover, the first lunar motor car, and only on the 2nd EVA, following advice from MC, were they able to get the front-wheel steering to work. Rear-wheel steering was available, however, and though they found that when driving in one-sixth G the vehicle was 'a real bucking bronco', making seat belts essential, Scott and Irwin were delighted with it. When they reached the rim of Hadley Rille, 4 km from the LM, the world shared their view when they turned on the TV camera mounted on the Rover, and deployed the umbrella-like

antenna which transmitted excellent pictures direct to Earth. On this first EVA, however, energy and thus oxygen consumption was 17% greater than expected, and MC warned that activities would have to be curtailed. Before it ended, however, the 3rd ALSEP was deployed—an exciting moment for the geologists, confident that, by using the Apollo 12, 14 and 15 ALSEPS, they could finally pinpoint, and even predict, the monthly Moonquakes occurring in 11 areas (mostly in the Copernicus region) each time the Moon passes at its closest point to Earth. MC gave much unavailing advice to Scott, who was watched by millions as he struggled with the percussion drill, leaning on it and panting heavily as he bored holes in the surface, both for bringing back core samples and inserting instruments for measuring the heat flow from the interior. One core 2·6 m long, was found to contain 57 separate layers of soil, illustrating 2400M yr of lunar history. The thickness of the layers ranged from 1·27 cm to 12·7 cm, each probably representing a different meteorite impact which deposited a new layer of loose top soil.

The 2nd EVA started badly. First there was a 11kg water leak to be mopped up; then Irwin had trouble with air bubbles in his portable life support system (PLSS), and had to get Scott to tape on his communications antenna, which had broken off. But their 12·5km tour of Hadley Mt's foothills, collecting rock samples, proved to be the most rewarding period so far on the Moon. Scott excitedly pronounced that he thought they had discovered what they went to find: a 10cm piece of crystal rock—geologists call it 'anorthosite'—believed to be part of the pristine Moon, and quickly dubbed the 'Genesis rock'. (Later, geologists decided it was 4150M yr old, and 150M yr older than any previous sample recovered; but it was *not* as old as the Moon itself.) Irwin took a spectacular tumble, and had to be helped up by Scott—who in turn fell during the 3rd EVA. Because they adapted themselves unexpectedly quickly, benefiting both from the Rover and from their improved and more flexible spacesuits, Scott and Irwin used much less energy on the last 2 EVAs, and they did not have to be shortened as much as anticipated. However, with MC insisting that 100% success had already been achieved, the planned 2nd visit to the rim of Hadley Rille was eliminated. The astronauts finally returned to Falcon after driving a total of 28 km, and having collected 78 kg of rock samples.

The TV camera on the Rover was carefully set by Scott so that, for the first time, lift-off from the Moon could be televised—an astonishing 2 sec of TV as the screen burst into red and green as the ascent stage rushed upwards, to rejoin Worden, who had had a busy 73 hr alone in Endeavour, conducting the orbital science experiments. There was a 3-hr drama in lunar orbit when jettisoning of Falcon had to be delayed because of doubts as to whether Endeavour's hatch was adequately sealed. The astronauts had to re-open the docking tunnel, check the seals, and delay LM jettison for one orbit; when it was achieved Falcon was in front instead of behind Endeavour, and there was anxiety—and, as Mission Control admitted, 'confusion' there—about how to ensure that the 2 craft did not collide. Finally the danger was averted, the LM impacted on the Moon, and the weary astronauts ignored advice to take sleeping pills at the end of a 20-hr day. Satellite ejection was followed by TEI on the 74th lunar orbit, then came Worden's 19-min spacewalk, 321,870 km from Earth, to retrieve 2 film cassettes from the SIMBAY, according to plan. It was the first spacewalk ever made for a practical, working purpose. Geologists were disappointed when the TV camera on the Rover failed, and

Artist's reconstruction of first televised lift-off from Moon

they were unable to view the Sun being eclipsed by the Earth; but interest in this was itself eclipsed by the drama of re-entry, when, for the first time, one of the 3 25·3m parachutes failed to open. The recovery carrier warned the crew to 'stand by for a hard landing'. The crew emerged none the worse for their dramatic splashdown.

The reason for the parachute failure was never conclusively established; most likely causes were either faulty

123

metal links connecting the suspension lines, or residual RCS fuel burning through the shroud lines while being dumped during the final descent. On following missions the links were strengthened and the fuel retained on board.

Apollo 16 lift-off

Apollo 16   Apr 1972   The 5th Moonlanding was a triumphant success, achieved despite the fact that at one period the mission appeared to be in an Apollo 13 situation, with the possibility of an emergency return to Earth. It lasted 11 days 1 hr 51 min, 24 hr 45 min less than the original flight plan. Launch on Apr 16 was on time, despite a threatened delay until 40 min before lift-off because of a gyroscope problem in the Saturn 5 instrument unit. Total lift-off weight was 2,914,000 kg; lift-off thrust was 3,503,405 kg. Spacecraft weight was 48,602 kg.

First of about a dozen technical problems which kept John Young, Commander, Thomas (Ken) Mattingly, CMP, and Charles Duke, LMP, troubleshooting for much of the outward journey, occurred $8\frac{1}{2}$ hr after lift-off: Young feared that particles and flakes flowing past the windows might indicate a nitrogen leak; Mission Control instruments showed no such leak, but Young and Duke were ordered to open up the docking tunnel and check out the Lunar Module (Orion) 24 hr earlier than planned. Orion was in good order; and meanwhile, Mattingly, operating the TV camera, enabled Mission Controllers to establish that thermal paint was flaking off some of Orion's panels. After some alarm about possible causes, it was decided it was not serious. Mattingly was soon busy with an electrical signal fault affecting the vital Guidance and Navigation System in the Command Module (Caspar); the Massachusetts Institute of Technology worked out a computer programme to ensure that the fault would not abort critical manoeuvres such as lunar orbit and descent orbit insertion. The crew were partially compensated for their extra work; only one of 4 scheduled midcourse corrections was required.

The major crisis came following undocking on the 12th lunar orbit. Mattingly, having delivered Orion into a 107 by

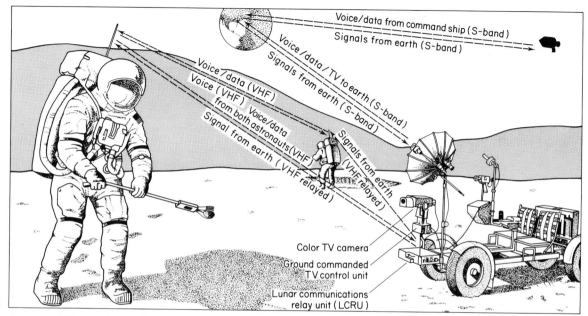

Voice/data from command ship (S-band)

Signals from earth (S-band)

Voice/data/TV to earth (S-band)

Signals from earth (S-band)

Voice/data (VHF) from both astronauts (VHF)

Voice (VHF)

Signals from earth (VHF relayed)

Signal from earth (VHF relayed)

Color TV camera

Ground commanded TV control unit

Lunar communications relay unit (LCRU)

Astronauts' Communication System with Earth and CM

19km orbit, had to fire Casper's main SPS engine to circularize his orbit and be available for a possible rescue manoeuvre before Orion attempted its lunar landing. He did not do so because of indications of yaw oscillations in the back-up system for firing the engine. With only minutes to go before starting their final descent, Young and Duke were given the first 'space wave off': both craft were ordered to continue orbiting, and to reduce the gap between them, ready to re-dock, while Mission Control studied the problem, and even talked of an Apollo 13 situation, in which the lunar module might once again have to be used as a life raft and tow home a disabled Command Module. But on the 15th

Young tests wheelgrip by skidding Rover

orbit Christopher Kraft, Houston's director, gave a 'go' for landing. Mattingly, using the primary system, successfully fired the SPS engine to circularize his orbit; Young and Duke finally settled 6 hr late on the Cayley Plains of the Descartes landing site at 02.23 GMT on Apr 21, 200 m N and 150 m W of their Double Spot landmark. They were only about 3 m from a crater some 7·6 m deep.

Young had become the first man to orbit the Moon twice;

he and Duke, the 9th and 10th men on the Moon, became the most enthusiastic, most vocal, and most efficient of them all; but after their exhausting experience during the approach, they elected to sleep before making the first EVA. Duke, who had complained earlier about his spacesuit being tight, had no difficulty at the start of the first 7-hr 11-min EVA. The one major disappointment on the Moon was that Young backed into, and broke, the cable connecting the top-priority heatflow experiment to the ALSEP power plant, thus destroying the scientists' hopes of checking the discovery by the Apollo 15 crew that heat was flowing from the lunar interior at a far higher rate than expected. But an improved drill this time enabled Duke to obtain 3m deep core samples without difficulty. Although at a later stage a rear mudguard (fender) fell off the Lunar Rover, causing the astronauts to be sprayed with soil, it performed well, and at the end of EVA 1, Young, filmed by Duke, carried out the planned 'Grand Prix'. He drove it flat out in circles, skidding it to test wheel grip.

EVA 2 dramatically demonstrated the improved quality of TV pictures; when they had driven 4·1 km to Stone Mountain, and 232 m up it, the astronauts could be seen working on ridges and crater rims; geologists on the ground used the TV camera completely independently to look around, and zoom in for close-ups of boulders and other features of interest. The geologists were embarrassed by the lack of evidence of volcanic activity.

EVA 3 was shortened to 5 hr 40 min to meet the revised lift-off time; it was devoted to exploring North Ray crater, a large depression 4·9 km from Orion. Before re-entering for the last time, Young and Duke competed to see how high they could jump; Duke, who had tumbled many times, fell heavily backwards causing much concern at MC, but escaped with no personal or spacesuit damage. Ascent and docking went perfectly; but a wrongly positioned switch caused Orion to start tumbling immediately after undocking. Casper had to make a hurried evasive manoeuvre, and Orion had to be left in lunar orbit instead of being impacted on to the lunar surface. In view of the slightly suspect back-up system for firing the SPS, the journey home started as soon as the 40kg subsatellite had been ejected. (The planned plane change, to place the subsatellite in a more suitable inclination, was cancelled, to avoid another SPS burn. As a result, the subsatellite lasted only a few weeks; after 425 revolutions it crashed into mountains on the farside on May 29 1972.) The highlight of a relatively quiet return flight was Mattingly's EVA 274,000 km from Earth to recover film cassettes from the SM SIMBAY. All 3 parachutes deployed perfectly after re-entry, and were recovered for examination because of the Apollo 15 failure. The CM, however, turned upside-down in the water until the buoyancy balloons were released. Residual propellants were retained on board for the first time, because of fears that they might have damaged Apollo 15's parachute lines; but the wisdom of venting them on previous missions was demonstrated when these propellants exploded after the spacecraft had been taken back to North American at Downie, California, causing damage and injuring a number of technicians.

A special diet of food and orange-juice laced with potassium enabled this crew to become the first to survive a mission without any irregular heartbeats, and removed doubts about the practicability of 56-day flights planned for Skylab. Young and Duke clocked up record totals of 20 hr 40 min on the lunar surface; drove 27 km; and brought back an estimated 97·5 kg of lunar samples.

Apollo 17 Dec 1972 The final Moonlanding proved to be the most perfect of the Apollo expeditions. The crew, Eugene Cernan, Commander, making his 3rd flight, Ronald Evans, CMP, and Dr Harrison (Jack) Schmitt, LMP, the first trained geologist to visit the Moon, stayed longer, travelled further and brought back more lunar samples, 113 kg, than any previous crew. But America's first night launch was both tense and exciting. Timed for 21.53, Cape Kennedy time, it came to an abrupt halt at T minus 30 sec when the automatic firing sequence failed to order pressurization of the Saturn 5's 3rd-stage LOX tank. The countdown clocks were put back to $T-22$ min, while technicians located a faulty computer circuit; but at $T-8$ there were further repeated holds; it seemed that the $3\frac{1}{2}$-hr launch window, all that was available if Apollo 17 was to arrive at the Moon with the Sun angle exactly right for a landing on the edge of the Sea of Serenity, would expire. But finally, 2 hr 40 min late, half a million people saw night turned into day by the pure gold flame as Saturn 5 completed its unbroken success record by placing the spacecraft accurately on course. Total lift-off weight was 2,923,383 kg; lift-off thrust was 3,476,794 kg. Spacecraft weight totalled 52,744 kg: CM America 30,320 kg, LM Challenger 16,448 kg.

The journey to the Moon was enlivened by a series of Master Alarms warning of a major problem that did not exist; by the need to remove the docking probe because 3 of the 12 docking latches failed to fasten during the transposition and docking manoeuvre; and by Jack Schmitt's continual scientific commentaries as he looked back to Earth. Not everyone appreciated his detailed weather forecasts and analyses of continental drift. But trajectory adjustments made up all the lost time, and Apollo 17 was back on schedule when, at 19.47 GMT on Dec 10, it went

HERE MAN COMPLETED HIS FIRST

EXPLORATIONS OF THE MOON

DECEMBER 1972, A.D.

MAY THE SPIRIT OF PEACE IN WHICH WE CAME

BE REFLECTED IN THE LIVES OF ALL MANKIND

EUGENE A. CERNAN
ASTRONAUT

RONALD E. EVANS
ASTRONAUT

HARRISON H. SCHMITT
ASTRONAUT

RICHARD NIXON
PRESIDENT, UNITED STATES OF AMERICA

Final EVA at Taurus-Littrow: Schmitt studying huge split-boulder

into lunar orbit. While Cernan was recovering from gastric pain, and the seismometers recording the impact, equivalent to 11 tonnes of dynamite, of the S4B, Schmitt reported seeing a light-flash near Crater Grimaldi, possibly caused by a meteorite strike. On their 12th revolution, Cernan and Schmitt separated the LM Challenger from the CM America; when Evans had safely established America for a possible rescue manoeuvre in a 125 × 100km orbit, Cernan swept Challenger down between 2 2130km high mountain ranges, to touch down between the Taurus Mountains and Littrow Crater, on the edge of the Sea of Serenity. Cernan lamented that, with nearly 3 min of hovering fuel left, he could have hovered over the scarp: 'No thank you,' replied Mission Control, 'we're happy right where you are.' The area had

been chosen by scientists because Alfred Worden, orbiting in the Apollo 15 CM, had seen conical mounds, similar to those formed on Earth by volcanic debris accumulating around a vent, with the surrounding area ranging from the most ancient highlands to young mantling material, or dust, overlaying the lowest areas. The 6th and final Apollo landing co-ordinates were 20°12′16″N and 30°45′0″E; just over 4 hr after the landing at 19.55 GMT on Dec 11, Cernan dedicated 'the first step of Apollo 17 to all who made it possible'; Schmitt had more difficulty than his predecessors in adjusting to one-sixth gravity, took some heavy falls and admitted it was embarrassing for a geologist. The US flag which had hung in Mission Control ever since Apollo 11 was somewhat emotionally deployed, and the 2 men settled down to the heavy task of setting out America's 5th lunar observatory; to save the weight of the camera, we had no TV of the descent from the LM to the surface; but during the ALSEP work, we had both panoramic and close-up views, showing with perfect clarity the smooth-topped mountains which surrounded them, and at one place, an avalanche of light material. Cernan, after some initial difficulty which used more of his oxygen than expected, succeeded in drilling both the 1m holes required for the Heat Flow Experiment; after the Apollo 16 failure, this had been given top priority to check Apollo 15's finding that twice as much heat was flowing up from the Moon's interior as had been expected. Because Cernan had inadvertently knocked off the Lunar Rover's rear right fender, the astronauts and their equipment were showered with Moondust during their first drive. It began 40 min late, and because of this, and high oxygen consumption, had to be shortened. They drove south to Steno Crater, and gathered 21·7 kg of samples. Back inside Challenger, they were advised by John Young, from Hous-

Schmitt working with lunar scoop: GNOMON in foreground

after many anxious urgings by Mission Control, because, if the Rover broke down, it would have taken them $2\frac{3}{4}$ hr to walk back. Then came the highlight of the whole mission at Crater Shorty, on the way back, which Mission Control had wanted to cut out. Cernan kicked up orange soil, and for 20 min, with Schmitt scrambling excitedly around the rim of the crater, scientists were riveted to the TV screens; the remotely operated zoom lens clearly showed the rust-coloured soil. There was a general agreement that the astronauts had found a volcanic vent, and that there had been, and perhaps still was, water on the Moon. Crater Shorty was re-named 'Volcano Shorty'—prematurely, as it turned out. For 3 months later, following analysis of the orange soil, the US Geological Survey announced that it consisted of tiny, orange-coloured glass beads, formed by the heat generated by an ancient meteor impact. By 1975, however, it was decided that the beads, unusually rich in lead, zinc and sulphur, were of volcanic origin, but had been thrown out from perhaps 300 km inside the lunar interior, during a period when the lunar mantle had partially melted. With this still in the future, Cernan and Schmitt re-entered the LM after travelling a record 19 km and setting up an EVA record of 7 hr 37 min 21 sec. From lunar orbit, Evans had been adding to the excitement with sightings of more orange-coloured areas and what appeared to be 2 volcanoes on the nearside, and one on the farside. The final EVA began with anti-climax, when Crater Van Serg, which geologists confidently expected would be another volcano, turned out to be an impact crater of little interest. But there was dramatic TV of Schmitt scaling boulders the size of London buses to collect samples; at times the operation looked more like an Everest climb. The Lunar Rover was taken up mountain sides so steep that its wire-mesh wheels were dented; and

ton, how to improvise a makeshift fender, or mudguard, by taping together 4 of their lunar maps, and using emergency lighting clips to fasten it to the Rover at the start of the 2nd EVA. Schmitt began this, somewhat depressed after being instructed to cancel some of his own activities. When he asked what he was supposed to do, Mission Control, in a rare moment of curtness, replied: 'Help Gene, I guess.' But then came a full hour's drive, taking them over 6·4 km W from the LM, to the foot of the 2130m high mountain range; after taking rockslide samples, Schmitt came away reluctantly, both astronauts, after struggling with a series of football-sized rocks (found on previous missions to be the most rewarding type of sample) showed signs of real exhaustion. Before they returned to Challenger, however, students from 75 countries assembled in Mission Control to watch Cernan and Schmitt, against the background of the US flag stretched on its wire frame, select and dedicate a Moonrock to be shared among them. To mark the completion of the 6 Apollo Moonlandings, Cernan then unveiled a plaque on the leg of the LM, signed by President Nixon and all 3 crew members, reading 'Here man completed his first exploration of the Moon, December 1972, A.D. May the spirit of peace in which we came be reflected in the lives of all mankind.' During their 3 EVAs, Cernan and Schmitt had travelled a total of 35 km in the Rover, and spent 22 hr 6 min outside Challenger. They had been on the Moon's surface for 74 hr 59 min.

Its record load of 113 kg of Moon rocks meant that the LM was 18 kg overweight; nevertheless, lift-off, vividly seen via the TV camera carefully pre-positioned on the abandoned Rover, went smoothly; the extra weight was counterbalanced by re-organizing the engine firing to burn off more fuel more quickly. During a communications failure just after lift-off, Mission Control hurriedly passed messages through America; and it took Evans nearly 20 min to complete docking after a too-gentle approach resulting in the 2 craft, half-joined, rotating against each other. Following transfer of men and materials, Challenger was jettisoned and deliberately crashed into the South Massif mountains just visited by Cernan and Schmitt; but it fell 14·5 km, instead of 9·6 km from the TV camera, which failed to pick up the expected picture; the crew, however, reported seeing a bright spot at the point of impact. 2 more days were spent in lunar orbit, mainly mapping the farside; on their 75th lunar orbit, the SM engine was fired to start the homeward journey, with the TV camera simultaneously sending back the first perfect views of Crater Tsiolkovsky, with close-ups of the mountain peak in the centre of the 200-km wide crater. The main concern during the homeward journey was repeated searches for a missing pair of scissors, which might inflict lethal wounds if they flew out of some recess during re-entry. They were never found, and must have drifted into space during Evans' 60-min cislunar

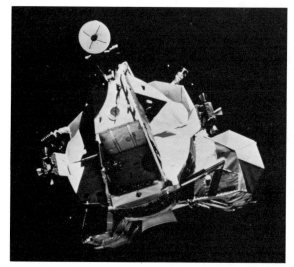

LM Ascent Stage about to RV with CSM

Apollo 17: Evans collects film cassettes during transearth EVA

spacewalk, 290,000 km from Earth, during which he retrieved film and data canisters from the SIMBAY without difficulty. Debriefings, during which Cernan gave Schmitt full credit for first spotting the orange soil (Schmitt had

always urged that astronauts should kick up the surface dust as they moved about to expose what was beneath) and a TV news conference, enlivened the rest of the homeward journey. Splashdown in the Pacific, with the CM settling right way up for once, was 6·4 km from the recovery ship. During the 12-day, 13-hr, 51-min flight, the CM (and Evans) had travelled 2,390,984 km.

Summary and Conclusions The total cost of Project Apollo, including the development and production of all the hardware, the facilities, tracking networks and launch costs, are estimated at £10,000 million ($24 billion). The main disappointment was that, 2 yr after Apollo 17, man seems as far as ever from understanding the origin of the Moon. None of the 3 alternatives has been ruled out: that it consists of material which broke away from Earth; that it formed from a cloud of debris (perhaps like the rings of Saturn) which once surrounded Earth; or that it was originally an independent body in solar orbit, captured by Earth's gravity. Lunar scientists think there is now most evidence for the last theory; and at the same time, following new information about Mars and Mercury gained by Mariners 9 and 10, there is growing evidence that all the terrestrial planets started from common material. The Apollo 15 and 17 Heat Flow Experiments show that the amount of heat escaping from the lunar interior is surprisingly high—about half that of Earth. While, by comparison, the Moon is seismically 'quiet' there are many small quakes, possibly triggered by Earth's tides, about 800 km below the surface. Its crust is more than 60 km thick and below 1000 km the interior is partially molten. The centuries-old arguments as to whether the Moon is hot or cold, alive or dead, are at an end; but so far, all that the Apollo data has done is to change the type of questions being asked. What is needed next is a polar orbital flight around the Moon; and of course, manned exploration of its polar surfaces and of the farside. Man's first Moonlandings, the technical mastery of space travel and the final scientific expeditions to the lunar surface, must inevitably rank as the most significant step so far in man's development.

Nevertheless, NASA themselves admit that, 'since the Apollo landing sites were highly restricted by operational and safety limitations, most of the Moon's surface, containing many intensely interesting and mysterious features, remain untouched, awaiting mankind's next series of contacts.' Since the US has no plans for further manned missions to the Moon until the 1980s, by which time the Space Shuttle should have been developed, it seems likely that manned inspection of these mysterious lunar features will be taken over during the next decade by Soviet explorers. But Project Apollo should enable the US, if it so desires, to be the first nation to establish permanent manned bases on Space Station Moon. They now have first-hand experience of the fact that the lunar material is firm enough to support structures, and yet can be easily dug for the construction of shelters and protective barriers. The examination of parts of the Surveyor 3 spacecraft, recovered by the Apollo 12 crew after Surveyor had been on the Moon's surface for 31 months, has established that there should be no unexpected difficulties with permanent installations set up on the Moon. Most important of all, the total of almost 385 kg of lunar materials amassed by the conclusion of Apollo 17, should enable techniques to be developed for making lunar bases self-supporting. By the end of Apollo 14, it had already been established that both water and oxygen could be produced from the lunar soil, since it contains a high percentage of iron oxide.

By using a solar furnace and introducing hydrogen, 6·35 kg of water could be produced from 45·4 kg of iron ixode, and the water then separated into oxygen and hydrogen. Thus life could be supported and fuel provided for space vehicles.

APPLICATIONS TECHNOLOGY SATELLITES

History Arthur C. Clarke, who first conceived the idea of synchronous satellites, described ATS-6 as 'the most ambitious and important educational experiment in history'. It was launched in 1974 to beam direct TV programmes into village homes in under-developed countries; its achievements since can be found below. Originally ATS was conceived as a 5-satellite series, starting in Dec 1966, to test components and techniques for communications, meteorological and navigation satellites. But ATS 1 and 3 proved that predictions made a century earlier that such vehicles would be able to provide 'fixes' to within 10 m of any point on the surface of the Earth were correct. Before the ATS programme, NASA tried out communications satellites at low, intermediate, and synchronous orbits; they included 'passive reflectors' such as the 33m dia Echo balloons and 'active repeaters' such as Relay and Syncom spacecraft. The ATS programme, based on the results of these experiments, included Syncom techniques, such as spin-stabilization and complicated synchronous orbit manoeuvres. From their geosynchronous orbits, 35,680 km above the equator, ATS satellites match Earth's rotational period and thus remain fixed over the same point; they view about 45% of the globe as compared to a spacecraft in a 320km orbit, which views only 3% of the globe at one time.

ATS-1 L Dec 7 1966 by Atlas Agena D from C Canaveral. Wt 352 kg. Soon after lift-off, this satellite, 1·42 m dia and 1·34 m high, with 23,870 solar cells, was successfully manoeuvred into its permanent position over the Pacific; at the end of 1976 it was still operational, though transmitting only on command. In addition to many scientific and technological experiments, it provided a wide variety of services. The capabilities of such satellites were vividly demonstrated by providing Australia with 10 hr of TV from Canada's Expo '67; on Oct 12 1967, the Japanese were able to see their Prime Minister's visit to Australia; and later that year the world saw President Johnson and other heads of State attending the funeral of Australia's Prime Minister Holt. Many lives and millions of dollars were saved in Alaska, during that same year, when ATS-1 provided emergency communications during the great flood there. In 1969, when the Apollo 11 astronauts returned from the first Moonlanding, ATS-1 provided the main communications link between President Nixon and the USS Hornet.

ATS-2 L Apr 6 1967 by Atlas Agena D from C Canaveral. Wt 370 kg. Intended to be gravity-gradient-stabilized, placed in a medium 11,100km equatorial orbit; but a failure in the Agena rocket prevented the booster from re-igniting at a crucial point in the trajectory. As a result, the satellite remained in a $185 \times 11,177$km orbit at $28°$ incl. Severe decelerations in this irregular orbit were too great for the gravity-gradient stabilization system to overcome; limited data was obtained from the systems and experiments as ATS-2 slowly tumbled and rotated. It re-entered on Sep 2 1969.

ATS-3 L Nov 6 1967 by Atlas Agena D from C Canaveral. Wt (after apogee motor burnout) 365 kg. A perfect launch

ATS-3 Earthview shows Americas, Africa and Greenland ice-cap

placed this satellite in a synchronous equatorial orbit of $35,765 \times 35,807$ km over the Atlantic Ocean, where it was still operational, and transmitting on command, at the end of 1975—8 yr later. The most notable of many achievements was that it sent back the first colour photograph of Earth from space; it also obtained many cloud pictures, and was used to monitor severe storms. On Nov 21 1967 it provided the first ground-to-spacecraft-to-aircraft communications link over the Atlantic during a Pan American flight; transmissions to and from the aircraft via the satellite were monitored not only in America but in London, Hamburg, Frankfurt, and Buenos Aires. Later ATS-3 provided support for the Apollo Moonlandings and TV coverage of Pope Paul's visit to Bogota, Colombia, in 1968. In a series of maritime experiments, involving ship location and ship-to-shore communications, it demonstrated that such satellites could bring major improvements in the management of shipping fleets. At the end of 1974 it was being used by the US National Oceanic and Atmospheric Administration (NOAA) in experiments to predict and measure rainfall in remote areas by comparing satellite pictures of the brightness and size of clouds with radar scans of their density. This system will provide flood warnings to farm populations in mountain areas.

ATS-4   L Aug 10 1968 by Atlas-Centaur from C Canaveral. Wt 392 kg. Intended to be a gravity-gradient spacecraft, but because of a Centaur failure, it was left in a 217×772km orbit. Although its systems were turned on, little data was obtained. Decay was on Oct 17 1968.

ATS-5   L Aug 12 1969 by Atlas-Centaur from C Canaveral. Wt (after apogee motor jettison) 433 kg. Another gravity-gradient spacecraft, still providing useful information on command at the end of 1976. From the start, however, only 9 of its 14 experiments were operational because the spacecraft began to wobble on its spin axis shortly after separation. The attitude control hydrazine jets had to fire 15 times faster than normal to reduce the wobble to required limits. The satellite was allowed to drift to its planned geostationary location at the equator, 1770 km W of Quito, Ecuador, taking 20 days, instead of 11 hr if the apogee motor had been fired a 2nd time. ATS-5 concluded its flight upright, but wrongly spinning counter-clockwise; this meant its gravity-gradient experiment booms could not be deployed. However, its auroral particle experiments sent back a large quantity of useful data until the end of 1974. The US tanker Manhattan was used to demonstrate the ability of ATS-5 to determine its exact position by L-Band ranging signals from the Mojave (Calif) ground station and to exchange teletype messages with it on unattended equipment on board the ship. Other tests included transmissions for the Federal Aviation Administration to a Boeing aircraft; and investigation of the effects of rain on signal 'fading'.

ATS-6   L May 30 1974 by Titan 3C from C Canaveral. Wt 1402 kg. Orbit $35,781 \times 35,791$. Incl $1 \cdot 6°$. The world's first educational satellite, intended to bring the benefits of space technology to backward areas by lessening ignorance and poverty. 14 hr after launch it was perfectly stationed over the equator above the Galapagos Islands, with a 40% global coverage, and capable of being re-located E or W. ATS-6 is $8 \cdot 51$ m high, topped by a 9m dia reflector antenna, with 2 combined VHF antennas and solar panels above that on the ends of a 16m boom. Radiating 200,000W of effective radio-energy, compared with 6400W produced by Intelsat 4, combined with the capability to point the antenna within one-tenth degree of arc, meant that it was powerful enough to beam colour TV direct to thousands of small ground receivers. It was essentially a full-size ground transmitter placed in orbit. It began its 20 experiments by providing the first satellite educational course on Jul 2 1974; more than 600 elementary school teachers in 8 Appalachian states took

ATS-6: Indian Technicians near Hyderabad prepare for Direct Satellite TV

a graduate-level course by means of its transmissions. After a year of providing remote schools in the Rocky Mountains with special courses, and 2-way communications for medical diagnoses in Alaska, the ATS Ground Station at Rosman, N Carolina, began the complicated operation of moving ATS-6 a distance of 12,800 km so that it could provide India with the promised year-long educational experiment. The 2-storey spacecraft was set in motion by 2 burns of its onboard rocket motor; the 2nd, lasting 5 hr 37 min 17 sec was the longest made so far of a chemical rocket in space. 3 separate braking manoeuvres, spread over a week, brought it to a halt at its new station above Lake Victoria in E Africa, 6 weeks later. There it played a major part in the Apollo-Soyuz flight of Jul 1975; for a 50-min period of each orbit, Apollo communications were passed to Houston via ATS-6. Its presence resolved a difficult issue between America and Russia as to where the joint docking should take place. Both countries wanted it to be within range of their own tracking stations; by using ATS-6, it was possible for it to take place over Europe, within range of Soviet stations, while at the same time direct transmissions to and from Houston were possible via the satellite. On Aug 2, only 9 days after the end of Apollo-Soyuz, the educational transmissions began to 5000 villages in 7 states in India. Half the villages were equipped with 3m dia antennas made of chicken wire, plus a TV set costing a total of $1000. India's Government-operated TV network (normally able to transmit to only 4 cities in India) had devised a series of simple programmes, demonstrating for instance, rice-planting, in which the distance between 2 plants is measured by the width of a man's fist; cooking lessons using only the primitive utensils available in an Indian village. And another lesson was aimed at persuading Indian mothers to feed solid-food (always,

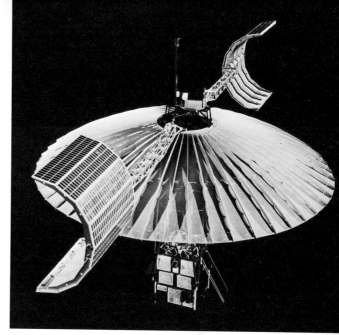

ATS-6 view shows 16m Solar 'Paddles', 9m dia reflector and Earth-viewing module below

again, food that was available) to their babies, instead of exclusively breast-feeding them until the age of 3. From a ground station at Ahmedabad, N of Bombay, the Indian Space Research Organization then transmitted the programmes via ATS-6; NASA made it available for 4 hr per day for 1 yr. In some villages there was anti-climax when their screens failed to work for the opening speech by the Prime Minister, Mrs Gandhi. But in many others hundreds of people, including naked children, crowded round single sets to watch the programmes.

One drawback to the experiment is that now it has been moved back again for further work over the US, it may be a decade before the Indian villages see TV again. Other 'firsts' achieved by ATS-6, called the 'L-band Experiment', included the first aircraft-to-ship communications relay via satellite; the first direct flight control of an aircraft by an ocean air traffic controller using a satellite; and the first search-and-rescue operation (simulated) directed by satellite. By the end of 1974, however, one of its experiments had failed: the Very High Resolution Radiometer, a camera-like instrument, trying out new ways of gathering meteorological information from synchronous orbits. It did, however, send back over 1000 images before the failure. It is expected to have sufficient fuel to continue operations until 1981: 7 years' work instead of the expected 2. As a result, proposals to launch the back-up ATS-6 were dropped as an economy measure. Cost of the mission was $180M for the spacecraft, plus $25M launch costs.

BIOSATELLITE

History   America's least successful space programme. It was originally intended to investigate the prolonged effects of weightlessness and radioactivity on living organisms. There were to be 6 flights, lasting 3, 21 and 30 days, the first 2 with plants and organisms; the next 2 were intended to

experiment with pig-tailed monkeys; the last 2 with rodents. As will be seen below, Bonny, the first of the pig-tailed monkeys, was sent up while the world's space correspondents were gathered at the Cape for the first Moon-landing. The principle of using animals in this way was heavily criticized before the flight. Astronauts themselves took the view that it was pointless to use animals, which were unable to cope with technical troubles, and that men were willing and able to take part in such experiments. Then, when Bonny was launched, and had such a bad time, there was so much world-wide criticism from the assembled correspondents that the last 3 flights were cancelled on the grounds of 'economy'. Apart from OFO-1 (Orbiting Frog Otolith) in 1970, in which 2 male bullfrogs were sent up to test effects on ears and balance, it was to be 6 yr later before NASA was able to resume such biological experiments—and then only by joining in a Soviet biological satellite, Cosmos 782. That was the first unmanned Soviet-American satellite—a follow-up to the successful first Apollo/Soyuz flight.

Bios 1   L Dec 14 1966 by TAD from C Canaveral. Wt 426 kg. Orbit 307 × 316 km. Incl 33·5°. Retro-rocket failure led to the recovery capsule, a covered aluminium bowl of 79cm dia with 0·1699 cu m of payload space, being stranded in orbit. Spacecraft re-entered 2 months later, on Feb 15 1967.

Bios 2   L Sep 7 1967 by TAD from C Canaveral. Wt 508 kg. Orbit 302 × 326 km. Incl 33·5°. Similar to Bios 1, this carried 13 experiments, with a Strontium 85 radiation source to subject the weightless plant and animal specimens to regular doses of radiation. Onboard watering systems, with lights and cameras, enabled the specimens to be

Astromonk Bonny before launch, wearing monitoring device

photographed every 10 min—268 times during the flight. The 3-day flight was cut short at 45 hr—17 orbits earlier than planned—because of communications problems and a tropical storm in the Pacific recovery area. Subsequent pictures of a pepper plant, looking healthy before launch, showed it looking very dejected indeed after 17 hr 40 min of weightlessness. Recovery, however, was successfully

achieved, and showed that animal cells were least affected by weightlessness.

Bios 3   L Jun 28 1969 by LTTA-Delta N from C Canaveral. Wt 697 kg. Orbit 355 × 386 km. Incl 33·5°. This mission cost £40 ($96m). To monitor his progress, Bonny, the pig-tailed monkey, had 24 sensors planted in his body, including 6 in the brain and one in each eye. It was these pre-flight preparations which led to criticism. He had to earn his food twice a day, by operating various push buttons for 15 min at a time. But though he had enjoyed doing the exercises on the ground, Bonny proved to be lonely and unhappy in space, and refused to work or drink. After 9 days of what was intended to be a 30-day flight, he was brought back, but died shortly after being recovered. The official report recorded: 'Inflight experiment data satisfactory for brief mission; post-flight experiments minimal due to premature post-recovery death of primate.'

COMSAT GENERAL   Sea and Air Satellites

Marisat   After several years' international controversy as to who should provide satellite communications for ships and planes, Comsat, America's Communications Satellite Corporation, which operates the 107-nation INTELSAT organization, set up a subsidiary called Comsat General. This made a start in Feb 1976 by providing the Marisat 2-satellite experimental system, costing $100M for commercial shipping and off-shore installations, with the help of a US Navy contract to use much of its capacity for up to 3 yr. Norwegian cruise ships and Esso tankers were among the 14 vessels which, by the start of the service, had installed antenna receivers and agreed to pay the minimum of $9600 for a year's service. Later the *Queen Elizabeth 2* installed the service. Telex messages cost $6 per min. Marisat 1, L Feb 19 1976 by Thor-Delta from C Canaveral. Wt 362 kg in 35,700km stationary orbit over Atlantic.

Comstar With approx 14,000 circuits, 2 Comstar satellites launched May 13 and Jul 22 1976 (a 3rd is planned in the late 70s) have been leased to American Telephone & Telegraph Co (A. T & T) for domestic US communications. Investment of $182M is expected to yield a return of $327M on the 7-yr operational life of 3 satellites.

Aerosat An experimental system to provide communications to transoceanic aircraft. Comsat General, the European Space Agency (*qv*) and Canada are collaborating. First launch is due 1979.

DISCOVERER

History An early programme of major importance. Its tests on orbital manoeuvring and re-entry techniques not only played a large part in enabling the first manned flights to be made in Project Mercury, but developed the 'spy satellite' systems now in regular use. Between 1959 and 1962, after which this type of work was classified, there were 38 launches; all were made by the US Air Force from the Western Test Range at Vandenberg. Project objectives were: 'Military space research; development of capsule recovery techniques; and biological research.' The series began 13 months after the successful launch of Explorer 1, America's first satellite, but for 18 months after that, as the list of the early flights illustrates, it was dogged by failures.

Discoverer spy satellite: Vandenberg launch

The plan was for a modified Thor IRBM to boost a 2nd-stage Agena A rocket to near-orbit altitude; after separation, the Agena's 6804kg engine was fired to place the whole stage in orbit. There its gas-jet orientation system turned the vehicle through 180°, so that the ballistic nose-cone containing a 136kg ejection capsule, was pointing rearwards and tilted down. It thus met the requirement that, for a satellite to drop a bomb or return an object to Earth, the object must separate and use retro-rockets fired in the opposite direction to the satellite's path, at a downward angle to ensure that it re-enters the atmosphere. At first ejections were often at the wrong angle, and sometimes upwards. This resulted in a false alarm in Feb 1960, when the US Navy's early detection system found an unknown satellite, believed to be Russian, in a near-Polar orbit. It was ultimately established that it was an early Discoverer capsule which had been ejected upwards instead of down. Ejection was achieved by explosive bolts, on either orbit 17 or 33. At 15,240 m a parachute pulled the recovery package clear of the heat-shield, which fell into the sea—in these early tests usually in the Pacific near Hawaii. Radio and light beacons, and even radar chaff, were used to assist location as the equipment package descended into the target area. At first C119 transport aircraft, towing trapeze-like frameworks at a height of 2438 m, patrolled the area in some numbers in vain efforts to develop the technique of 'snatching' the parachute harness with the package attached, before it reached the sea. A sea recovery was finally achieved on Explorer 13, and the first mid-air capture on the following mission. Beginning with Discoverer 30 in Sep 1961, C130s were used in this role. Amid much public concern that America's military capacity in space was probably far behind Russia's, the programme was pursued with almost frantic haste. There were 12

Line-up of Agenas, ready for Discovery launches

launches in 1961, with the last 2 officially designated Discoverers being launched in 1962. Looking back, and noting the large percentage of failures, it is easy to forget that space computer techniques, as well as hardware such as re-entry nose-cones, were still being developed. But in the 21 launches between the first mid-air success and Discoverer 38, there were only 7 more successful mid-air recoveries; 3 sea recoveries could be counted partial successes; 11 launches were admitted failures. The other 3 could probably be counted as experimental. The high proportion of failures inevitably attracted much public comment; Russia meanwhile was able to pursue her parallel programme in complete secrecy, merely announcing the Cosmos series numbers of her launches. Inevitably, from Nov 22 1961 the US Department of Defense decided to classify all military spaceflights. Since then it has been possible to follow such activities only indirectly by studying the Tables of Earth Satellites, compiled by Britain's Royal Aircraft Establishment. These show 27 capsule ejections in orbit between 1963–71; undesignated launchings are listed under the name of their rocket, Atlas-Agenas B and D until 1967, when more powerful Thor-Agenas and Titan-Agenas began to come into use. These launches have long since passed from the experimental to the operational stage; further details can be found under the heading MILITARY SATELLITES of the regular passes now being made over Russia, China, Middle East and other countries, whether or not they happen to be US allies, by satellites able to return data either directly or by means of ejected film packs. Below is a detailed list of the early Discoverers and, for comparison, of the last of them:

Discoverer 1 L Feb 28 1959 by Thor-Agena A from Vandenberg. Wt 590 kg (including 111 kg of instruments). Orbit 159×974 km. Incl $90°$. First satellite in polar orbit; objective was to test propulsion, guidance, staging and communications, but accurate tracking was prevented by tumbling. Re-entered after 5 days.

Discoverer 2 L Apr 13 1959 by Thor-Agena A from Vandenberg. Wt 730 kg. Orbit 229×354 km. Incl $90°$. Objective was to eject and recover a 88kg hemispherical capsule, inside which temperature and oxygen would be sufficient to maintain life. Capsule ejected on orbit 17, but was lost in Arctic. Satellite re-entered after 13 days.

Discoverer 3 L Jun 3 1959 by Thor-Agena A from Vandenberg. Similar to Discoverer 2, but carrying 4 black mice in recoverable capsule. Failed to orbit after 2nd-stage failure.

Discoverer 4 L Jun 25 1959. Similar to Discoverer 2. Failed to orbit, again because of 2nd-stage failure.

Discoverer 5 L Aug 13 1959. Similar to Discoverer 2. Orbit 219×724 km. Incl $80°$. Capsule ejected, but not recovered. Satellite dec. Feb 11 1961.

Discoverer 6 L Aug 19 1959. Orbit 210×850 km. Incl $84°$. Capsule again ejected on orbit 17 but recovery failed.

Discoverer 7 L Nov 7 1959. Orbit 159×835 km. Incl $81°$. Due to poor stabilization, capsule was not ejected. Dec. after 19 days.

Discoverer 8 L Nov 20 1959. Orbit 193×1660 km. Incl $80°$. Capsule ejected on orbit 15, but overshot recovery area. Dec. Mar 8 1960.

Discoverer 9 L Feb 4 1960. Failed to orbit; premature 1st stage cut-off.

Discoverer 10 L Feb 19 1960. Failed to orbit; destroyed by Range Safety Officer when it went off-course.

Discoverer 11 L Apr 15 1960. Orbit 165×603 km. Incl $80°$. Capsule was ejected on orbit 17, but recovery failed. Dec. after 11 days.

Discoverer 12 L Jun 29 1960. Failed to orbit following 2nd-stage attitude instability.

Discoverer 13 L Aug 10 1960. Orbit 253×694 km. Incl $82°$. Success at last; capsule ejected on orbit 17, and recovered from sea. Satellite dec. Dec 14 1960.

Discoverer 14 L Aug 18 1960. Orbit 182×808 km. Incl $79°$. After ejection on orbit 17, capsule was captured for the first time in mid-air by patrolling aircraft. Dec. after 28 days.

Discoverer 38   L Feb 27 1962 by Thor-Agena B from Vandenberg. Wt 952 kg. Orbit 334 × 495 km. Incl 82°. Last in the series to be officially announced; capsule was successfully ejected and recovered in mid-air after 65 orbits. Satellite dec. Mar 21 1962.

EXPLORER   Scientific Research

History   Explorer 1 became the United States' first satellite, and the world's 3rd. The Explorer series, which still continues, is somewhat comparable to Russia's Cosmos in that it embraces a varied series of experimental, research and scientific satellites.

The series began in a flurry of haste, following the successful Soviet launch of Sputnik 1 on Oct 4 1957. While the Russians had fulfilled their undertaking, made 2 yr earlier, to launch a satellite for meteorological purposes as part of the International Geophysical Year of 1957–58, America's own efforts, concentrated on the Vanguard project, had been a dismal failure. It was at this point that Washington turned at last to Dr Wernher von Braun's group, at the Army Ballistic Missile Agency at Huntsville, whose satellite proposals had been repeatedly turned down. They had evolved a rocket called Juno 1—a 4-stage development of Jupiter C, which itself had been developed from Redstone, the 21·3m rocket which von Braun had evolved for America from his wartime German V2. In a plan submitted in April 1957, ABMA had recommended a programme to launch 6 7·7kg satellites, the first of which would orbit in Sep 1957. Although of course it could not have been foreseen, this would have given America a 1-month lead over Russia in what was to become the major event of the 20th century—the 'Space Race'. 3 weeks after Sputnik 1, the ABMA group was given authority to go ahead with plans to launch 2 satellites, with a target date of Jan 30 1958 for the first. Explorer 1 was successfully launched 1 day behind this schedule. To its technical success can be added its major contribution to the International Geophysical Year—confirmation of the existence of radiation belts around the Earth, forecast and named after Dr James Van Allen. While the early Explorers were tiny compared with Russia's Sputniks, their miniaturized instruments gathered data of extreme scientific value; it was a trend which ultimately put America far ahead in space techniques, and ensured, in spite of Russia's apparent lead in the early years, that the first men on the Moon were Americans. Details of some of the Explorer flights are listed below:

Explorer 1   L Jan 31 1958 by Jupiter-C from C Canaveral. Wt (including integral last-stage motor) 14 kg. Orbit 360 × 2532 km. Incl 65°. Total length, with rocket case, 2 m. Explorer 1 carried 8 kg of instruments, designed to gather and transmit data on cosmic rays, meteorites and orbital temperatures. Confirmed the existence of belt of intense radiation beginning 965 km above the Earth. Continued transmissions until May 23 1958. Remained in orbit over 12 yr, re-entering over S Pacific after 58,000 revolutions on Mar 31 1970.

Explorer 2   L Mar 5 1958 by Jupiter C from C Canaveral. Wt 14 kg. Failed to orbit due to unsuccessful 4th-stage ignition.

Explorer 3   L Mar 26 1958 by Jupiter C from C Canaveral. Wt 14 kg. Orbit 195 × 2810 km. Incl 31°. Similar to Explorer

1, but with addition of small magnetic tape recorder able to release 2 hr of stored data on cosmic-ray bombardment in 5 sec as satellite passed over ground station. Transmitted until Jun 16 1958; dec. Jun 28 1958.

Explorer 4   L Jul 26 1958 by Jupiter C from C Canaveral. Wt 17 kg. Orbit 262 × 2210 km. Incl 50°. Mapped Project Argus radiation until Oct 6 1958; dec. Oct 23 1959.

Explorer 5   L Aug 24 1958 by Jupiter C from C Canaveral. Wt 17 kg. Failed to orbit; upper stage fired in wrong direction, leading to collision of rocket and instrument section.

Explorer S1   L Jul 16 1959 by Juno 2 from C Canaveral. Wt 42 kg. Failed to orbit; destroyed by range safety officer. The last of 5 successive failures (others were 2 Vanguards and 2 Discoverers).

Explorer 6   L Aug 7 1959 by Thor-Able from C Canaveral. Wt 64·4 kg. Incl 47°. Similar to Pioneer 1, but used to send 'Paddlewheel' satellite, 0·66 m dia, including 4 solar panels, or paddles, 0·5 m sq for recharging the batteries. Instruments measured behaviour of radio waves in ionosphere, mapped Earth's magnetic field, cloud cover etc. It also sent back first photograph of Earth. Transmitted until Oct 6 1959; dec. Jul 1961.

Explorer 7   L Oct 13 1959 by Juno 2 from C Canaveral. Wt 41 kg. Orbit 557 × 1088 km. Incl 50°. Returned data on Earth's magnetic field and solar flares until Jul 24 1961. Orbital life 70 yr.

Explorer S-46   L Mar 23 1960 by Juno 2 from C Canaveral. Wt 16 kg. Failed to orbit; upper stage apparently failed to ignite.

Explorer 8   L Nov 3 1960 by Juno 2 from C Canaveral. Wt 41 kg.

Explorer Development   Begun by the US Defense Department's Advanced Research Project Agency (ARPA), the series was transferred to NASA (National Aeronautics and Space Administration) when that was formed in Oct 1958. Since then Explorer has become a multi-purpose research programme studying interplanetary geophysics, astronomy and geodesy, with a total of 55 launches by the end of 1976. The basic spacecraft, which in practice has many variations, weighs 59 kg; there are series of satellites within the series, and some of them are listed below:

Explorer 18, 21, 28, 33, 34, 35, 41, 43, 47, 50   These 10 Interplanetary Monitoring Platforms (IMPs) investigated Earth's radiation environment during a complete 11-yr cycle of solar activity; they provided a vast range of new information about what goes on in space between the Earth and Moon and beyond the Moon. A notable discovery was that the solar wind includes a double ion stream; the 2 streams penetrate one another and have a speed difference of over 300,000 kph. While the first Imp (Exp 18, L Nov 26 1963; orbit 202,000 × 125,000 km) weighed only 62 kg with a 35W power supply, Exp 50 (L Oct 26 1973; orbit 238,989 × 94,697 km) weighed 81 kg with double power supply for spacecraft and experiments of 120W and 42W. Exp 35 (L Jul 19 1967) was placed in lunar orbit and studied the Moon's magnetic field and radiation belt, as well as Earth's magnetospheric 'tail'. IMP spacecraft provided continuous warning of solar flare radiations during the Apollo and Skylab missions, and were used as 'survey baselines' for the Pioneer 10 and 11 missions to Jupiter, and the Mariner missions to Venus and Mercury in 1973–74. Their scientific achievements include defining the nature and extent of the magnetosphere—the enormous, teardrop-shaped envelope surrounding Earth which is formed by the solar wind, which is a supersonic stream of particles entering Earth's magnetic field. Launches were by Delta; orbital life is mostly over 1M yr. Follow-on projects in 1977, called ISEE will involve the European Space Agency (qv).

Explorer 22, 27   Beacon Explorers, to study gravimetric geodesy and electron density in ionosphere.

Explorer 24 & 25, 39 & 40   Dual launches of Air Density/Injun Explorers; Air Density satellites, 3·6m balloons, measure atmospheric density. Injuns measure radiation data.

Explorer 29, 30, 36   GEOS Explorers, a US Naval Research programme to monitor solar ultraviolet X-rays, and geometric geodesy (the size and shape of Earth).

Explorer 37   Solar Explorer, to study solar X-rays and ultraviolet radiation.

Explorer 38 & 49   Radio Astronomy Explorers. Exp 38 (L Jul 4 1968; orbit 642 × 5862 km), spent 4 yr measuring galactic radio sources, but for about one-third of its observing time its receivers were swamped with noise (radio broadcasting, storms etc) coming from Earth. For that reason Exp 49 (L Jun 10 1973) was placed in a 1109 × 1120 km lunar orbit so that when it was on the farside, the Moon would shield its radio experiments from Earthly interference. Once in lunar orbit, its 4 radio antennas, 458 m between tips, formed a huge 'X'. At 1975-end it was still transmitting on command.

Explorer 42 & 48   Small Astronomy Satellites, which are

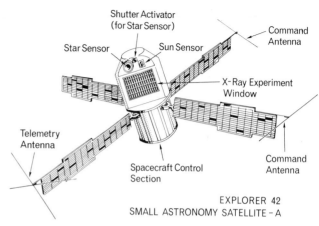

EXPLORER 42
SMALL ASTRONOMY SATELLITE – A

building up a sky map of gamma rays—electromagnetic radiations which cannot be detected on Earth because they are absorbed in its atmosphere. Exp 42 was launched Dec 12 1970, by Scout from Italy's equatorial platform at San Marco, wt 306 kg, into 521 × 563 km orbit with 3° incl. Its data on X-ray sources and their location was followed up by Exp 48, launched Nov 15 1972 from San Marco, into a 442 × 632 km orbit with 1·9° incl. This 186kg satellite was equipped with a spark chamber gamma ray telescope to study gamma rays and X-ray sources emanating from the galactic plane and the Crab Nebula. It is hoped this series, to be completed with a 3rd launch, will lead to some understanding of the dynamics of the Milky Way.

Explorer 51, 54, 55   Atmosphere Explorers. This series, following up the aeronomy satellites, Exp 17 & 32 of 1963 & 1966, continued to study Earth's atmosphere, and fears about pollution of the 'ozone' in the upper layers (22–25km altitude) by, for example, freon gas emitted by aerosols. Exp 51 (L Dec 16 1973, orbit 158 × 4303 km) had its orbit lowered to 120 km every few weeks to take low-altitude samples, then was raised again after a few days to prevent decay as a result of atmospheric drag. Exp 54 (L Oct 6 1975; orbit 155 × 3816) sampled nitric oxide, one of the controlling agents of ozone production and depletion. Exp 55 (L Nov 20 1975; orbit 156 × 2983 km) measured changes in the total ozone field, and Earth's heat-balance and energy conservation mechanisms.

GEMINI Manned Spacecraft

GEMINI 12 FIGURES

Launch Weight:	3792 kg
Length:	5·60 m
Base Diameter:	3·05 m
Height:	5·80 m
Insertion Velocity:	28,221 kph
Apogee:	279·4 km
Perigee:	161 km
Period of Revolution:	95 min

History   Successor to Mercury, the Gemini spacecraft was twice the weight, and drew heavily on proven Mercury technology. But it was also far more advanced, complex and versatile, and achieved far more than its original aim, which was to bridge the gap in US manned spaceflight between the

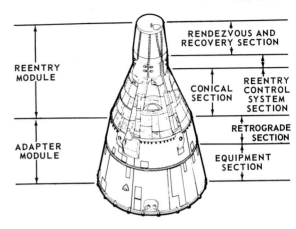

Gemini: Two-man spacecraft

end of Project Mercury and the start of Project Apollo. As it was, 2 yr elapsed between the last Mercury mission and the first Gemini flight; but once the programme did begin, 10 missions, Gemini 3–12, were flown at a breathtaking rate of 5 a year in 1965–6. Despite some technical problems, invaluable experience was obtained in long-duration flight, rendezvous, docking, use of target-vehicle propulsion for orbital manoeuvres, extravehicular activity (EVA, or spacewalking), and guided re-entry. In those 2 yr, the US put 20 men into space, while the Soviet Union achieved only the 2-man Voshkod 2 flight. It was unfortunate for America that that flight was the occasion of Leonov's historic first spacewalk—12 min which enabled Russia to retain her international prestige as leader of the space race despite the solid American progress. It also completely obscured the fact that Gemini 8 achieved the first space docking, and established the technical lead that finally enabled the US to land the first men on the Moon years ahead of Russia.

Spacecraft Description Unlike Mercury, which had nearly all its systems inside the pressure shell, Gemini had most systems' components located in unpressurized equipment bays. This enabled launch crews to remove a hatch, and quickly replace any malfunctioning component, as well as making possible a larger crew compartment—1·558 m³ of accessible volume. Titanium and magnesium were the principal metals used. The spacecraft consisted of 2 main parts, a re-entry module and an adapter module.

Re-entry Module This included the crew compartment, with side-by-side seating and emergency ejection seats, replacing the Mercury escape tower. (These could be used not only for escape during a launchpad or lift-off emergency, but for a final parachute descent if the spacecraft descended over land, instead of sea.) Also part of the re-entry module were electronics for guidance, instrumentation and communications; 2 hatches; heat shield; a re-entry control section, with propellant tanks and 16 thrusters in 2 independent systems for re-entry control; and a rendezvous and recovery section, with radar, drogue and main parachutes.

Adapter Module With a maximum diameter of 3·05 m, this provided the bays for rendezvous systems and equipment, and the mating structure between re-entry module and launch vehicle. It contained the life support and electrical systems, with 2 fuel cells providing the primary source of power from launch to re-entry. Gemini was the first spacecraft to use fuel cells (a British development) providing electrical power through the chemical reaction of oxygen and hydrogen, with each cell producing 0·57 litre of drinking water per hour as a by-product.

Altitude and Manoeuvring System This consisted of 16 thrusters mounted in fixed positions on the Adapter Module. The desired amount of impulse or control was obtained by varying the firing time: 8 of the engines provided 11·3 kg thrust; 2 produced 38·6 kg; and 6 produced 45·4 kg thrust. An illustration of the amount of attitude and manoeuvring control required was that during the relatively short Gemini 6 flight more than 35,000 individual thruster firings occurred.

GEMINI FLIGHTS

The first 2 Gemini flights were unmanned tests; the first 7 in the series, launched by Titan 2 rockets, were code-named 'GT'; the last 5 which involved dual launches, with an Agena rocket being placed in orbit to act as a docking target, were code-named 'GTA' launches.

Gemini 3 First manned flight in the series—a 3-orbit mission in March 1965. The first computer was carried into space—22·7 kg of miniaturized equipment capable of making 7000 calculations a sec. It was a feature hardly noticed at the time; but it had much greater long-term significance than Leonov's spacewalk. With it Grissom, the Commander, was able to compute the thrust needed to change his orbit. From that time, instead of being carried helplessly round

Ed White making first US Spacewalk

and round the world on a fixed orbital path, man had a genuine ability to 'fly' in space.

Gemini 4   On this flight, 3 months later, the computer failed and Jim McDivitt had to make a manual re-entry. He was 1 sec late punching the button to fire the retro-rockets, and the spacecraft came down 64·4 km off course. All the same, it was a triumphant flight, lasting 4 days—the first US long-duration flight. During it, 3 months after Leonov had first tried it, Ed White ventured outside. His spacewalk lasted 21 min; experimentally, and somewhat uncertainly, he manoeuvred himself around with an oxygen-powered space gun. The colour pictures brought back have been used in books, magazines and newspapers all over the world ever since to symbolize man in space (*See* Voskhod Spacecraft—History.)

Gemini 5   An 8-day flight by Cooper and Conrad in Aug 1965, which proved that men could withstand weightlessness for as long as it would take to fly to the Moon, visit it briefly and return.

Gemini 7   A 14-day flight in Dec 1965 by Borman and Lovell, which held the long-duration record for 5 yr until finally overtaken by Russia's 18-day Soyuz 9 flight in 1970.

Gemini 6   This mission, postponed in Oct 1965, at T − 42 min, when the Agena rocket intended as a rendezvous target was lost, used Gemini 7 as the target instead. On Gemini 7's 11th day in orbit Schirra and Stafford went up to join them, and Schirra brought his craft within 2 m of Gemini 7—the first real 'rendezvous' of men in space. The 1-day Gemini 6 flight should have taken place 3 days earlier, but once again had to be postponed when a fault occurred in the launch vehicle 2·2 sec before lift-off.

Gemini 8 On Mar 16 1966 Armstrong and Scott achieved the first space docking, by linking their spacecraft with an Agena target rocket, placed in orbit 101 min earlier. Shortly afterwards the linked vehicles began tumbling and spinning, out of control, as a result of a jammed thruster. The astronauts escaped only by firing their retro-rockets, and had to return to Earth 2 days early. They survived man's first space emergency, and the drama completely obscured the fact that it was on this mission that the US overtook the Soviet Union's long-held lead in space technology.

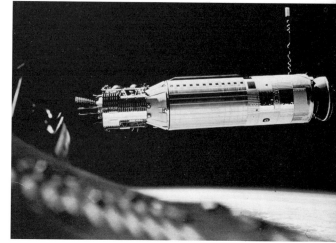

Gemini 7 as seen from 6 m by Gemini 6: Man's first rendezvous in Space

Agena Docking Target: photo by David Scott from Gemini 8

Red Sea and Gulf of Aden seen from 660 km by Gemini 11

Gemini 9 The unlucky flight. Its original crew, See and Bassett, were killed when trying to land a jet fighter in bad weather at the McDonnell works at St Louis, crashing through the factory roof only yards from the spacecraft. The mission was finally flown by the back-up crew, Stafford and Cernan, in Jun 1966, after 2 delays: first because the Agena target was lost; then, when a reserve target was launched, Gemini 9 missed its window. But the flight, when it did take place, included a 2-hr spacewalk by Cernan, and for the first time splashdown was less than 0·80 km from the recovery ship.

Gemini 10 A 3-day mission in Jul 1966 during which Young and Collins first docked with an Agena target rocket, then fired its engine to boost themselves into a 761 km orbit. There they separated and rendezvoused with Gemini 8's Agena target, which had been left in a 'parking orbit' for this purpose. Collins spacewalked across to it and retrieved a dust-collecting device from the outside of the Agena.

Gemini 11 Conrad and Gordon, in Sep 1966, made a direct-ascent rendezvous and docked with an Agena target while still on their first revolution. During a 44-min spacewalk, Gordon connected the 2 craft with a 30 m tether, and when undocked, Gemini's thrusters were used to put the craft into a cartwheel motion—the first experiment in creating artificial gravity even though it was only 0·00015 of normal Earth gravity. Automatic, computer-steered re-entry was achieved for the first time.

Gemini 12 The last mission, flown by Lovell and Aldrin in Nov 1966, included 3 spacewalks totalling $5\frac{1}{2}$ hr by Aldrin—the first fully successful attempts at EVA. Once again the spacecraft docked with an Agena target, artificial-gravity experiments were made, and a fully automatic re-entry performed.

Summary When Project Gemini ended 9 dockings with a target had been achieved and 7 different ways of doing it had been worked out. The 10 missions were completed without the loss of a single life. The total cost was £534,750,000 or $1,283,400,000.

HEAO   Astronomy Observatory

History   3   High   Energy   Astronomy   Observatories

HEAO-1 successfully launched 12 Aug 1977 to search for 'black holes'

(HEAO), planned for launch in 1977, 1978 and 1979, are the result of a project several times postponed since it was conceived in the late 1960s. HEAO-1 for launch in mid 1977 is to make a complete map of X-ray sources throughout the celestial sphere in its 1-yr lifetime, in a 444km circular orbit with $22.75°$ incl. This places it above the atmosphere but below the radiation belts which would interfere with its measurements of stellar radiation. It is hoped to obtain clues about the creation of supernovae, and the distribution of energy in neutron stars. Attempts are being made to locate the fast-rotating neutron stars called pulsars, and to discover the contents of so-called 'black holes' from which no light escapes; and to investigate quasars, thought by some to be the furthest and oldest objects in the universe.

Spacecraft Description With 6.10m length, 2.35m dia, and 2721 kg wt, HEAO has 4 highly specialized experiments to survey and map X-ray and gamma ray sources. Atlas-Centaur launchers are being used instead of the Titan 3Cs for the much bigger spacecraft originally planned. The aim was to halve the original cost of $275M.

LAGEOS   Earthquake Predictor

History With the ability, it was hoped, to predict earthquakes, the Laser Geodynamics Satellite was NASA's first satellite wholly devoted to laser ranging. With a practical life of 50 yr, and an orbital life of about 8M yr, the idea was to establish a permanent reference point, so that Earth's progress could be tracked relative to the satellite, in contrast to the traditional system of tracking satellites, relative to the Earth. Launched on May 4 1976 by a Delta 2913 rocket from Vandenberg, into a near circular 5900km orbit with $110°$ incl, it should ultimately make it possible to track movements as small as 2 cm on Earth's surface. The US Geological Survey, responsible for earthquake research and prediction, will use LAGEOS to measure continental drift (large land masses known as 'tectonic plates'); by paying particular attention to earthquake areas such as the San Andreas fault in California, it should be possible to provide early warning of impending earthquakes.

In the first 4 yr, while laser-ranging techniques were being worked out, and a global network of 14 Earth stations (some of them mobile) was built up, the aim was to achieve measurement accuracies of 10 cm, improving these to less than 5 cm in the 1980s. Lageos is the first step in NASA's Earth and Ocean Dynamics Application Programme. Cost of satellite and launch vehicle, $8.5M.

large number of retroflectors, but small enough not to be affected by solar radiation drag. Aluminium construction would have been too light, and brass too heavy; finally 2 aluminium hemispheres were bolted together around a brass core to provide a large mass/surface ratio. Materials were also selected to reduce magnetic effects between the satellite and Earth's magnetic field. Each end of the bolt connecting the hemispheres carries a copy of a message prepared by Dr Carl Sagan. It includes 3 maps of Earth's surface, showing the period 225M yr ago when it is believed the land masses were one 'super-continent' (sometimes called 'Pangaea'); the position of the continents as they are now; and their estimated position 8·4M yr from now when the satellite, so solid it will survive re-entry, falls back to Earth.

LANDSAT   Earth Resources

History   Although it was only fully operational for 8 months, instead of the planned minimum of 1 yr, Landsat 1 provided the US with the most remarkable success so far achieved with this type of satellite. Landsats, originally known as ERTS, are Earth Resources Technology Satellites—improved and enlarged versions of the Nimbus weather satellites, which they resemble in appearance. When Landsat 2 was launched in 1975, NASA administrator, Dr James Fletcher said if he had to pick one space-age development to save the world, he would pick Landsat and the spacecraft which would be developed from it. The original plan was to launch 2 in successive years (1972–3) as an experiment in systematically surveying the Earth's surface to study the health of its crops, and the potential use and

Lageos: permanent Earth reference point

Spacecraft Description   With 60 cm dia and 411 kg wt, Lageos is an aluminium sphere with brass core. Its 426 prisms, called cube-corner reflectors, give it the dimpled appearance of a golf ball. The 3-dimensional prisms reflect laser beams back to their source, regardless of the angle from which they come. The satellite had to be big enough for a

development of its land and oceans. The dramatic flow of vivid, revealing photographs (actually 'false-colour images') sent back from the moment it became operational, were regarded as sensational by 300 investigators in the 50 countries participating in the experiment; their quality, clarity and visibility led to the information being put to immediate practical use. An example of the pictures can be found in the colour section of this book. During 1975 countries as far apart, both geographically and politically, as Chile, Zaïre and Iran decided to build and operate their own ground stations to receive Landsat data.

Before launch, NASA decreed that all ERTS data, including 9000 pictures per week, should be unclassified and made available to the public; the photographs can be bought from $1.25. An unexpected and dramatic result was that for the first time the public were able to obtain and study detailed pictures of Soviet and Chinese launch centres and missile sites—pictures previously highly classified when obtained by 'spy' satellites. The immediate success of the project led to the processing and distribution systems being swamped with requests. Just as Nimbus experience was used for the development of Landsat, knowledge gained from Landsat was used less than a year later, by the rotating crews of astronauts operating even more advanced equipment in Skylab. Total Landsat costs were £72·5M ($174M) including £11·6M ($28M) for data-handling computers and processing facilities, and £14M ($34M) for investigations.

Spacecraft Description Butterfly-shaped. Ht 3 m. Width 3·4 m with solar paddles deployed. Sensors and electronics are housed inside a 1·5m dia sensory ring below the paddles. Sensors as follows. MSS (Multispectral Scanner Subsystem): this collects data by continually scanning the ground

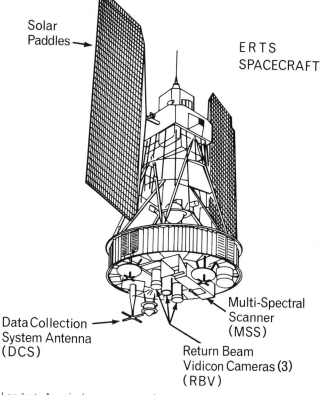

ERTS SPACECRAFT

Solar Paddles

Data Collection System Antenna (DCS)

Multi-Spectral Scanner (MSS)

Return Beam Vidicon Cameras (3) (RBV)

Landsat: America's most successful Unmanned Satellites

directly below in 4 spectral bands, 2 in the visible spectrum and 2 in the infrared. RBV (Return Beam Vidicon Subsystem): this views the same 185km swathe as the MSS. 3 cameras of 4125 lines are reshuttered simultaneously every 25 sec to produce 185 × 185 km overlapping images of the ground along the direction of the satellite motion. WBVTR (Wide Band Video Tape Recorder): because Landsat is not within range of a ground station for much of its time, 2 tape-recorder systems can store images, and later transmit them, simultaneously. Data Collection System: this collects measurements as it passes over 'remote platforms' installed on icebergs, etc. to send up readings of soil, water and other conditions for later retransmission and comparison with data collected from the other sensors.

Landsat 1   L Jul 23 1972 by Thor-Delta from Point Arguello. Wt 891 kg. Orbit 901 × 920 km. Incl 99°. The near-polar orbit enables the satellite to circle Earth 14 times a day, and to view any point on Earth except for small areas around the Poles. Because the orbit is also Sun-synchronous, every 18 days the Landsat cameras can view the same spot at the same time of day; each pass enables the cameras to view a swathe 185 km wide with some overlap on each pass; it is back in its original position after 252 passes. By Mar 30 1973, when faults occurred in the video tape-recorder, the N American continent had been photographed 10 times; the total of over 34,000 images taken since launch included all the world's major land masses at least once. The tape-recorder fault meant that images could no longer be stored for transmission; but when the satellite was over one of NASA's 3 ground receiving stations in California, Alaska and Maryland, it was still possible to transmit 'live' pictures of the entire N American continent. An example of the ability

Landsat-1 view of Japan from 914 km; Nagasaki City, lower left corner

of such satellites to make available to the public pictures of launch sites was one of southern Russia showing 7 major ABM complexes near the Chinese border. (Landsat No E-1049-05260-7.) Orbital life of Landsats is 70–100 yr.

Landsat 2   L Jan 22 1975 by Thor-Delta from Vandenberg. Wt 816 kg. Orbit 907 × 918 km. Incl 99°. Intended to supplement and later replace Landsat 1, but at the end of 1975 the 2 were still working together, having by then logged 22,345 orbits. Their data was being used to measure crop yields, so that shortages and gluts could be forecast and avoided; to map mountain snows, so that the amount of 'run-off' for irrigation and generating electric power would be known in advance; to help cities, regional authorities and developing nations to plan wiser use of their land resources; to develop potential earthquake zones and monitor off-shore sewage and industrial wastes; and for many other purposes. Landsat 2 also provided a picture (E-2101-07085-5) giving the public full details for the first time of Russia's Kapustin Yar launch centre (*qv*).

Landsat 3   Due for launch in 1977, this will have improved sensors, and tasks will include water management and a census of population density.

Remote Sensing   The basis of Landsat is that all objects, living or inanimate, transmit or reflect visible and invisible light, and thus have their own 'signature' or individual 'fingerprint'. All energy coming to Earth from the Sun is either reflected, transmitted, or absorbed in objects on Earth, each in its own way. 'Remote sensing'—measuring objects from a distance—goes back to aerial photography in the 1930s. It began to develop more rapidly with military reconnaissance in World War II, followed by the use of

Landsat-1 view of Dead Sea area. River Jordan is top centre

sounding rockets and satellites after that. A camera is a remote sensor in that it records the shape and colour of an object by its reflected light without touching the object. The human eye is the simplest example of remote sensing; but most of the information reflected or radiated by the Earth, cannot be detected by the human eye. Near infrared light lies just beyond our vision. Healthy green vegetation is even brighter in the infrared than in the visible, and this information is particularly valuable to farmers, because it can provide early warning if their crops are sick. Just as the eye can cover only a minute portion of the total electromagnetic spectrum, so no single instrument is capable of sensing and measuring the radiations from objects with different physical and chemical properties. One of the objects of Landsat is to determine what sort of sensors, and what combination of them, will yield the most useful information.

Picture Reading and Results The energy reflections described above are converted by the Landsat scanners into electrical signals in 4 selected bands. (In this case, Bands 1 and 2 are in the visible wavelengths of 0·5–0·6 and 0·6–0·7 micrometres; Bands 3 and 4, which are not visible to the human eye, are in the near infrared portion of the spectrum with wavelengths of 0·7–0·8 and 0·8–1·1 micrometres). These reflections are processed into 'digital bits', transmitted to a receiving station, and then reprocessed into either black-and-white reproductions of what was seen on Earth, or as 'false colour images', by projecting the data of 3 of the 4 bands through blue, red and green filters. The colours assigned are in the same order as the primary colours of the visible spectrum, but result in an 'exchanged' colour: what we seen as green in Band 1 is shown as blue; what we see as red in Band 2 appears as green; and Band 3, which normally we cannot see, appears as red. Band 4 may be used instead of Band 3, and also appears red.

The result of this is that clear water will appear black in Bands 3 and 4, because water almost totally absorbs radiant energy—in other words sends back hardly any reflections; water carrying silt, or otherwise polluted, will appear blue.

Trees and plants appear bright red because of the very high reflectivity of chlorophyll-bearing leaves in the near infra-red; vegetation brightness depends on such things as the size of leaves, big leaves showing up as brighter than small leaves, with the effect that hardwood trees will register as brighter than pine trees. The big leaves of tobacco plants will be brighter than wheat. And, the vital factor in the whole exercise, crop brightness depends on plant health; thus, healthy crops, shown in the infrared Bands 3 and 4, will be much brighter than diseased vegetation. But crop disease would be difficult or impossible to detect on a single photograph; abnormal changes, suggesting disease, would show up when successive pictures were compared.

The amount of knowledge which can be extracted from the data and pictures will increase as the scientists learn the technique of reading them. But at an early state, observers were able to detect geologic faults and water-bearing rock areas in Nebraska, Illinois, and New York State, which had been unknown before; areas of clear and polluted waters in Chesapeake Bay were readily discernible. From Brazil came reports that the Landsat pictures had revealed that villages and towns were sometimes wrongly located on their maps by tens of kilometres, and that lagoons shown as 20 km long were in reality over 100 km long. Ghana reported that locust-control was being attempted because pictures had shown vegetation at the edge of deserts which attracted locusts for breeding. Iran reported that it had become apparent that lowering of the Caspian Sea by evaporation had apparently changed the shape of the Bandar Shah peninsula. Pictures of Britain are acclaimed as 'remarkable', and disclose, among other things, that a linear feature or 'fault' starts near Harwich on the E coast and runs right through London to Land's End.

LUNAR ORBITER

History The second of 3 unmanned exploration projects, carried out in parallel with the 3 manned programmes aimed at getting men on the Moon before 1970. Following the successful Project Ranger flights, which yielded the world's first TV pictures of the Moon's surface, 5 Lunar Orbiters were launched within a year, starting on Aug 10 1966 to help select Apollo landing sites in equatorial regions from 43°E to 56°W. Other objectives were to study variations in lunar gravity, radiation and micrometeoroid data. Orbiter 1 was placed in a 191 × 1867km lunar orbit, with 12° inclination. Its pictures covering 5·18 million sq km, of the Moon, included 41,440 sq km of potential Apollo landing areas. Perturbations in its orbit provided the first knowledge of what became known as 'mascons'—at least a dozen mass concentrations of materials, usually associated with the mare, which have a powerful gravitational effect on spacecraft remaining for lengthy periods in lunar orbit. All 5 Orbiters were immensely successful; it proved possible to manoeuvre them by Earth commands into orbits descending as low as 40 km. Objects as small as 1 m across were photographed, and their pictures provided the first lunar atlas including the farside, to be built up. The first 4 Orbiters provided between them experience of several thousand lunar orbits before each was deliberately crashed on to the surface with the last of its attitude control gas, to ensure than there was no radio frequency interference with later missions. Orbiter 4 provided the first pictures of the lunar south pole. Orbiter 5, launched on Aug 1 1967, after it had completed its photography was retained as a target for NASA's Manned Spaceflight Network while it had sufficient

Copernicus Area, taken by Lunar Orbiter 2. Such views made manned landings possible

fuel, until its final controlled impact on Jan 31 1968. By then the 3rd unmanned lunar exploration project, Surveyor, was also nearing completion.

Spacecraft Description A truncated cone structure. Lunar Orbiters were folded for launch. When deployed, the 4 windmill-like solar panels and antennas provided a maximum span of 5·6 m, and 1·6 m ht. Total wt of 390 kg included a photographic laboratory weighing only 65·8 kg, but carrying 2 cameras for wide-angle and telephoto coverage, film processing and photo readout (scanning) systems. These viewed the Moon through a quartz window protected by a mechanical flap. Launcher: Atlas-Agena D.

MARINER/VOYAGER   Planetary Explorers

Mariner 1:	1962	Venus	Failure
Mariner 2:	1962	Venus	Success
Mariner 3:	1964	Mars	Failure
Mariner 4:	1964	Mars	Success
Mariner 5:	1967	Venus	Success
Mariner 6:	1969	Mars	Success
Mariner 7:	1969	Mars	Success
Mariner 8:	1971	Mars	Failure
Mariner 9:	1971	Mars	Success
Mariner 10:	1973	Venus/ Mercury	Success
Voyager 1:	1977	Jupiter/ Saturn Uranus	en route
Voyager 2:	1977		

History One of NASA's 3 planetary exploration programmes. By the end of the remarkable Mariner 10 mission, the secrets of our smallest planet, Mercury, had been revealed, and over half of its surface photographed; much had been learned about Mars and Venus, with the promise of more revelations ahead. The Mariner programme had complemented and leapfrogged Pioneer, and built up the technology for the Viking landings in 1976. The first 9 Mariner launches, spread over 10 yr, included 3 intended for Venus and 6 for Mars; their success enabled the 10th to be aimed at Mercury, passing Venus on the way, with the following pair intended to fly past Jupiter and Saturn. The latter replaced the more expensive 'Grand Tour' missions intended to use the period 1977–79 when all 5 outer planets will be lined up in such a way that their gravity could be used to swing spacecraft past each in turn; this occurs only once in 180 yr. The revised plan is costing only £133·3 million ($320 million) compared with the Grand Tour's £375 million ($900 million). The 2-yearly launch 'windows' (when Earth and Mars come within about 56 million km of each other) were used by Mariner 4 in 1967 to obtain man's first close look at another planet; and by Mariners 6 and 7 in 1969 to follow up with much better pictures. In 1971 Mariner 9 became man's first planetary orbiter, and provided us with a complete map of the Red Planet. The secrets of Venus, still largely concealed beneath her dense cloud cover, despite Russia's landings, will be explored by both Mariner and Pioneer missions during the remainder of this decade. Whatever their success, the place of Mariner in space history is secure.

Spacecraft Descriptions These will be found at the end of the Mariner section, together with a full description of the

Mariner 10 mission. This is given in detail, as an illustration of the techniques employed in planetary exploration.

Mariner 1   L Jul 22 1962 by Atlas-Agena B from Cape Kennedy. Wt 202 kg. This first attempt at a Venus fly-by failed because of an error in the flight guidance equation; the rocket went off course immediately after launch, and had to be blown up. The object had been to obtain details of the Venusian atmosphere, cloud cover, magnetic field etc.

Mariner 2   L Aug 27 1962 by Atlas-Agena B from Cape Kennedy. Wt 202 kg. The first successful planetary fly-by, it was fired into a Venusian trajectory from Earth-parking orbit; after a 109-day journey it flew past the planet at a distance of 34,830 km, providing 35 min of instrument scanning. Surface temperatures registered at 428°C, above the melting point of lead, and far higher than expected. The atmosphere appeared to contain no water vapour. The cloud layer was unbroken, with one spot near the southern end of the terminator 11°C cooler than the rest, possibly due to a mountain range. It was also established that, unlike Earth, Venus did not have a strong magnetic field and radiation belt.

Mariner 3   L Nov 5 1964 by Atlas-Agena D from Cape Kennedy. Wt 261 kg. Intended to take 21 TV pictures as it passed Mars at a distance of 13,840 km, but failed to achieve the necessary speed of 41,228 kph when fired from Earth-parking orbit. Although it went into solar orbit, Mars was missed by a wide margin.

Mariner 4   L Nov 28 1964 by Atlas-Agena D from Cape Kennedy. Wt 261 kg. Following Mariner 3's failure, launch of the 2nd of the pair of vehicles was delayed till the end of the Martian window. Injection was successful, and prob-lems resulting from the instruments locking on to the wrong stars were overcome: 2 days after launch, Canopus was acquired, and after 228 days and a flight of 523M km, Mars was passed at a distance of 9844 km, on Jul 14 1965. During the next 10 days 21 TV pictures, and 22 lines of a 22nd photograph, were received at NASA's Jet Propulsion Laboratory at Pasadena, California. Man's first close-range pictures of another planet showed that Mars was heavily cratered, more Moon-like than Earth-like, very dry, with no trace of surface water, and certainly not possessing any of the canals theorized by astronomers. Although the possibility of some form of life could not be ruled out, the very thin atmosphere, coupled with the evidence that there might never have been enough water for oceans or streams, made any advanced life forms seem most unlikely. Long after passing Mars, Mariner 4, in solar orbit, provided convincing evidence that such vehicles could be operated for many years; $2\frac{1}{2}$ yr later, 90M km from Earth, a JPL command again turned on the TV equipment, and fired the spacecraft engine for 70 sec.

Mariner 5   L Jun 14 1967 by Atlas-Agena D from Cape Kennedy. Wt 245 kg. Originally the back-up vehicle for Mariner 4, this was modified for flight towards the Sun and Venus, instead of away from the Sun to Mars. Solar panels were reversed and reduced in size, and a thermal shield added. A flight of 349M km resulted in Mariner 5 passing only 3990 km ahead of Venus in its orbit around the Sun, on Oct 19 1967. Using more advanced instruments, surface temperatures of about 267°C were recorded; measurements of the magnetic field ranged between zero and 1/300th of Earth's; an electrified ionosphere was identified at the top of the atmosphere.

Mariner 6 & 7. The 4 panels have 17,472 solar cells and gas jets at the ends

Mariner 6 and 7 L Feb 24 and Mar 27 1969, by Atlas-Centaur from Cape Kennedy. Wt 413 kg. Intended to study the atmosphere and surface of Mars as part of the search for extraterrestrial life, and to develop technology for later Mars missions, these flights were immensely successful. The author, watching the 201 TV pictures flowing back to JPL at Pasadena, found it even more exciting and dramatic than covering the first Apollo Moonlanding a few days earlier. They passed Mars at distances of 3412 and 3524 km. Mariner 6 had flown 387·8M km in 156 days, to arrive on Jul 31 for encounter at 95·7M km, about $5\frac{1}{2}$ light min from Earth.

Atlas-Centaur launches Mariner 7

Mariner 7 flew 316·9M km in 130 days for encounter on Aug 5 at 99·4M km. Mariner 7 was probably struck by a small meteoroid a few days before arriving; after loss of signal, commands sent ordering it to switch antennas were successful both in restoring communications and establishing that it had been damaged, losing some of its telemetry channels; a slight velocity change caused it to arrive 10 sec late. The spacecraft began sending back far-encounter pictures from

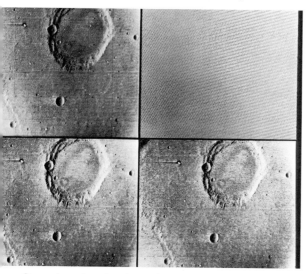

Computer enhancing showed extra detail on Mariner pictures. *Top right*: amount of 'Clutter' removed

distances of up to 1,126,540 km; but greatest interest naturally lay in the 24 near-encounter pictures sent back by Mariner 6 during its 68 min of closest approach, and in the 33 near-encounter pictures provided by Mariner 7's 74 min of closest approach. Mariner 6, concentrating on the equatorial region, dramatically established that Nix Olympica, at first thought to be a gigantic crater, was a 24km high volcano, with a 64km wide crater at the top. Mariner 7, concentrating on the southern hemisphere and part of the south polar ice cap, confirmed that this was largely solid carbon dioxide (dry ice), with perhaps a little water content. Mars emerged at the end of the fly-bys as heavily cratered, with a thin atmosphere consisting of at least 98% carbon dioxide, its craters differing from those on the Moon as a result of being worn down by winds and dust. It seemed that any advanced form of life could be ruled out, but one scientist speculated on the possibility of some form of life evolving which by-passed the need for liquid water. One of the Mariner 7 pictures showed a minute, potato-shaped speck which proved to be one of the 2 Martian moons, Phobos; but full details of the moons and of the Red Planet itself were to be finally established by Mariner 9 during the next launch window. Cost of this twin-mission was £61·6M ($148M).

Mariner 8 L May 8 1971 by Atlas-Centaur from Cape Kennedy. Wt 1031 kg. Intended to be the first of a pair of Martian orbiters, but an autopilot fault sent the 2nd stage off course and it fell into the Atlantic 1450 km SE of Cape Kennedy.

Mariner 9 L May 30 1971 by Atlas-Centaur from Cape Kennedy. Wt 1031 kg. Intended to map 70% of Mars during 90 days in orbit around it, Mariner remained operational 349

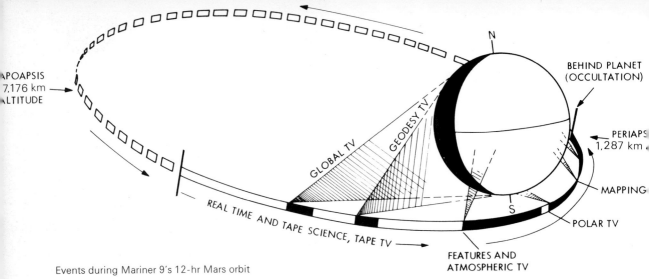

APOAPSIS
7,176 km
ALTITUDE

N

BEHIND PLANET
(OCCULTATION)

GEODESY TV

GLOBAL TV

PERIAPS
1,287 km

MAPPING

S

REAL TIME AND TAPE SCIENCE, TAPE TV

POLAR TV

FEATURES AND
ATMOSPHERIC TV

Events during Mariner 9's 12-hr Mars orbit

days before it was shut down on Oct 27 following exhaustion of its attitude control nitrogen gas; by then it had circled the Red Planet 698 times, mapped the whole of it, and transmitted 7329 TV pictures, including detailed photographs of both Phobos and Deimos. Following the loss of Mariner 8, plans were revised so that Mariner 9 could cover both missions. The spacecraft arrived at Mars on Nov 13 1971, at the end of a 167-day flight covering 397M km. A 15-min firing of its 136 kg thrust liquid propellant engine reduced the approach speed, relative to Mars, from 18,000 kph to 12,500 kph, and placed it (after a later trim manoeuvre) in a 12-hr Martian orbit, 17,140 × 1387 km. It

thus became the first man-made object to orbit another planet. (Russia's Mars 2 and 3 followed later in 1971.) The braking burn reduced the spacecraft's wt to 590 kg. As the spacecraft was approaching Mars in mid Nov, it took 3 series of pictures of a violent dust-storm which astronomers had been watching envelop the entire Martian globe during a 2-month period. While this delayed Mariner 9's mapping sequences for 6 weeks, it provided a unique opportunity for its instruments to peer down into the most extensive dust storm to occur on Mars since 1924. It reached an altitude of 50–60 km. Only the bright, waning ice cap at the S pole, and 4 dark mountain peaks (one of them Nix Olympica) were visible through the haze, which had the effect of cooling the surface, and warming the atmosphere. When the dust storm subsided Mariner 9 was able to maintain an instrument surveillance of the changing seasons below for more than half a Martian year. By the end of its mission Mars was known to be a geologically active planet, different from both Earth and Moon with volcanic mountains and calderas (craters) larger than any on Earth; there is a vast equatorial crevasse which would dwarf America's Grand Canyon, 4000 km long, and plunging to a depth of 6096 m. The 'Martian canals' were an illusion, yet this gigantic rift was never suspected, showing up on Earth-based telescopes only as dark markings. One theory is that the dark trough is warmed by the Sun at one end, while it is still dark at the other, with the effect that violent winds are set up each day. Contrary to earlier conclusions it is now thought that free-flowing water may have existed on Mars at one time; and that dust storms and cloudiness account for much of the variability of appearance that has puzzled astronomers for centuries. Because of a previously unknown gravity-field variation in Mars' equatorial plane, Mariner's orbital period was found to be too short in relation to its Earth tracking stations; so, after the dust storm cleared, its engine was fired to raise the periapsis, or low point, to 1625 km, to co-ordinate the orbit with the Goldstone, California, station. With the Martian surface clear at last, the mapping cameras looked down on a shrinking S polar cap; sinuous channels which appeared to be dried-up river beds cut by water; chaotic, bouldery terrain first glimpsed by Mariners 6 and 7; and huge impact craters, their floors covered with wind-blown dunes. Surface temperatures ranged from 81°F on the equator to −189°F at the poles; the N pole was much colder than Earth's coldest spot, which is in Antarctica, at −125°F. Several

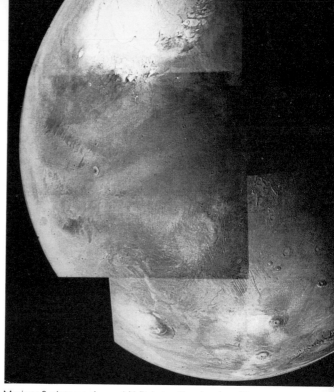

Mariner 9 pictures showed N Pole Ice Cap (*top*) and huge volcanic mountains. Nix Olympica, higher than Everest (*bottom left*)

localized dust storms were seen after the main storm cleared. Variable cloud patterns were observed, mainly in the north, but also over large volcanoes, and were believed to contain water ice; though if large quantities of water exist, they seem certain to be locked in the permanent polar ice caps. Atmospheric winds were measured up to 185 kph. Nix Olympica 500 km across its base, is the Red Planet's highest spot, the peak reaching at least 17 km above the surrounding plain, making it far higher than Everest on a planet half the size of Earth. The tiny Martian moons were both studied: Deimos, orbiting at 20,070 km, has a 16km dia equator, and is 9·6 km from N to S. Phobos, orbiting at only 5986 km, has a 27·3km dia equator and is 19·3 km from N to S. Both moons are heavily cratered, apparently from meteorite impacts; gravity is so low on Phobos that a man could throw a cricket ball into orbit around it. From Apr 2 to Jun 4 1972, Mariner's instruments were turned off while its orbit took it into Mars' shadow during each twice-a-day revolution. After they had been successfully turned on again it became possible to study the N polar region, and to look for potential landing sites for the Viking 1975 project. The last of 45,960 commands to Mariner 9 was to turn off its radio transmitter; it is expected to remain in Martian orbit for at least 50 yr. The costs of Mariners 8 and 9 totalled £56·7M ($136·4M).

Mariner 10 L Nov 3 1973 by Atlas-Centaur from C Canaveral. Wt 503 kg. This was the first dual-planet mission, and the first designed to use the gravitational attraction of one planet to reach another; and despite a series of technical failures soon after launch—such as TV heaters failing to turn on—its remarkable success provided man with his first detailed knowledge of Mercury. Course corrections were successfully performed on Nov 13 and Jan

Mariner 10: TV cameras (*top*). Steerable High-Gain Antenna (*lower left*). Rocket Nozzle visible in sunshade (*bottom*)

21 1974 (the latter a 3·8 sec burn) to ensure passing Venus at a distance of 5760 km; thus obtaining the gravity-assisted bending of the trajectory needed to enable the onboard rocket engine to achieve a Mercury encounter. Closest approach to Venus was on Feb 5, and during the fly-by 3500 pictures were sent back by its twin TV cameras. Although both US and Soviet spacecraft had visited Venus before, this was the first time cameras had been carried, equipped with ultraviolet filters able to send back pictures of the global circulation of the Venusian atmosphere. These pictures confirmed radar measurements made by NASA's Pasadena

laboratory in 1962, suggesting that the upper atmosphere hurtles round in 4 days compared with the 243-day rotation of Venus itself. The pictures showed striking details of circulation patterns swirling from the equator; those of the edge of the planet's disc showed dense, lower layers below the fast-moving upper cloud deck. Because Venus has no detectable magnetic field, but a very dense atmosphere, the solar wind acts directly upon the atmosphere, forming an ionosphere which in turn generates a bow shock as the planet orbits the Sun. (*See* Venus entry in Soviet section for more information.)

On Mar 16 1974 a 3rd trajectory correction placed Mariner 10 on a course which took it past Mercury at a distance of only 271 km on Mar 29. In an 11-day period—6 days before closest approach and 5 days after—a total of 2300 TV pictures of truly remarkable clarity came back to Earth. They revealed a highly cratered, lunar-like surface; impact craters, valleys, and features resembling the mare regions of the Moon, were recorded down to resolutions as small as 100 m. Numerous impact craters had central peaks, like those on the Moon; overlapping craters suggested a formation of meteorites striking the surface in quick succession. Lava-filled craters suggested there was surface activity after the main impact patterns had been formed. The first Mercury feature to be given a name—a large, multiple-rayed crater in the centre of a half-disc picture built up from 18 photographs—was named after the late Gerard Kuiper, principal investigator on Ranger 7, which sent back the first lunar photos. Kuiper was a member of the Mariner 10 team until his death in 1973. The spacecraft's infrared radiometer recorded temperatures ranging from 370°F on the planet's day side to minus 280°F on the night side at local midnight. It was estimated that surface temperatures on the

Venus from Mariner 10, passing at 720,000 km en route to Mercury

165

Man's first look at Mercury: Mariner 10 picture from 77,800 km

day side could reach 560–800°F, depending on the planet's distance from the Sun, providing a total temperature range as great as 1000°F, and leading to the conclusion that it was covered by a thin blanket of porous material which has a very low thermal conductivity. But while Mercury's crust is Moonlike, the interior is much more like Earth's than the Moon's. A minor disappointment was that what was thought on some photographs to be a small moon, perhaps 26 km in dia, was later found to be a star in the background. On May 9 and 10 Mariner 10 was placed on course for its 2nd Mercury fly-by with a 2-part trajectory burn, performed that way to avoid overheating of the rocket engine. A coincidence of celestial mechanics—the fact that Mariner's 176-day orbit round the Sun was in phase with Mercury's 88-day orbit—made a 3rd, as well as a 2nd encounter possible; but correction of overheating problems, and getting the spacecraft out of a 'search-roll mode', when it drifted off Canopus, caused greater consumption than expected of the nitrogen gas used by the reaction control system.

Mariner 10 completed its 2nd sweep past Mercury on Sep 21 1974, with closest approach at 48,000 km. With the spacecraft 170M km from Earth (19M km further than on the first pass), Mariner 10 sent back about 1000 TV pictures. The closest, over the S pole, with resolutions down to 1 km showed that Mercury was once struck with such force in that area that a crater was left 1290 km wide, and the opposite side of the planet was scarred by the vibrations. The collision could have been with an object 96 km in dia. The crater was named the Caloris Basin, since it becomes extremely hot because of the planet's proximity to the Sun. Scientists studying the pictures decided that extensive thrust faults and scarp formations were due to buckling effects, caused by 2 slabs of the surface moving towards each other. This was an opposite process to the tension features on both Mars and the Moon, formed when sections of the surface crust pulled apart. It became clear, too, that Mercury's interior differs from that of the Moon; it has an iron, or high-iron-content core, accounting for about half the planet's volume. As a result, although Mercury is only one-third the size of Earth, its density, or weight, is similar to Earth's. Mariner 10's 3rd and final flight past Mercury, with nearest approach 319 km on Mar 16 1975, and another 1000 pictures of the surface, completed a triumphant mission for the Jet Propulsion Laboratory's guidance and control team. With its original 3·6 kg of nitrogen attitude control gas down to only 0·13 kg, the controllers developed a new 'solar sailing' technique to conserve its fuel during its 6-month flight around the Sun, following the 2nd encounter. The pressure of solar-radiation on the movable high-gain antenna and the solar panels was used to stabilize the spacecraft and to steer it on the designed course between trajectory correction manoeuvres. The last of 8 course corrections—the most ever made on a single mission—was a 3-sec burst of the orbital propulsion system. This directed it towards Mercury from the sunlit side, so that it then looped behind it in relation to the Sun. TV pictures were taken during the inbound and outbound periods; despite a technical fault at Canberra, Australia, which put the big 612m dia receiving dish out of action, pictures showing surface details as small as 450–900 m were obtained. The final batch of pictures meant that 57% of Mercury's visible surface features had been photographed. The pictures suffered from some blurring, however, because of the spacecraft's speed of 40,000 kph as it swept past. Altogether Mariner 10 sent back more than 10,000 pictures of

Mercury. The primary aim of the 3rd fly-by, however, was not pictures, but to probe the planet's magnetosphere, streaming out into space on the anti-Sun side. The 3rd pass confirmed the findings of the spacecraft's magnetometers a year earlier; that Mercury is one of the few magnetized planets in the solar system. Its magnetic field, scientists decided, was not just the effect of the solar wind on Mercury's surface; it was caused by the planet's interior, though they could not decide immediately whether this was due to permanently magnetized rocks, or to 'an active dynamo mechanism' in a fluid core. On Mar 24 1975, following a signal that the control gas was exhausted, a command was sent turning off the transmitter before this historic spacecraft drifted out of communication with Earth. Having penetrated the mysteries of Mercury in just one flight, Mariner 10 is now tumbling or orbiting round the Sun, and will eventually fall into it and burn up. The cost of the basic mission—the first fly-by, had been strictly limited to £41M ($98M); in fact, excluding the booster, launch costs, and NASA tracking charges, that mission cost £40·4M ($96·87M); the 2 subsequent flights cost £1·24M ($2·98M)—a total of £41·64M ($99·85M).

Voyagers 1 and 2 (formerly **Mariners 11 and 12**) L Sep 5 and Aug 20 1977 by Titan Centaur from Cape Canaveral (wt 825 kg), are due to visit Jupiter, Saturn, several moons of both planets, and possibly Uranus in the next decade. Voyager 2, launched first, should be overtaken by Voyager 1 to reach Jupiter in Mar 1979 and use Jovian gravity for 'slingshot' arrival at Saturn in 1980, 3·2 yr after launch, a journey of 2200M km. The Voyagers, advanced Mariners with nuclear power units and 3·7m dia antennas, should send TV pictures of Jupiter and its 4 Galilean satellites for 8 months, then go on to Saturn to study both its rings and satellites, including Titan, with an atmosphere as thick as (though different from) Earth's. Voyager 2 should be re-targeted to reach Uranus (with its newly discovered rings) in 1986. (More details inside back cover.)

Spacecraft Description The early Mariners, 3·04 m long, and 1·52 m across the base, were very similar to the Ranger spacecraft used for impacting on the Moon; a tubular centre was attached to an hexagonal base, from which a dish antenna and solar cell panels were extended. The weight of Mariners 3 and 4, as a result of a more powerful launcher, was increased from 202 kg to 260 kg. An octagonal magnesium centrebody had 4 rectangular solar cell panels to power its computer and sequencer, TV camera, cosmic-ray telescope etc, and a hydrazine-fuelled main engine. Mariners 6 and 7, twice as large again, 3·35 m high, had an 8-sided magnesium framework with 8 compartments containing electronics, TV assembly etc; and 4 rectangular solar panels 213 cm long, with attitude control jets on the tips of the panels. Each had 2 TV cameras (wide and narrow angle) mounted on a rotating platform, able to resolve objects down to 275 m. Mariners 8 and 9 retained the same basic design, but the need for a 136kg thrust retro-engine to inject them into Mars orbit again increased total wt to 1031 kg; this included 454 kg of fuel. The narrow-angle TV camera could resolve features down to 100 m. Other instruments for investigating the atmosphere and surface included an infra-red radiometer, ultraviolet spectrometer, and an infrared interferometer spectrometer.

MERCURY   Manned Spacecraft

History   Project Mercury, the first US manned spaceflight programme, was initiated in Nov 1958; its object was to establish, with a 1-man vehicle, that men could be sent into space and returned safely to Earth. 6 flights started with Alan Shepard's suborbital of 15 min on May 5 1961, and ended on May 16 1963, with Gordon Cooper's 22-orbit mission lasting $34\frac{1}{4}$ hr. Redstone rockets were used for the sub-orbital, and Atlas rockets for the orbital missions.

Spacecraft Description

Launch Weight:	1935 kg	
Orbit Weight:	1355 kg	MA 6
Retrofire Weight:	1347 kg	Weights
Splashdown Weight:	1131 kg	
Recovery Weight:	1099 kg	
Height:	2·90 m	
Base Diameter:	1·89 m	
Escape Tower:	5·18 m	

The bell-shape was determined primarily because of heating conditions during re-entry; limited launch capability at that time—Redstone's thrust was 35,380 kg—necessitated a design just large enough to contain the astronaut and essential equipment: hence, with some justification, it is usually referred to as 'the capsule'. The main conical portion contains the pilot, the life-support system, the electrical power system, and the systems' controls and displays. The cylindrical section contains the main and reserve parachute landing system. The topmost (antenna canister) section contains RF transmission and reception equipment, and the

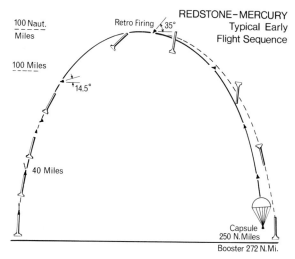

America's first Space 'Lob'

drogue parachute. The cabin section consists of an inner pressure vessel and an outer heat-resisting skin, fashioned from titanium and nickel alloy, with ceramic fibre insulation. Front and rear pressure bulkheads are fabricated from titanium, the rear bulkhead supporting, among other things, the astronaut's individual, form-fitting couch. This is designed so that the gravity forces during launch and re-entry are evenly distributed throughout his body.

Heat Shield   For the suborbital missions, blunt end construction was of beryllium; for the 4 orbital missions, an ablative shield mixture of glass fibres and resins, so that during re-entry (max temp 3000°F; 1649°C), the resin vapourized and boiled off at low temperatures into the hot boundary layer of air. The upper part of the spacecraft (max temp 1300°F; 718°C) was protected with an insulated double-wall structure. The heat shield is detachable, so that following re-entry and main parachute deployment, the shield can be dropped 1·22 m below the spacecraft, pulling out with it a circular collar of rubberized glass fibre. The heat shield thus strikes the water first, absorbing the initial shock; the collar acts as a cushion, and then, as it fills with water, serves as a sea 'anchor'.

Control System Originally designed so that, in normal circumstances, control should be exercised entirely from the ground, though with the astronaut able to take over manually in the event of failure of such things as retro-rocket firings, life-support systems, air conditioning, communications and attitude control. A red light on the panel indicates failure of the retro-rockets to fire at the correct time. Attitude control is by 2 independent thruster systems: 1) Twelve H_2O_2 pitch, yaw and roll thrust nozzles which can be operated automatically, or by ground control or by the astronaut. In the last mode, 'fly-by-wire', the astronaut operates the nozzles by movements of his control stick.   2) 6 more nozzles in a separate system, used either mechanically or manually. The 18 nozzles provide thrust from 0·45–10·9 kg. The astronaut is provided with an instrument panel periscope for observing Earth's surface, to provide a datum for manual control of the spacecraft's attitude with the reaction jets in case of failure of the automatic system; it uses infrared horizon scanners.

Communications Duplicated systems for voice, radar, command, and telemetry links.

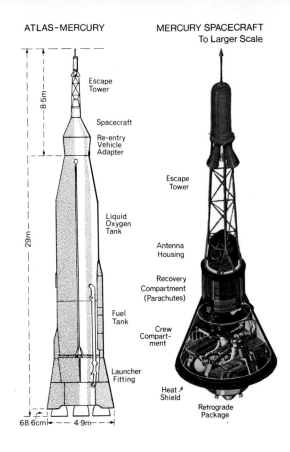

ATLAS–MERCURY

MERCURY SPACECRAFT
To Larger Scale

Escape Tower

Spacecraft

Re-entry Vehicle Adapter

8·5m

29m

Liquid Oxygen Tank

Fuel Tank

Launcher Fitting

68·6cm ← 4·9m →

Escape Tower

Antenna Housing

Recovery Compartment (Parachutes)

Crew Compartment

Heat Shield

Retrograde Package

Life Support 2 control circuits for cabin and suit, providing breathing, ventilation and pressurization, with 100% oxygen atmosphere at 0·36 kg/cm².

Escape Tower Operated by one solid-propellant rocket of 22,680 kg thrust with 3 canted nozzles, and able to lift the spacecraft clear of the rocket for recovery even from ground level. Separation of tower in normal operation is by 3 solid-propellant rockets of 159kg thrust.

Retro-rockets 3 ripple-firing (which means that burn times overlap) solid propellant rockets of 526kg thrust, operated with the spacecraft oriented at 34°, heat-shield forward and up. They reduce the orbital speed of about 28,165 kph by 563 kph, thus starting re-entry.

Parachutes One 1·83m dia conical ribbon-type drogue housed in antenna cone and opening at 6400 m; 19·2m dia ringsail main, and reserve, opening at 3050 m, and stowed in spacecraft's cylindrical neck.

MERCURY FLIGHTS

The first research and development spacecraft was successfully launched by an Atlas-D rocket on Sep 9 1959. Various test-firings of boosters, spacecraft and escape systems followed. The main flights were named and numbered according to the booster employed: Mercury-Atlas 1, a ballistic trajectory test of an unmanned spacecraft in Jul 1960, was unsuccessful because of rocket failure. Mercury-Redstone 1, in Nov 1960, also failed because the rocket engines cut out after it had lifted 2·54 cm above the pad, but was successfully repeated 1 month later. Mercury-Redstone 2 (MR 2) is described below; Mercury-Atlas 2–4, further unmanned tests, were completed by Sep 1961. Details of chimpanzee and manned flights are given below.

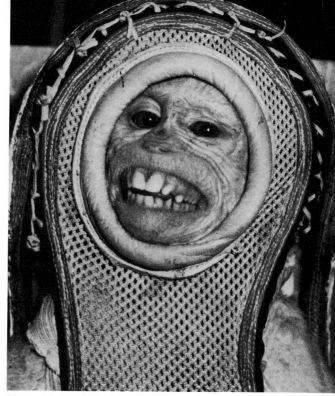

Rhesus Monkey 'Miss Sam', fired 14 km high by Little Joe rocket, successfully tested Mercury's escape tower

Mercury-Redstone 3 takes first US astronaut into space

MR 2   On Jan 31 1961 Ham, the chimpanzee, was launched from Cape Canaveral, and endured 17 G during lift-off, and loss of cabin pressure; after being weightless for 6 min, he again endured heavy re-entry loads. Owing to a series of malfunctions, the capsule reached a speed of 9335 kph instead of 7080 kph, and a height of 253 km instead of 185 km. It landed 212·4 km from the target, and was shipping water and submerging by the time helicopters recovered it, partially tearing off the heatshield as they did so. But Ham worked throughout the flight, was little the worse, and accepted an apple with a large grin on being released. His survival in spite of the setbacks gave the 7 Mercury astronauts confidence in the system.

MR 3   On May 5 1961 at 9.34 am Alan Shepard, 37, became the first US man in space in Freedom 7, with a flight lasting 15 min 22 sec. He was weightless one-third of that time, rose to an altitude of 187·5 km, attained a maximum speed of 8336·2 kph and landed 486 km downrange from the Cape. He experienced 6 G during acceleration, and 11 G on re-entry. Recovery operations went perfectly, the spacecraft was undamaged, and Shepard exuberant.

MR 4   On Jul 21 1961 'Gus' Grissom, in Liberty Bell 7, admitted that he was 'a bit scared' at lift-off on the 2nd suborbital mission. He was weightless for 5 min 18 sec of the 15 min 37 sec flight, reached a height of 190·4 km and endured 11·1 G on re-entry. Recovery did not go according to plan. The new explosive hatch-cover, incorporated for quick rescue of an injured astronaut, blew off after Grissom had removed the pin from the detonator and was awaiting arrival of the helicopter. Liberty Bell began shipping water, so Grissom hurriedly climbed out and swam away. Recovery efforts failed to prevent the spacecraft sinking in 4570 m;

but a rather waterlogged Grissom was hauled into the helicopter none the worse for his experience. The planned MR 5 and 6 flights were cancelled as unnecessary.

MA 5 On Nov 29 1961 chimpanzee Enos went into a nearly perfect orbit of 159 by 237 km, after holds totalling 2 hr 38 min. Because of environmental control system troubles, Enos' temperature rose to over 100°F (37·8°C), and re-entry was ordered after only 2 instead of the planned 3 orbits. But recovery was perfectly executed after a 3-hr 16-min flight, during which Enos was weightless for 2 hr 41 min. He survived over 7 G during acceleration and 7·8 G during re-entry, and 79 undeserved electric shocks when one of the control levers, which he had been trained to operate to avoid shocks, malfunctioned. Nevertheless Enos continued to operate it; and it was decided that, had an astronaut been aboard, he would have been able to correct the ECS problem.

MA 6 On Feb 20 1962, after delays and postponements starting on Jan 23, John Glenn, 41, 3 hr 44 min after entering Friendship 7, and patiently enduring a whole series of last minute problems, became the first US man in orbit—11 months after Gagarin. His apogee was 261 km, perigee 161 km, and velocity 28,234·3 kph; he was weightless for 4 hr 48 min of the 4-hr 55-min flight. Glenn was intrigued, as was the world below, with the fact that, at sunrise on each orbit, his spacecraft was surrounded with brilliant specks— 'fireflies' he dubbed them—which disappeared in bright sunlight. A sticking valve in a reaction jet, similar to the trouble that had necessitated termination of MA 5 after 2 orbits, enabled Glenn to demonstrate that an astronaut could overcome such problems. Then a telemetry fault started a major alarm by indicating that the heatshield was no longer locked in position. A difference of opinion on the

Atlas rocket almost ready to launch Glenn in Friendship 7

ground was resolved with the decision to order Glenn not to jettison the retropack before re-entry; it was hoped this would help retain the heatshield in position. As the retropack burned and broke away, Glenn commented: 'That's a real fireball outside'; and later, as chunks flew past his window, he thought his heatshield was disintegrating. But all was well, and Friendship 7 splashed into the Atlantic 64·5 km short of the predicted area; retrofire calculations had not taken into account the spacecraft's weight loss in consumables. Glenn's only injury was knuckle abrasions when he punched the detonator to blow off the hatch after Friendship 7, with Glenn still inside, had been hoisted on to the deck of the recovery destroyer. Recovery forces totalled 24 ships, 126 aircraft, and 26,000 personnel.

MA 7 On May 24 1962 Scott Carpenter, in Aurora 7, became the 2nd US man in orbit, following a near-perfect countdown, delayed only by 3 15-min weather holds. The successful Glenn flight led to a decision to make this a scientific, rather than an engineering flight. Carpenter carried out a number of experiments, including the release of a multi-coloured balloon on a 30·5m nylon line, which failed to inflate correctly; he manoeuvred the spacecraft with such enthusiasm, yawing and rolling and discovering that even when flying head-down to Earth he was not disorientated, that by the end of 2 orbits he had used more than half his fuel. When eating, he discovered that weightless crumbs floating in the spacecraft could be a danger to breathing. Problems during preparations for re-entry, and a hurried switch to fly-by-wire control with the result that he forgot to switch off the manual system, resulted in the spacecraft running out of both manual and automatic fuel; the retro-rockets, which Carpenter operated by pushing a

Salt-wind corrosion necessitated blowing up of famous Mercury launch tower on 1 Dec 1976

174

button, fired 3 sec late; other errors contributed to Aurora 7 overshooting the Atlantic splashdown target by 420 km. It was 50 min after splashdown before it was established that Carpenter had struggled out of the spacecraft and was safe in his liferaft. He had been weightless for 4 hr 39 min out of 4 hr 56 min, had reached a velocity of 28,241·6 kph and endured 7·8 G.

MA 8   On Oct 3 1962 with Walter Schirra in Sigma 7, this was 'the textbook flight'. Apart from an alarming clockwise roll 10 sec after lift-off, and temperature problems with his spacesuit which had Ground Control discussing possible termination after the first orbit until Schirra succeeded in adjusting it, the 6-orbit flight—double that of previous missions—went well from beginning to end. Schirra carried out disorientation tests, and practised conserving his fuel so effectively, that a final, 1-day MA 9 mission at last became possible. He splashed down only 7·24 km from the recovery ship in full view of the crew and waiting newsmen. He had been weightless for 8 hr 56 min of the 9-hr 13-min flight, and post-medical checks showed unusual symptoms. His heart-beat, 70 when lying down, rose to 100 when standing; and his legs and feet turned reddish purple, indicating pooling of blood in the legs. But the condition lasted only 6 hr, and next day his heart and blood pressure were normal.

MA 9   On May 15 1963 Gordon Cooper in Faith 7 started a 22-orbit flight which lasted 34 hr 20 min, of which, as usual, all but 17 min was spent in weightlessness. He described the spacecraft as a 'flying camera' since it had a slow-scan TV Monitor and various other photographic systems. Extra equipment, fuel, oxygen, water etc., for the longest Mercury mission had raised the in-orbit weight to 1373 kg. Experiments included the release of a 15·24 cm sphere

Project Mercury parachute and capsule recovery test

containing a flashing light which he was able to spot—a significant step towards rendezvous manoeuvres. Cooper also carried out experiments aimed at developing a guidance and navigation system for Apollo; and caused scientific scepticism and interest by claiming to see individual houses and smoke from their chimneys while passing over Tibet at an altitude of 161 km. On the 19th orbit a warning light suggested the craft was decelerating and beginning to re-enter; by the 21st his automatic stabilization and control system had short-circuited and the carbon dioxide level was rising in the spacecraft. Despite his dry comment, 'Things are beginning to stack up a little', Cooper manually conducted his re-entry so efficiently he splashed down only 6·44 km from the recovery carrier.

Performance Summary   A 3-day MA 10 mission desired by the astronauts was finally turned down. So project Mercury lasted 55 months from authorization. While John Glenn's first orbital flight was 22 months later than originally targeted, the final 1-day flight took place only 4 months after the decision to make it. Only 1 of the 7 Mercury astronauts got no flight; that was Donald Slayton, originally selected for MA 7; he was found to have an erratic heart rate, and was replaced by Scott Carpenter. The total cost of Project Mercury was £163,375,000 ($392,100,000).

MILITARY SATELLITES   Past and Future

History   The Soviet-American 'space race' originally began, not so much because of the international prestige involved, but because each feared the other might achieve complete military dominance if it became the first to master space technology. By 1960 the US Air Force's Ballistic Missile Division at Patrick Air Force Base, 32 km S of Cape Kennedy, had 14 major programmes aimed at 'searching out' the possible advantages of space activity. The design and use of military satellites was far advanced long before the first 2 Sputniks were orbited in 1957; and this section should be read in conjunction with the Discoverer and NOAA entries. Tiros 1 was the first US reconnaissance satellite to be successfully orbited. Projects Midas (Missile Defence Alarm System) and Samos (Satellite and Missile Observation System) were the first efforts to use space to double the warning time given by the ground-based BMEWS (Ballistic Missile Early Warning Stations) of an enemy missile attack. The BMEWS radar stations in Alaska, Greenland, and England are able to detect missiles the moment they rise above the horizon, and give 4 min warning to Europe and 15 min to America before they strike their targets. Midas doubles that by giving warning the moment they lift-off.

For convenience I have divided the costly variety of military and reconnaissance ('spy') satellites which have been developed since then, into 5 broad categories:

1 Early Warning
2 Nuclear Explosion Detection
3 Photo Surveillance
4 Electronic Surveillance
5 Communications and Navigation

This entry seeks to penetrate (so far as it is proper to do so) the veil of secrecy surrounding US military activities in space, by co-ordinating the very limited amount of information officially announced, with expert deductions made from the orbits and weights worked out by independent observers in Britain and the US. By end 1975 America's military launches numbered 326 compared with 586 by Russia; but many US Defense Department launches are classified as civil rather than military; these are announced launches of small research satellites like Radcat and Radsat (L Oct 2 1972 from Vandenberg) as radar calibration targets, and to measure radiation effects on the lifetimes of instruments.

In the same way, satellites providing telephone communications can hardly be classified as 'military' in the strict sense. Since it is not so easy to differentiate in this way with Soviet Cosmos launches, this may account, to a limited extent, for the apparently much higher Russian launch rate; but it will be noted that Russia's total of both military and civil launchings overtook America's in 1967, and has remained more than double the American figure ever since. This has led to fears among some of America's military space experts about the vulnerability of their much thinner space systems. They fear that one day Russia might attempt to 'blind' American defences by intercepting and destroying her early-warning satellites, and by putting her navigation and communication satellites out of action by attacking them with laser beams. To counter this possibility, much work is being done to increase the survivability of America's military satellites by shielding them against laser beams, and by keeping a close watch with the IMEWS system, not only on Russian launches, but on the performance and behaviour of the rival satellites when they are in orbit; and by providing multiple communications paths, in various frequencies, by means of multi-purpose systems— 'satellite buses', as they are called. The results of that process are dramatically apparent in the elaborate, 24-satellite Global Positional System (GPS), which Dr Malcolm Currie, Director of Defense Research and Engineering for the

USAF, said 'will have a revolutionary effect on both strategic and tactical warfare' when it is fully operational in the mid 80s.

EARLY WARNING SATELLITES

Midas Equipped with highly sophisticated infrared sensors capable of picking up the exhaust heat from a ballistic missile as it leaves the ground, this was originally intended to increase from 15 to 30 min the advance warning for the USA of a missile attack, provided by BMEWS. The plan was to have 12 to 15 satellites in polar orbits. The 'readout stations', built to receive their information, and feed it into the BMEWS system, included one at RAF Kirkbride, Cumberland.

Midas 1 (L Feb 25 1960) failed to reach orbit as a result of a faulty stage separation; Midas 2 (L May 24 1960) achieved orbit, but telemetry failure prevented transmission of infrared data. Midas 3 (L Jul 12 1961) was successfully launched from Vandenberg into a circular polar orbit of 3428 km, and 91° incl. Orbital life was 100,000 yr; though it was known to be fully operational, there is no record of how long it lasted. Midas 4 (L Oct 21 1961) and Midas 6 (L May 9 1963) aroused an international furore, when details of an experiment called Project West Ford became known. This was to eject a 35kg canister, containing 350 million hair-like, copper dipoles 21 mm long. The idea was that, after separation, the spinning canister should slowly dispense the dipoles in an orbital belt 3220 km high, 8 km wide and 40 km deep, to test whether they would act as passive reflectors for relaying military communications. For over a year the project was violently attacked by the world's scientists, particularly in Britain and Russia, because they felt the dipoles might interfere with astronomical observations, especially with radar tele-

Atlas-Agena launch of Midas satellite to give early warning of enemy ICBM attack

scopes. Professor Keldysh, President of the Soviet Academy of Sciences, said the experiment could result in 'serious contamination of near-terrestrial space and greatly hamper both manned spaceflights and astronomical observations'. Midas 4 successfully ejected its canister, but the dipoles failed to disperse; despite the protests, the US Air Force insisted on repeating the experiment with Midas 6, which was said to be successful; but after that, talk of dipoles was dropped. Apart from the dipole incident, Midas 4 was credited with detecting the launch of a Titan missile from Cape Kennedy 90 sec after lift-off. This incident inevitably led to increased secrecy being applied to military space experiments; but the space logs identified 9 Midas launches in circular 3220km polar orbits, up to Oct 5 1966, carried out by Atlas-Agena D rockets.

IMEWS Integrated Missile Early Warning Satellite, a USAF project now operational as successor to Midas. From a geostationary orbit, it employs an infrared 'telescope' to detect exhaust plume emissions from missiles as soon as they are launched. Immediate warning of hostile launches is believed to be transmitted to ground stations in Guam and Woomera in Australia, and from there via military communications satellites to NORAD (North American Defense Command headquarters) at Colorado Springs. In addition to the early warning role, such satellites can obviously monitor test launches, and are believed to have provided immediate information when Russia tested missiles carrying 3 separate warheads which spread out to make interception more difficult. They may have taken over the early warning role of the Vela satellites. IMEWS 1, L Nov 6 1970, by Titan 3C from Cape Kennedy, wt 820 kg was not completely successful. After being placed in a $26,070 \times 36,050$km orbit at $7\cdot8°$

incl, for checkout of the systems over the USA, fuel was exhausted before it could be moved into position to observe missile tests in China and firings along Russia's Pacific test range. IMEWS 2, L May 5 1971, from Cape Kennedy into a 42,124 × 35,651km orbit with 0·87° incl, was successfully placed over the Indian Ocean; shortly afterwards the US Senate Armed Services' Committee was told there was a satellite 'capable of immediately reporting ICBM launches from the Soviet-Sino area.' IMEWS 3, L Mar 1 1972, was stationed in a 42,067 × 35,962km orbit, with 0·2° incl, over the Panama Canal, where it can view both the Atlantic and E Pacific oceans to detect a submarine-launched missile attack. It seems likely that IMEWS, powered by cruciform solar panels spanning about 7 m, have an operational life of over 5 yr; unless removed by future space tug operations, they will remain in orbit for around 1M yr.

Calsphere/Thorburner   One of the most mysterious series, but possibly concerned with establishing a US ability to intercept and destroy enemy satellites. 3 tiny satellites, 2 later identified as Calspheres 1 and 2, were launched in a joint USAF/USN operation, on Oct 6 1964, into 1046 × 1078km orbits at 90° incl, and a successful interception manoeuvre is believed to have been carried out. Thorburner 2 launches (named after the rocket combination) have taken place twice a year into identical orbits since 1969; the 4th Thorburner 2, L Feb 17 1971, placed Calspheres 3, 4 and 5 into precisely similar orbits. The US Defense Department has said that while it has not matched Russia's satellite-intercept capability, it could be developed.

NUCLEAR EXPLOSION DETECTION

Vela/ERS   Launched in pairs, and planned before Ame-rica's first satellite had been launched, Vela's main task is to detect and identify nuclear explosions in space. They will provide instant warning of any violation of the 1963 treaty which prohibits the testing of nuclear weapons either in the atmosphere or distant space. The initial pair, launched by an Agena D, on Oct 17 1963, were manoeuvred by onboard rocket motors into circular 107,825km orbits on opposite sides of the Earth, well beyond the Van Allen radiation belts. Velas 3 and 4 were launched on Jul 17 1964; and Velas 5 and 6 on Jul 20 1965, by Atlas-Agena D launchers. Uprated Velas 7 and 8 were launched on Apr 28 1967, and uprated Velas 9 and 10 on May 23 1969, by Titan 3C launchers. Advanced Velas 11 and 12 were launched on Apr 8 1970, also by Titan 3C. Similar orbits were employed in each case; transmission life of these satellites, which nowadays also includes solar flare and other observations, is probably about 3 yr. Orbital life of the satellites is in all cases estimated at 1M yr.

Spacecraft Description   The original Vela was 20-sided, 1·42m wide, wt 231 kg. Its 18 detectors could identify X-ray, gamma ray and neutron emissions and would have detected nuclear explosions as far away as Mars and Venus. The latest Velas are 26-sided polyhedrons, weighing 263 kg, after burnout of the onboard solid-propellant apogee motor. Approx 22,500 solar cells cover 24 sides to generate 120W of electric power; the satellites are continuously oriented to look down into the Earth's atmosphere from opposite sides. Vela launches invariably include piggy-back auxiliary payloads, such as ERS (Environmental Research Satellites). These range in weight from 0·7–20 kg, and usually carry a single scientific or engineering research experiment. 29 ERS launches had been made by the end of 1972.

PHOTO SURVEILLANCE

Samos A USAF programme, providing operational versions of the satellites first developed by the early Discoverers, and able to photograph all parts of the world from polar orbits, tape-recording TV pictures while over potentially hostile territory and transmitting them when passing over US territory; able to supplement and improve on these pictures by periodically dropping off capsules containing film.

Samos 1, L Oct 11 1960 by Atlas-Agena A from Point Arguello, wt 1860 kg, failed to achieve orbit. Samos 2, L Jan 31 1961, into a 554 × 475km orbit, and 97° incl, successfully returned experimental data. Samos 3, L Sep 9 1961, exploded on the launchpad. Subsequent launches came after the decision that details of military satellites should be classified; but the series, operational since 1963, progressed through steadily more advanced versions. These are placed in polar orbits with perigee, or low points, of less than 161 km, and on occasions specially launched to take a close look at points of interest picked out by routine surveillance satellites in much higher orbits. There is some evidence that Samos 87 was launched on Mar 17 1972, by a Titan 3B/Agena D booster from Vandenberg into a 131 × 409km orbit with 110° incl. It weighed about 3000 kg, was a long cylinder with 1·5m dia; typically, it remained in orbit 25 days.

Big Bird A huge USAF multi-function satellite, designed to perform both the 'search-and-find' and 'close-look' functions which required 2 different spacecraft until it came into operation in 1970. Also known as LASP, for Low Altitude Surveillance Platform. Weighing over 12,000 kg in orbit, it probably consists of a modified Agena rocket casing, 15·2 m long and 3·05 m dia, fitted with a high-

Titan 3D launches Big Bird spy satellite

resolution Perkin-Elmer camera capable of identifying objects as small as 0·3 m across from heights of more than 160 km. Operational techniques are similar to those employed by the ERTS satellites: they are placed in Sun-synchronous orbits so that they pass regularly over the targets at the same time of day. A series of pictures with identical Sun angles is thus obtained, and changes occurring, such as the construction of new missile sites, and the number and types of missile being installed, are easily read. Film is processed on board, then scanned by a laser device, and converted into electronic signals transmitted back to Earth to at least 7 receiving stations at the USAF's global bases. Drag encountered by such a large vehicle at such low altitude would mean that it would re-enter in 7–10 days; the Agena engine is therefore fired periodically to raise the orbit and extend its life. Big Bird is almost certainly capable of carrying out some Elint, or electromagnetic surveillance as well, and sometimes carries with it a small 60kg piggy-back capsule, placed in a higher orbit for such 'ferret' operations. Because both East and West had adopted the practice of collecting and examining each other's space debris whenever possible, America adopted Soviet policy with Big Bird of exploding them in orbit, to ensure debris fell in their own area or in the sea, rather than allowing them to decay naturally.

Big Bird 1, L Jun 15 1971 from Vandenberg, the first known launch by a Titan 3D/Agena, with wt of 11,400 kg, was placed in a 114×186km orbit, with $96°$ incl, and re-entered after 52 days. By the end of 1973, a regular pattern of launches had developed, with slightly higher orbits, and lifetimes extended to 90 days. There was a 3-month gap, however, between Big Bird 8 (L Apr 10 1974) which re-entered on Jul 28, and Big Bird 9 (L Oct 29 1974; orbit 162×271 km). This was probably due to modifications being made to its sensors, to counter Soviet efforts to camouflage missile silo and control centre constructions. Big Bird 9 provided surveillance at a time of increased tension in the Middle East, and took pictures of 16 Soviet ships unloading crated materials, believed to include SAM-6 spare parts, and components for Scud surface-to-surface missiles,

at the Syrian port of Latakia. By the end of 1975, Big Bird 11 (L Dec 4 1975; orbit 157×234 km), had a 5-month life; as usual it carried a small pick-a-back capsule, possibly for radiation or atmospheric research, which was ejected by its own motor at Big Bird apogee into a 236×1558km orbit with 2-yr life. Big Bird 12 (L Jul 8 1976) and Big Bird 13 (L Dec 19 1976) had lives of 157 and 161 days, and also ejected capsules.

ELECTRONIC SURVEILLANCE

Elint A general name for electronic intelligence, or 'ferret' satellites. Little is known about their numbers or effectiveness. They are usually launched into higher orbits—about 500 km—than photographic satellites. They record electromagnetic radiations being transmitted from areas of military activity, and replay them to ground stations for study and identification. By the end of 1972 nearly 30 Elint satellites had been identified, more than 20 having been launched with a photo satellite in lower orbit. The 'radar signatures', or characteristics, such as pulse repetition frequency, pulse width, transmitter frequency, modulation, and so on, enable the likely function and method of operation of a particular centre to be identified; the numbers and types of electronic systems at the site, and subsequent changes will give a valuable indication of its purpose and capability. Electronic Warfare, Electronic Countermeasures, and Electronic Counter Countermeasures, carried on at present by a wide variety of aircraft and ground-based stations, are likely to be increasingly taken over by satellites in the next decade. The ability to intercept and decode an enemy's satellite and other communications, and to interfere with them by rival satellite activities, is expected to be a decisive factor in any large-scale future hostilities.

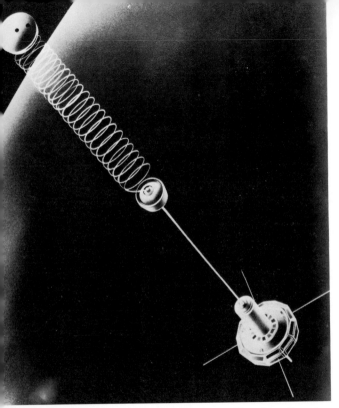

US Navy 'Yo-Yo' communications satellite, aligned by wire spring and 30m boom

COMMUNICATIONS AND NAVIGATION

A military satellite operating in synchronous orbit (where it revolves at the same speed as Earth, and therefore remains over the same point on Earth), at an altitude of 35,890 km, overlooks 163M sq km of Earth's surface, compared with an aircraft 8 km high, overlooking 284,900 sq km. The development of these satellites, therefore, for military communications and navigation has formed a major part of the US space effort.

Communications   The world's first communications satellite, the 21st in the world log to be orbited, was Score, L Dec 18 1958 by Atlas B from Cape Canaveral, into a 185×1470km orbit and $32°$ incl. Wt 70 kg. It transmitted taped messages for 13 days, and re-entered 34 days later. Next came Courier, L Oct 4 1960, the first active-repeater Comsat, which operated for 17 days. Lack of rocket power, together with political and economic argument, delayed further developments until Jun 16 1966, when a Titan 3C successfully orbited 8 satellites, including the first 7 of the Initial Defence Satellite Communications System (IDSCS). 26-sided polygons, 86cm dia, covered with solar cells, and with no moving parts each weighed 45 kg. Dispensed over a period of 6 hr at slightly different orbital velocities to give them global coverage, they were placed just below synchronous altitude, at 33,915 km. Drifting about $30°$ relative to Earth, each stayed in view of an equatorial station for $4\frac{1}{2}$ days, so that even if one malfunctioned there was always another drifting into position. Spin stabilized, with a service life of 3 yr, their electronic components were programmed to shut off automatically at the end of 6 yr. The system, totalling 26 satellites, was completed by 3 more Titan 3C launches, the last on Jun 13 1968. The satellites

DSCS-2 provides communication links between US Military Commands

were capable of linking ground points 16,090 km apart, and from 1967 provided a S Vietnam–Hawaii–Washington link for transmitting, among other things, high quality reconnaissance photographs. 5 were still operational after 8 yr in orbit.

Improved versions weighing 242 kg at launch, 129 kg after onboard rocket motors had manoeuvred them into stationary, equatorial orbits, were NATO 1, L Mar 20 1970, NATO 2, L Feb 2 1971 and NATO 3A, L Apr 28 1976; with 5-yr operational life, they provide regular communications between operational ground stations. As the 5-yr planned life of the first IDSCS satellites was nearing its end, launch of more advanced replacements began. Launched in pairs, DSCS 1 & 2 (L Nov 3 1971 by Titan 3C from C Kennedy; wt 520 kg ea) and DSCS 3 & 4 (L Dec 14 1973) were placed in stationary orbit. A 3rd pair (L May 20 1975) failed to achieve orbit after a Titan 3C malfunction, and fell into the Pacific 6 days after launch; 3 such pairs could provide complete world coverage, except for the polar areas, between US military installations. A 3rd series of DSCS satellites, with a 10-yr life cycle, is due to begin work in 1980. Equipped with both multi-beam antennas and a jam-resistant command and control system, they will be able to handle 1300 2-way phone conversations for both tactical and strategic users.

Meanwhile, development of tactical communications satellites (Tacsat), needing much greater onboard power, so that they could be received by small, low-powered ground terminals carried by ships, tanks, jeeps and aircraft, were also under development. LES 5 (Lincoln Experiment Satellite) L Jul 1 1967 from C Kennedy, wt 102 kg, as part of a 6-satellite payload (which included IDSCS 16, 17 and 18) was America's first. 2 days after it had been manoeuvred into a

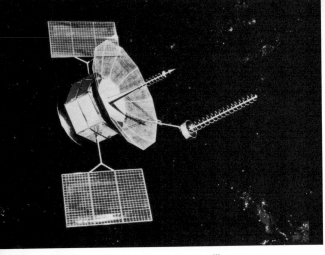

FLTSATCOM: US Navy's latest satellites

33,360km, near-synchronous orbit, the first satellite communications between US aircraft, a US Navy submarine and surface vessel, and Army ground units, had been carried out. Tacsat 1, L Feb 9 1969, 7·6 m tall, 2·7 m dia, wt 725 kg, gyrostat-stabilized so that the antennas and telescopes could be continuously pointed while the major part of the satellite spun within them, was a follow-up project; its object was to communicate with tiny land-based receivers 0·3 m in dia.

Another communications satellite system being developed is FLTSATCOM (Fleet Satellite Communications), consisting of 4 large geosynchronous satellites in equatorial orbit. Starting in 1977, they will cover the US Navy, Air Force and other Defense Department users. Closely linked to this will be AFSATCOM (Air Force Satellite Communications

System), linking airborne command posts, the bomber fleet, strategic support aircraft, missile control centres, and other ground sites. LES 8 & 9, launched Mar 15 1976, into synchronous orbits, were test flights for AFSATCOM. They carried nuclear-power generators, which it is hoped can be used to replace the highly vulnerable solar panels of conventional spacecraft, and other techniques intended to improve the spacecraft's chances of survival if they are attacked by either conventional or nuclear armed 'space interceptors'.

Navigation The first navigation satellite, Transit, was developed primarily to provide Polaris missile submarines with the ability to fix their positions within 150 m; it soon became evident that all-weather navigation could be provided in this way for all types of shipping. After an initial launch failure, Transit 1B, L Apr 13 1960, wt 120 kg, in a 373×745 km orbit at $51°$ incl, transmitted information including time signals for 3 months which enabled navigators to fix their position. By 1968, 3 Transit series, totalling 23 satellites, had been placed in circular orbits of about 805 km; in mid 1972, 5 were operational, 2 of these having been launched 5 yr earlier. A successor, DNSS (Defense Satellite System), providing continuous position-fixing for all 3 US Forces, is under development. The US Navy has been experimenting with Timation, requiring multiple satellites in high polar orbits of about $16,000 \times 20,000$ km. The only named launch, Timation 2, on Sep 30 1969, placed 6 Timations, together with 2 other satellites in 906×940km orbits, with $70°$ incl. An unnamed launch on Dec 14 1971, placed 4 more satellites in similar orbits. By the time Timation 3 was launched (Jul 14 1974; wt 293 kg; orbit $13,445 \times 13,767$ km; incl $125°$), the programme had been merged with Global Positioning System, and Timation 3 was renamed NTS-1 (Navigation Technology Satellite). Details of GPS below.

GPS

The Global positioning System, also known as NAVSTAR (for Navigation System using Time and Ranging), is due to be operational in 1984. It will be able to fix any point on or near the Earth with an accuracy, horizontally and vertically, of 10 m; it will be able to measure velocity (for instance of ICBMs) with an accuracy of 0·03 m per sec. It will consist of 24 satellites, generating continuous navigation signals 'using broad spectrum techniques and very long pseudo-noise codes to provide secure, jam-resistant operation'. Users of the system, ranging from foot soldiers carrying 5·4kg manpacks, to aircraft and ballistic missiles, will remain passive, so they cannot be electronically located. The first 6 satellites should be in subsynchronous, 20,000km orbits by mid 1977; when fully deployed, 8–9 satellites should be in view of any point on Earth at a given time, carrying their precise atomic clocks, synchronized to a common system time. In addition to improving the accuracy of both ICBMs and SLBMs (submarine-launched ballistic missiles), one of its most important applications will be accurate all-weather, day-and-night weapon delivery—an extension of the so-called 'smart' weapon technology first developed in Vietnam and during the 1973 Middle East war. Cost of this USAF-managed programme will be about $2000M over a 10-yr period.

SPACE SHUTTLE

Starting in 1983, the USAF expects to use a military version of NASA's Space Shuttle for about 20 launches per year from Vandenberg. With its ability to launch 29,500kg payloads into 278km circular orbits—twice as much as a Titan 3C at less cost—the Shuttle is expected to offer solutions to some of America's major military problems, such as the loss of foreign bases. It will be used to bring back to Earth for repair satellites which have gone wrong. As one of their contributions to the Space Transportation System, the USAF is developing for NASA and itself the Shuttle's Interim Upper Stage (IUS). Since the Shuttle Orbiter can operate only up to about 700 km, an additional propulsion stage will be needed to take spacecraft to higher orbits, including stationary ones, until a more permanent reusable Space Tug has been developed.

NIMBUS   Meteorology and Oceanography

History   The US Navy now regards Nimbus as indispensable to shipping operations in the Arctic and Antarctic. Satellite pictures showing the location and movement of ice masses now enable Naval ships to move in these areas for several extra months, and ultimately are expected to make operations possible right through the 6-month polar night. (Russia is using special spring and autumn Cosmos launches for the same purpose). The Nimbus series was originally conceived (as the Latin title word, meaning 'raincloud', implies) as meteorological satellites to provide atmospheric data for improved weather forecasting; but as increasingly sophisticated sensing devices were added to successive spacecraft, the series grew into a major programme studying Earth sciences. From it, too, sprang ERTS (now renamed Landsat), the results of which proved to be quite sensational and are described separately. By mid 1973, data provided by Nimbus 1–5, covered oceanography (the geography of the oceans), hydrology (the study of water in the atmosphere, and on land surfaces, in the soil and underlying rocks), geology (the history of Earth, especially as recorded in rocks), geomorphology (the study of the Earth, its distribution of land and water and evolution of land forms), geography (description of land, sea and air, and distribution of life, including man and his industries); cartography (chart and map making), and agriculture (data on moisture and vegetation patterns over various land surfaces).

By 1975 all 7 of the Nimbus spacecraft originally planned, had been launched, at an estimated cost of £142M ($340M). They were lettered A–F before launch, and given numbers only if successful. All except B, lost as a result of a launch vehicle failure, exceeded their mission objectives; and the 8th, to be launched in 1978, should become NASA's first satellite mainly devoted to studying atmospheric pollution.

Spacecraft Description The basic Nimbus spacecraft is butterfly-shaped when deployed in orbit, 3 m long, with a span across the solar panels of 3·4 m. The panels, each $2·4 \times 0·9$ m provide more than 200W of power, supplemented by 2 SNAP-19 nuclear generators. The body of the spacecraft consists of 2 main components, separated by struts; the larger carries the sensors and other equipment; the smaller, between the solar panels, houses the stabilization and control system; working on the principle that the Earth is warm and space is cold, the infrared sensors keep the satellite's cameras pointed towards Earth at all times. The system is controlled by a computer which maintains the correct attitude, to within 1 degree in each axis, by means of cold gas jets and inertia wheels. Pitch-and-roll stability is sensed by horizon scanners, while a gyroscope controls stability in the yaw axis.

Nimbus 1 L Aug 28 1963 by Thor-Agena B from Vandenberg. Wt 376 kg. Orbit 932×422 km. Incl 98°. This was fully operational for 1 month until failure of the solar array power system; a short 2nd burn of the Agena rocket resulted in an eccentric orbit instead of the intended circular 1110km orbit. It returned 27,000 cloud-cover photos, providing the first high-resolution TV and infrared weather photos, and proving to meteorologists that such pictures could be received at small, inexpensive portable stations as the satellite passed overhead. Hurricane Cleo's 'portrait' was taken during Nimbus 1's first day in orbit; subsequently many other hurricanes and Pacific typhoons were tracked; inaccuracies on relief maps were corrected, and the Antarctic ice front more accurately defined as a result of its pictures. Orbital life 15 yr.

Nimbus 2 L May 15 1960, by TAT-Agena B from Vandenberg. Wt 413 kg. Orbit 1100×1181 km. Incl 100°. A nearly perfect orbit, designed for 6 months (2500 orbits), actually provided 33 months of operations and terminated in orbit 13,029 on Jan 17 1969. Following the success of Nimbus 1, over 300 APT (Automatic Picture Transmission) stations in 43 countries were able to receive its high-resolution infrared pictures. Temperature patterns were obtained of lakes and ocean currents for shipping and fishing industries, and thermal pollution could be identified. An additional Medium Resolution Infrared Radiometer, which measured electromagnetic radiation emitted and reflected from Earth in 5 wavelength intervals from visible to infrared, permitted detailed study of the effect of water vapour, CO_2 and ozone on the Earth's heat balance. Orbital life 800 yr.

Nimbus 3 L Apr 14 1969, by Thorad-Agena D from Vandenberg. Wt 575 kg. Orbit 1070 × 1131 km. Incl 99°. The 3rd success in the series following the launch failure of Nimbus B on May 18 1968. Nimbus 3 carried 7 meteorological experiments, plus SNAP-19, a nuclear-power unit for generating electricity. Vertical temperature measurements of the atmosphere by SIRS (Satellite Infrared Spectrometer) were acclaimed as one of the most significant developments in the history of meteorology. Previously, because data over the oceans had been scanty, only 20% of the world had detailed weather information; SIRS made it possible to obtain temperature data over the entire Earth with an accuracy of 2°F above 6100 m, and within 4°F below that level. A 10·4kg electronic collar fitted to a wild elk in the National Elk Refuge in Wyoming, was interrogated twice daily to study the migratory habits of large animals. The Nimbus 3 rocket also placed Secor 13, a US army geodetic satellite, weighing 20·4 kg into an identical orbit. Life of Nimbus 3 is 800 yr; Secor 2000 yr.

Nimbus 4 L Apr 8 1970 by Thorad-Agena D from Vandenberg. Wt 675 kg. Orbit 1093 × 1107 km. Incl 107°. In its circular, near-polar orbit, Nimbus 4 was still making world-wide weather observations on a twice daily basis (once in daylight and once in darkness) more than 2 yr later. Of its 9 experiments, 4 were new and 5 improved versions of experiments on earlier flights. New experiments included IRLS (Interrogation Recording and Location System); examples of its tracking activities included weather balloons floating around the world, floating ocean buoys, a wild animal, and Miss Sheila Scott on a solo flight over the N Pole. One task was to measure the thickness of ice on floating islands in the Arctic by interrogating buoys placed

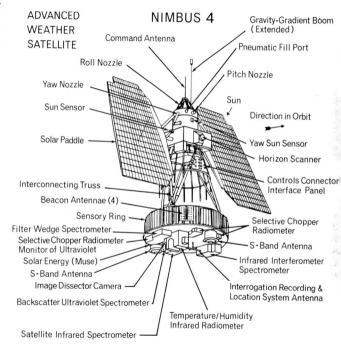

ADVANCED WEATHER SATELLITE

NIMBUS 4

Gravity-Gradient Boom (Extended)
Command Antenna
Pneumatic Fill Port
Roll Nozzle
Pitch Nozzle
Yaw Nozzle
Sun
Sun Sensor
Direction in Orbit
Solar Paddle
Yaw Sun Sensor
Horizon Scanner
Controls Connector Interface Panel
Interconnecting Truss
Beacon Antennae (4)
Sensory Ring
Selective Chopper Radiometer
Filter Wedge Spectrometer
Selective Chopper Radiometer
Monitor of Ultraviolet
S-Band Antenna
Solar Energy (Muse)
Infrared Interferometer Spectrometer
S-Band Antenna
Image Dissector Camera
Interrogation Recording & Location System Antenna
Backscatter Ultraviolet Spectrometer
Temperature/Humidity Infrared Radiometer
Satellite Infrared Spectrometer

in the water; one such island, T3, was 11 × 6 km long and 30 m thick. These islands melt in summer in brackish swamps covered by fog and are impossible to reach. Details of their behaviour are providing information on the 'cradle' of much of the weather affecting the US and Europe. Other Nimbus 4 tasks included the analysis of water qualities and sewage pollution of the Great Miami River near Cincinnati and of Lakes Erie and Ontario. Low oxygen content indicates raw effluent, the oxygen decrease being caused by bacterial decomposition. Other IRLS 'platforms' were also placed in Mt Kilauea, Hawaii, probably the world's most active volcano, to measure the relationship between temperature rises and eruptions, and in a bear's den in Montana to monitor its environment during hibernation. Each 'platform' will only respond when the satellite interrogates it by its individual 16-bit digital 'telephone number'; the address of the bear den in Montana is 0111100001111001. The Nimbus 4 rocket also placed TOPO 1, a US Army Secor-type satellite, wt 21·7 kg, into an identical orbit for use in space-ground tactical exercises. Orbital life: Nimbus 4 1200 yr; Topo 1 2000 yr.

Nimbus 5   L Dec 11 1972 by Delta from Vandenberg. Wt 768 kg. Orbit 1089 × 1102 km. Incl 99°. Carrying 6 new instruments, Nimbus 5 was designed to take the first vertical temperature and water readings of Earth's atmosphere through cloud—a major step forward since many parts of the world are under cloud-cover for more than 50% of the time. False colour pictures enabled investigators to say that rain was falling in certain areas 'at a rate of five-hundredths of an inch per hour'. Measurements were also made of water vapour which evaporated from the oceans, changed to water and fell back to the surface. Sensors were also carried to map the Gulf Stream off the US E Coast and the Humboldt Current off S America's W Coast. Plotting the daily position of the Gulf Stream enables southbound ships to avoid it, while northbound ships, by riding in the stream, get the benefit of several extra knots. Operational life 1 yr. Orbital life 1600 yr.

Nimbus 6   L Jun 12 1975 by Thor-Delta from Pt Arguello. Wt 827 kg. Orbit 1092 × 1104 km. Incl 99·9°. 1600-yr life. In addition to maintaining the regular weather, rainfall and icepack watch, Nimbus 6 was allotted the task of investigating the dangers involved in recovering the huge oil and gas deposits already located in the Arctic in the Prudhoe Bay and Beaufort Sea areas. A series of data collection platforms were dropped to enable Nimbus 6 to track the ice-pack movements, the amount of summer melting, etc. This should enable estimates to be made of possible locations for oilrigs, and whether pipelines from them would have to be buried under the ocean floor, laid on the sea bottom or over the icepack.

NOAA   Weather Programme

History   The US National Oceanic and Atmospheric Administration manages and operates the basic US weather satellites, called NOAAs in polar orbit, which provide daily cloud-cover pictures and other information for world-wide weather forecasting. The pictures are seen by millions when used by TV weathermen. By 1976 about 25 weather satellites had been launched; from 1960–65, 10 TIROS satellites were launched; from 1966–69 when 9 more were orbited, they were called TOS, for Tiros Operational Satellite, or ESSA, for the Environmental Science Services Administration, which managed the programme in that period. In 1970, a 2nd-generation satellite, called ITOS, for Improved TOS was introduced. Currently, the satellites are given an ITOS letter designation before launch and a NOAA number when in orbit. In 1974 came the SMSs (Synchronous Meteorological Satellites), which aided short-term weather forecasting by providing almost continuous observation of short-duration weather features. By then America was working towards

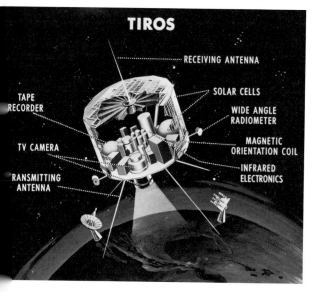

GOES, a Geostationary Operational Environmental Satellite System, to provide continuous day-and-night weather observation of N and S America and its oceans; and the World Meteorological Organization was planning, with Russia, the European Space Agency and Japan, to launch their own geostationary satellites in the late 70s to complement the SMS/GOES system for global use.

Spacecraft Description Tiros satellites had a 'hatbox' shape; 18-sided polygons, 1·07 m dia, and 0·55 m high. Solar cells covered the sides and top, with apertures for 2 TV cameras on opposite sides; each camera could take 16 pictures per orbit at 128-sec intervals, though the intervals could be decreased to 32 sec. 2 tape recorders could store up to 48 pictures when ground stations were out of range. The weight of 119 kg for Tiros 1 had risen to 138 kg for Tiros 9 and 10. Essa satellites, similar but more advanced, had 2 APT (Automatic Picture Transmission) cameras, able to photograph a 3000km wide area, with 3km resolution at picture centre. Pictures were taken and transmitted every 352 sec, allowing a typical APT station to receive 8–10 per day. ITOS/NOAA satellites remained box-shaped 1 m × 1 m × 1·24 m, with 3 solar panels, with momentum flywheel providing spacecraft stabilization, and 340 kg wt.

Tiros An acronym for Television and Infrared Observation Satellite, this began as a joint NASA/Defense Department project to develop a meteorological satellite. As soon as Tiros 1, L Apr 1 1960, began orbiting at 692 × 740 km, it was clear that the US had successfully established both a meteorological survey and military reconnaissance satellite. During the 78 days that its batteries lasted, it sent back 22,952 cloud-cover photographs; as the first of them came

Tiros 6 gave warning of Autumn storms in Atlantic and Pacific

in, a meteorologist at a ground station observed that the programme had gone 'from rags to riches overnight'. It is believed that its photographs included some of the Soviet Union and China, so detailed that aircraft runways and missile sites could be readily identified. By the time 10 had been launched, the more advanced Nimbus and ESSA satellites were taking over; the first 8, all operating in similar orbits, sent back several hundred thousand photographs, together with information about the flow of heat the Earth was reflecting back into space—vital meteorological information unobtainable until then. Tiros 9, L Jan 22 1965, was the first attempt to reach polar orbit from Cape Kennedy; the series of 3 Delta 'dog-leg manoeuvres' duly placed it in the planned 82° Sun-synchronous orbit; but due to a 2nd-stage failure to cut-off, the orbit, instead of being 644 km circular, was 700×2578 km. In the event, the higher apogee provided more Earth cover than planned; on Feb 13, the first 'photomosaic' of the entire world's cloud cover was provided by 450 excellent pictures. By the time Tiros 10, L Jul 2 1965, was shut down on Jul 3 1967, more than 500,000 cloud-cover pictures had been returned.

TOS This system, based upon Tiros technology, began operating with the launch of ESSA 1 on Feb 3 1966 (orbit 702×845 km) and ended with ESSA 9 (L Feb 26 1969, orbit 1427×1508 km). By that time 400 receiving stations were in operation around the world, and weather services of 45 countries, as well as 26 universities, up to 30 US TV stations, and an unknown number of private citizens who had built their own receivers, were receiving and using their weather photographs each day. In 1969 a picture from ESSA 7 made history by revealing that the snow cover over America's mid-west, in Minnesota and the Dakotas, was 3 times thicker than normal. Measurements showed that it was equivalent to 15–25 cm of water covering thousands of square miles. A disaster area was declared before it happened; and when the floods came much had been done to control the situation.

ITOS/NOAA The Improved Tiros Operational System more than doubled ITOS capability because its cameras were able to take night cloud cover pictures as well as daytime pictures. They were designated ITOS before launch and once in orbit given a new series of NOAA designations because NOAA had then taken over ESSA. Thus, NOAA 1 was launched Dec 11 1970 from Vandenberg in a 1429×1472km orbit with $101°$ incl; but ITOS B, L Oct 21 1971, was never given a NOAA designation because it failed to achieve a satisfactory orbit. NOAAs 3 & 4 were launched in Nov 1973 and 1974, and NOAA 5 (wt 340 kg) in Jul 1976 into similar orbits.

SMS-1 L May 17 1974 by Delta by C Canaveral (wt 627 kg) into synchronous orbit over the equator off S America, provided the first day-and-night stormwatch, with either weather or infrared pictures every 30 min. It was also used from Sep 1974 to receive and transmit data provided by 20 balloons released in French Guiana. Drifting round the world at 30 km, the balloons each carried 64 instrumented packages of 400g wt. When dropped on command by parachute, these packages transmitted wind, temperature and humidity data back to their carrier balloon, which sent the information on via SMS or Nimbus 6. **SMS-2** L Feb 6 1975, and placed over the equator $15°$ SE of Hawaii, meant that the 2 spacecraft together could cover the W Hemisphere. One of its tasks was to keep watch on California's forest areas, including the famous redwoods, to give warning within 90 min of fire outbreaks.

GOES-1 view of N and S America

GOES First of 3 launches was Oct 16 1975 by Delta from C Canaveral. Wt 293 kg. Incl $70°$W. Identical with SMSs, and renamed once more because of their operational status.

OAO   Astronomical Observatories

History   A programme of 4 Orbiting Astronomical Observatories, begun in 1959, and completed with the launch of the 4th (on Aug 21 1972) at a total cost of £154·6M ($364·4M). Although only 2 of the launches were successful, their achievements ranged from the first observation of ultraviolet (UV) emissions from the planet Uranus, to new insights into the structure and composition of Earth's upper atmosphere. OAO 2's success in studying young, hot stars which emit most of their energy or light in the UV—the blue portion of the spectrum not visible to the human eye or ground observatories because of Earth's atmosphere— proved that with such unmanned spacecraft astronomers could conduct sustained viewing of the universe. It enabled the 82cm UV telescope installed on OAO 3 (renamed Copernicus once in orbit, to celebrate the 500th anniversary of 'the father of modern astronomy') to be pointed within 3 one-hundredths of an arc second. In 1975 Copernicus began the study of 3 Sun-like stars, 11 light years from Earth, to see if it was possible that intelligent civilizations on their planets were trying to communicate with us. Astronomers hope to continue this work with the Large Space Telescope project, postponed in 1976 because of NASA budget cuts.

OAO-1   L Apr 8 1966 by Atlas-Agena D from C Kennedy; wt 1776 kg. Orbit 792 × 805 km. Incl 35°. The battery failed after only 3 days in orbit, but it provided engineering data showing that the concept of astronomical investigation from space was feasible, and led to improvements in later craft.

OAO-2   L Dec 7 1968 by Atlas-Centaur from C Kennedy. Wt 2016 kg. Orbit 770 × 780 km. Incl 34·99°. When it was

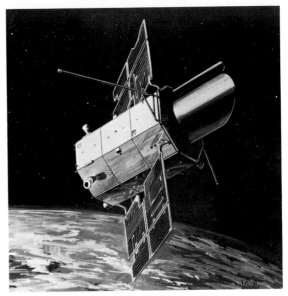

OAO-2 spacecraft discovered hydrogen clouds (artist's rendering)

shut down on Feb 13 1973, because it had developed an electrical fault, it had operated for over 4 yr instead of 1. During its 22,000 orbits, its 11 telescopes had viewed 1930 celestial objects and made 22,560 observations. These included the detection of a huge hydrogen cloud around comet Tago-Sato-Kosaka—the first evidence that such

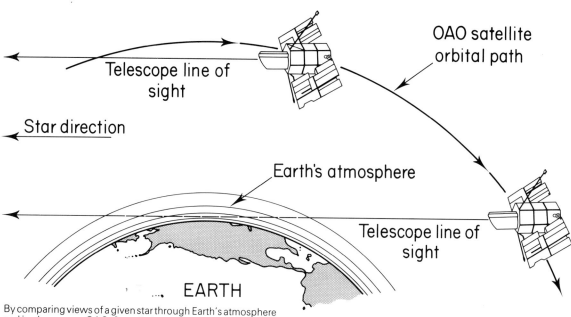

Telescope line of sight

Star direction

OAO satellite orbital path

Earth's atmosphere

Telescope line of sight

EARTH

By comparing views of a given star through Earth's atmosphere and in clear space, OAO-3 studied atmospheric pollution

clouds exist; observations of stars with magnetic fields over 10,000 times stronger than that of the Sun; in May 1972 the first UV observations above the atmosphere of a supernova—the momentary outburst of a star to a brightness millions of times greater than the Sun; as well as observing UV emissions from Uranus.

OAO-B   L Nov 30 1970 by Atlas-Centaur from C Kennedy. It failed to achieve orbit when a protective shroud could not be jettisoned, and fell back to Earth.

OAO-3   L Aug 21 1972 by Atlas-Centaur from C Kennedy. Wt 2220 kg; orbit 748 × 740 km. Incl 35°. Still transmitting on command at the end of 1975, Copernicus, as OAO-3 was renamed in orbit, had been declared successful within 5 months. By then, Princeton University's 80cm UV telescope, with a 47·6kg mirror made from thin fused silica ribs, had made 1780 observations of 37 different sources. It had detected large quantities (more than 10%) of molecular hydrogen in the denser interstellar dust clouds; and surprisingly large amounts of deuterium (heavy hydrogen) in interstellar dust clouds. (Deuterium is a basic element for fusion in the formation of stars, and current theories had suggested most of it should already have been used up; so those theories may have to be revised). These observations are made by collecting UV light from a star and directing it to a spectrometer, which then sends digital readings to Earth. The 2nd experiment, consisting of 3 smaller X-ray telescopes, developed at University College, London, and provided by the Science Research Council, successfully followed up the work of earlier satellites (notably Explorer 42, L Dec 1970) which established that X-rays were present in the universe in much larger quantities than were generated by the Sun. The object was to chart about 200 other X-ray sources already identified, as well as finding new sources. The most striking finding in observations of 55 unique objects was that the rotation rate of the Cygnus X-3 binary system increased perceptibly over a period of only one month. Clusters of galaxies in Perseus, Coma and Virgo, as well as the supernova remnants in Puppis were also studied. Copernicus, as mentioned above, also started work on trying to establish whether other civilizations were trying to contact us with UV laser beams; and US scientists, by studying a bright star, first when it was high in the sky, and then through Earth's atmosphere as it sank below the horizon, also used Copernicus to study how far Earth's ozone layer is being broken up by chlorine, resulting from freon, the gas used in millions of aerosol spray cans. Orbital life 500 yr.

OGO   Geophysical Observatory

Details of vast hydrogen clouds, the most abundant element in the universe, were provided by the 6-satellite Orbiting Geophysical Observatory series. OGO 1 was launched Sep 5 1964 by Atlas-Agena from C Kennedy. Wt 487 kg; orbit 35,743 × 114,040 km; incl 57°. OGOs 2–6 were launched between 1965–69. Each was launched into a different elliptical orbit and into a different sector of the cislunar space quadrant. The last (L Jun 5 1969; wt 620 kg) transmitted until Jun 23 1969. The 130 experiments they carried sent back 1·5M hr of data and added a number of notable 'firsts' to space history. They included: first observation of protons responsible for a ring of current surrounding the Earth during magnetic storms, at a distance of several Earth radii; first satellite global survey of Earth's magnetic field; first observation of daylight aurorae; first world-wide map of airglow distribution; and new knowledge about Earth's 'bow shock' as it sweeps around the Sun. In Apr 1970 OGO-5 measured a huge hydrogen gas envelope, 12M km across and 10 times larger than the Sun, surrounding Comet Bennett. Not visible from Earth, the hydrogen cloud was measured by a French measuring device while the comet was 104M km from Earth, and OGO-5 was operating in a special spin-scan mode, 22,500 × 107,800 km above the Earth. The existence of large amounts of hydrogen around comets was first discovered in Jan 1970 by OAO-2. Satellites' orbital life 10–16 yr.

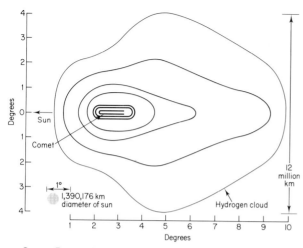

Comet Bennett's hydrogen cloud as measured by OGO-5

OSCAR   For Radio Amateurs

History   A series of small satellites enabling radio 'hams' all over the world to talk to each other. Oscars 1–4 were launched by the US Defense Department between 1961–5; Oscars 5–7 were launched between 1970–74, always 'piggy-back', first with Transtage military launches, then on NASA Itos launches. They transmitted low-power signals to enable radio amateurs to obtain training in satellite tracking and for experiments in radio propagation. By Oscar 7, emphasis was on transmission of educational broadcasts to schools. Although the satellites, battery-operated at first, had only a short life—from 2 months to about 1 yr—they were a great success. After Oscar 5, which operated 1½ months, Melbourne University in Australia had compiled tracking reports from hundreds of stations in 27 countries.

Spacecraft Description   Oscar 7, L Nov 15 1974 by Itos/Delta from Vandenberg, wt 29 kg, orbit 1444 × 1462 km, 101° incl, was built by a multinational team of radio 'hams' under direction of the Radio Amateur Satellite Corporation (AMSAT) of PO Box 27, Washington DC 20044. With 4 radio masts at 90° intervals from the base and 2 experimental repeater systems, it was able to receive and store messages, to be forwarded later as it moved around the world.

OSO   Solar Observatory

History   The Sun has now been kept under almost continuous observation by a series of 8 orbiting Solar Observatories since Mar 1962. This period more than covers a complete 11-yr cycle of solar activity, and astronomers are now well-placed, with observations added by Skylab's Apollo Telescope Mount in 1973, to study the next peak of solar activity in 1980–82. The spin-stabilized, Sun-oriented OSOs were NASA's first standardized observatory spacecraft. Among discoveries so far are that solar flares—the sudden release of bursts of energy and material from the Sun—have temperatures of over 30M°C; and a single flare, which may last minutes or hours, can release as much energy as the whole Earth uses in 100,000 yr. Solar radiation takes 8½ min to reach Earth, and it has been found that several

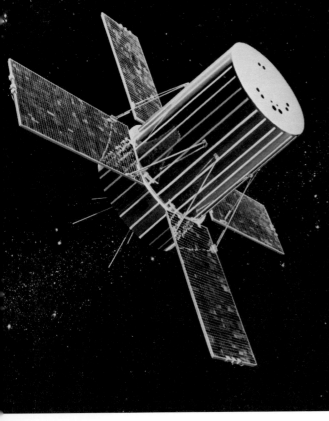

minutes before that there is an increase in soft X-ray emissions—suggesting that it may be possible to predict major flares. 'Holes' in the Sun's corona, where temperatures are much lower than average, is another discovery. The coronal holes are somewhat similar to the solar polar caps which have temperatures of about $1M^{\circ}C$ ($1\cdot8M^{\circ}F$) compared with $2M^{\circ}C$ ($3\cdot6M^{\circ}F$) in other parts of the solar corona. A follow-up programme, the Solar Maximum Mission (SMM) will use the Space Shuttle to launch spacecraft able to observe the next solar peak.

OSO 1 (L Mar 7 1962 by Delta from C Kennedy; wt 208 kg, orbit 553×595 km). With 30-yr orbital life, this returned data on 75 solar flares over 5-month period. **OSO 2** (L Feb 3 1965) operated 10 months and brought total data hours to 6000. **OSO 3 & 4** (L Mar and Oct 1967) and **OSO 5 & 6** (L Jan and Aug 1969) continued observations in the standard 550km orbit with $32\cdot8^{\circ}$ incl. **OSO 7** (L Sep 29 1971), much more sophisticated with 571kg wt, failed to achieve correct orbit; when it re-entered in Jul 1974, OSO 5 was successfully reactivated to monitor solar eruptions. **OSO 8** (L Jun 21 1975, wt 1064 kg), carried US and French UV telescopes to investigate the transfer of solar energy between layers of the Sun's atmosphere. It was also investigating the mysterious X-ray radiation arriving at the Earth from all directions in space, as well as X-ray 'binaries'—a visible star with what appears to be an invisible companion, or 'black hole'.

Spacecraft Description OSO 7, similar in appearance to earlier craft, but nearly double the weight of any previous one, has a bottom, rotating 'wheel' section with 152·4 cm dia, and 71·6 cm ht. The wheel spins at 6 rpm, automatically controlled by a pneumatic system. The upper, 'sail' section is 234 cm high and 209 cm wide; non-spinning, it provides power by means of solar cells, and a platform for the US and French spectrometers, which always look at the Sun. A gyroscope in the sail acts as a 'Sun-position memory', so that the Sun-pointing experiments re-acquire the Sun as quickly as possible when the vehicle emerges from Earth's shadow. Gas jets and magnetic torque coils keep the spin axis perpendicular to the Sun.

PAGEOS   Balloon Satellite

History   This was part of the Geodetic Satellite Programme (*see* Explorers 22–36) to create a world survey network with an accuracy of 10 m. Intended to provide an orbiting point source of light which could be photographed for 5 yr to determine the size and shape of Earth to a degree never before possible, it was brightly visible—a source of unfailing interest to professional and amateur astronomer alike—for 10 yr. Pageos (Passive Geodetic Earth-Orbiting Satellite) was an aluminium-coated Mylar balloon of the Echo 1 type. As it reflected the Sun, it was as bright as the star Polaris, and was observed simultaneously from 41 portable camera stations around the world, which were used to construct a 3-dimensional geodetic reference system. The resulting triangulation network made it possible to obtain the distance between 2 surface points on Earth, 5000 km apart, to an accuracy of 10 m.

Pageos 1   L Jun 23 1966 by Thrust Augmented Thor from Vandenberg. Wt 111 kg. Incl 85°. Orbit 5342 × 3016 km. The balloon was folded and packed inside a spherical canister, which was separated into halves by an explosive device after orbital insertion. The balloon then automatically inflated. Intended purely to reflect sunlight, it carried no instruments. Orbital life was 50 yr, but on Jan 20 1976, it mysteriously broke up into 14 pieces. Speculation about the cause of its explosion included a build-up of static electricity, or some deliberate interference by either America or Russia.

PIONEER   Interplanetary Explorers

History   With the latest in this series being used to reveal the secrets of Jupiter and Saturn, this has become the most exciting of all the unmanned projects. Pioneer 1, launched in 1958, was the first NASA spacecraft, the project having been handed over by the USAF. The series is complementary to the Mariner and Viking flights. Missions are given letters before launch, changed to the next number in the series only if successfully placed on course. In 15 yr, 11 out of 14 launches added immensely to man's knowledge of Earth's surroundings. The first 5 were used to study solar energy, and also to provide up to 15 days' warning of solar flares, so that astronauts could be protected during Moonflights. Pioneers 6–9 were used to study space from widely separated points during an entire solar cycle of 11 yr; the first of these, designed for a 6-month lifetime, operated over 10 yr. A major achievement of this series was to discover Earth's long magnetic 'tail', about 5·6M km, on the side away from the Sun. In 1971 the alignment of Pioneers 6 and 8, at points more than 161M km apart, enabled the solar wind's density to be measured more accurately than ever before. In Sep 1972 NASA's Ames Research Center in California, succeeded in locating and reviving Pioneer 7, which had turned its transmitters off, although it was on the far side of the Sun, 312M km from Earth. Then came the triumphant flights to Jupiter covering 1972–4. Pioneer 10 was the first spacecraft placed on a trajectory to escape from the solar system into interstellar space; the first to fly beyond Mars; the first to enter the Asteroid Belt; and the first to fly to Jupiter. It is hoped it will be the first to sense the interstellar gas beyond the Sun's atmosphere.

In achieving all this Pioneer 10 established that the much-feared asteroid belt contained less material than at one time believed, and presented little hazard either to unmanned or manned spacecraft.

Spacecraft Description
Pioneer 10

Lift-Off Weight Overall:	146,673 kg
Lift-Off Height:	40·3 m
Spacecraft Weight Overall:	260 kg
Spacecraft Payload:	27·8 kg
Spacecraft Radius:	6·4 m
Spacecraft Antenna Dia:	2·7 m
Launch Vehicle:	Atlas-Centaur-TE-M-364-4

Pioneer spacecraft are designed to carry a variety of instruments tailored to the study of either individual or a series of planets. Nos 10 and 11 were the first craft designed to travel into the outer solar system, to operate there for 7 yr, and as far from the Sun as 2400M km. Since launch energy requirements to reach such distances are far higher than for shorter missions, the spacecraft must be very light: the 258 kg wt includes 30 kg of scientific instruments, and 27 kg of propellant for attitude changes and midcourse corrections. 6 thrusters each provide 0·2–0·6 kg thrust. They can adjust the place and time of arrival at Jupiter by changing velocity, or merely adjust the attitude. This is done by pulse thrusts, timed by a signal from a star sensor which 'sees' Canopus once per rotation, or by one of the 2 Sun sensors which 'see' the Sun once per rotation. Velocity changes totalling 670 kph can be made during the mission. The spacecraft are spin-stabilized, giving the instruments a full circle scan 5 times a minute. Because solar radiation at Jupiter is too weak to operate an efficient solar-powered system, 4 nuclear units provide electric power; they are carried on 2·7m booms so that their radiation will not affect the scientific experiments (13 on Pioneer 10; 14 on Pioneer 11). Controllers use 222 different commands to operate the spacecraft; during the 4 days it takes to pass Jupiter commands take 45 min to reach it. The heart of the communications system is the fixed, 2·7m dia dish antenna, which focuses the radio signals in a narrow beam. The onboard experiments are intended to return data on the solar atmosphere from Earth to beyond Jupiter; to measure the numbers and characteristics of asteroids; to measure Jupiter's atmosphere, heat balance, and internal energy sources; its magnetic fields and radiation belts; and, by means of the imaging photo polarimeter, to return images of Jupiter. Pioneers 10 and 11 both carry a 15 × 23cm plaque showing the origin of the spacecraft in the solar system, and drawings of a man and woman related to the spacecraft's size in case they should one day be seen by another intelligent species; but such a species would need to have eyes like our own to be able to understand the plaque's message of goodwill.

Pioneer 1   L Oct 11 1958 by Thor-Able from C Canaveral. Wt 38 kg. Failed to reach the Moon, but looped 113,854 km into space. In 43 hr 17 min of flight, it discovered the extent of Earth's radiation bands.

Pioneer 2   L Nov 8 1958, by Thor-Able from C Canaveral. This 39·5kg probe, intended to reach the Moon, failed as a result of unsuccessful 3rd-stage ignition.

Pioneer 3   L Dec 6 1958, by Juno 2 from C Canaveral. The 5·9kg probe again failed to reach the Moon, but reached height of 102,333 km, and discovered Earth's 2nd radiation belt.

Pioneer 4   L Mar 3 1959, by Juno 2 from C Canaveral. 5·9kg lunar probe, passed within 59,983 km of Moon, and then into solar orbit 0·9871 × 1·142 AU.

Pioneer 5   L Mar 11 1960 by Thor-Able from C Canaveral. 43kg probe, sent into solar orbit 0·8061 × 0·995 AU, and sent back solar flare and wind data until Jun 26 1960, at a distance of 37M km.

Pioneer 6   L Dec 16 1965 by TAD from C Kennedy. 63kg cylinder, 1 m dia and 0·9 m high, launched towards Sun into 0·814 × 0·985 AU solar orbit. It sent back first detailed description of the tenuous solar atmosphere; with Pioneer 7, gathered continuous data on events on a strip of solar surface extending nearly halfway round the Sun. It also measured the tail of Comet Kohoutek. In Dec 1975, still working after 10 yr it set an operational record for an interplanetary spacecraft. With Pioneers 7, 8 and 9, it was then providing a network of solar weather stations, often millions of km apart, being used to predict solar storms for airlines, power companies, military and other organizations.

Pioneer-1, first NASA satellite, mapped the Van Allen radiation belts

Pioneer 7   L Aug 17 1966 by TAD from C Kennedy. 63kg cylinder, launched away from Sun into 1·010 × 1·125 AU solar orbit. (*See* Pioneer 6.)

Pioneer 8   L Dec 13 1967 by TAD from C Kennedy. 65·3kg cylinder, placed in 1·0 × 1·1 AU solar orbit, to join previous 2 Pioneers in obtaining data on solar wind, magnetic field and cosmic rays. Also defined tail of Earth's magnetosphere. Launched piggyback with Pioneer 8, was NASA's first Test and Training Satellite, used to exercise the Apollo communications network.

Pioneer 9 L Nov 8 1968 with 2nd Test and Training Satellite, by TAD from C Kennedy. 66·6kg Pioneer placed in 0·75 × 1·0 AU solar orbit. This was the 4th of what was to be a series of 5 solar probes; but Pioneer E, intended to be Pioneer 10, carrying 3rd piggyback Test Satellite, failed to orbit on Aug 27 1969.

Pioneer 10 L Mar 3 1972 by Atlas-Centaur from C Kennedy. Wt 270 kg. Initial speed 51,800 kph, faster than any previous man-made object, placed it on a trajectory which took it past Jupiter's cloud-tops at a distance of 130,300 km on Dec 4 1973. Initial speed was achieved by adding for the first time to an Atlas-Centaur booster (lift-off thrust 186,590 kg), a solid-fuelled 3rd-stage TE-M-364-4, developing 6800kg thrust. (This is an uprated version of the retromotor used for the Surveyor Moon lander.)

Pioneer 10 swept past the Moon's orbit in just over 11 hr, compared with the 89 hr taken by Apollo 17. 4 days after launch, 2 brief firings of the thrusters (8 min 7 sec and 4 min 16 sec) increased velocity by 4 m per sec, shortened the flight time to Jupiter by 9 hr and adjusted the arrival point to 14° below the Jovian equator. In Jul the asteroid belt, believed to be 270M km wide, was entered amid tense speculation as to whether Pioneer 10 would be damaged or destroyed by a 48,000kph collision with an asteroid fragment. It was estimated that a particle only 0·05 cm in size could penetrate vital spacecraft areas. But its closest pass to any known body was 8·8M km from Palomar-Leyden, which has a 1km dia. Pioneer 10 emerged unscathed 7

months later, having established that the asteroid belt could also be penetrated by men without too much difficulty. On Aug 7 it was teamed with Pioneer 9 to send details of one of the most violent solar storms ever known; one estimate was that, during it, the Sun produced as much energy as the whole Earth would use in 70M yr. When Pioneer 11 was launched on Apr 6 1973, Pioneer 10 was within 190M km of Jupiter. In Aug, when the Earth passed between the spacecraft and the Sun, the spacecraft's attitude had to be adjusted slightly to ensure that the alignment sensors would not lose count of the spin rate and confuse the automatic orientation system. From Nov 26 1973, when command-and-return time had increased to 92 min, a total of about 300 pictures were obtained of the approach and fly-by; early ones were mainly for calibration, but about 40 yielded much information after lengthy computer processing. In addition to the Great Red Spot, pictures were obtained of the moons Ganymede, Callisto and Europa. The biggest disappointment was failure to obtain good pictures of the orange moon, Io, but much was learned about that, as detailed below. 26 days before closest encounter, Pioneer 10 began threading its way through the orbits of the Jovian moons.

On Nov 8 it entered the Jovian environment by passing the orbital path of Hades, the outermost moon, at 23·6M km. 3 days later the path of Andrastea, the tiny, 16km dia, innermost of Jupiter's 4 outer satellites, was passed at 20·5M km. Accelerating steadily under the pull of Jovian gravity, Pioneer 10 travelled another 9M km in the next 11 days, to pass the 3 inner moons, Demeter, Hera and Hestia between 11·5M and 11·4M km from the planet. By now the velocity was 450 km/sec and on Nov 26, it encountered Jupiter's bow-shock wave, crossing the magnetopause 24 hr before the earliest predicted encounter, re-entering the magnetic

PIONEER / JUPITER SPACECRAFT

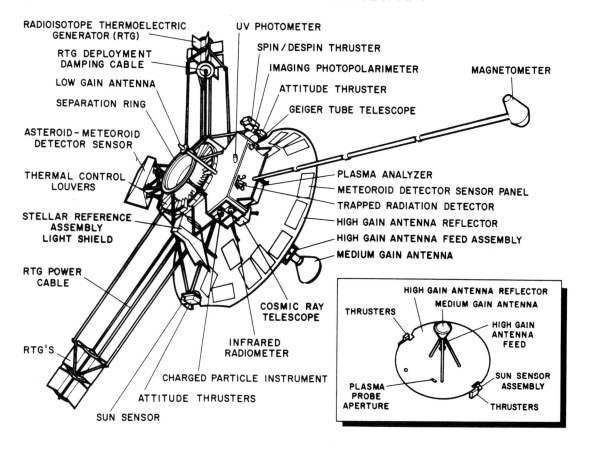

RADIOISOTOPE THERMOELECTRIC GENERATOR (RTG)

RTG DEPLOYMENT DAMPING CABLE

LOW GAIN ANTENNA

SEPARATION RING

ASTEROID – METEOROID DETECTOR SENSOR

THERMAL CONTROL LOUVERS

STELLAR REFERENCE ASSEMBLY LIGHT SHIELD

RTG POWER CABLE

RTG'S

UV PHOTOMETER

SPIN / DESPIN THRUSTER

IMAGING PHOTOPOLARIMETER

ATTITUDE THRUSTER

GEIGER TUBE TELESCOPE

MAGNETOMETER

PLASMA ANALYZER

METEOROID DETECTOR SENSOR PANEL

TRAPPED RADIATION DETECTOR

HIGH GAIN ANTENNA REFLECTOR

HIGH GAIN ANTENNA FEED ASSEMBLY

MEDIUM GAIN ANTENNA

COSMIC RAY TELESCOPE

INFRARED RADIOMETER

CHARGED PARTICLE INSTRUMENT

ATTITUDE THRUSTERS

SUN SENSOR

THRUSTERS

HIGH GAIN ANTENNA REFLECTOR

MEDIUM GAIN ANTENNA

HIGH GAIN ANTENNA FEED

SUN SENSOR ASSEMBLY

THRUSTERS

PLASMA PROBE APERTURE

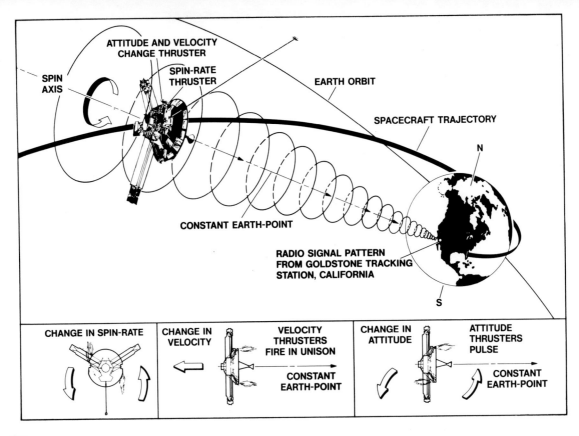

SPIN AXIS

ATTITUDE AND VELOCITY CHANGE THRUSTER

SPIN-RATE THRUSTER

EARTH ORBIT

SPACECRAFT TRAJECTORY

N

CONSTANT EARTH-POINT

RADIO SIGNAL PATTERN FROM GOLDSTONE TRACKING STATION, CALIFORNIA

S

CHANGE IN SPIN-RATE

CHANGE IN VELOCITY

VELOCITY THRUSTERS FIRE IN UNISON

CONSTANT EARTH-POINT

CHANGE IN ATTITUDE

ATTITUDE THRUSTERS PULSE

CONSTANT EARTH-POINT

field on Dec 1, and crossing the magnetopause for the 2nd time, slightly more than 3·2M km from Jupiter. The 4 Galilean moons still lay ahead: on Dec 2 the orbital path of Callisto, as big as Mercury, circling Jupiter at 1·8M km was crossed; 16 hr later, Ganymede, the largest moon, was passed 798,000 km from the planet. On Dec 3, 6 hr after that, Europa, similar in size to Earth's Moon and 583,260 km out, was passed. Finally, Io, 356,000 km from Jupiter, the most interesting of the Jovian moons because it is the most reflective object in the whole solar system, was passed. The big disappointment was that although pictures were obtained of Callisto, Ganymede, and Europa, none was obtained of Io, which brightens for about 15 min when it reappears from behind Jupiter. It is believed that this is because Io has an atmosphere of methane or molecular nitrogen which freezes on the surface during occultation, and evaporates to gas when heated by the Sun.

Closest approach came at 02.25 GMT on Dec 4 and 16 min later Pioneer 10 passed behind Io, and then behind Jupiter itself. The spacecraft reached a speed of 132,000 kph as it passed within 130,300 km of the cloud tops, crossing the Jovian equator at an angle of 14°. 15,000 commands were sent to Pioneer 10 during the flyby; 6 of its 11 instruments were operated continuously. They were the UV photometer, the magnetometer, 2 of the 4 high-energy particle detectors, the asteroid/meteoroid telescope and the meteoroid detector. The solar wind instrument and the 2 remaining high-energy particle detectors were calibrated several times each day and operated to meet the changing approach conditions; the infrared radiometer was sent several hundred commands each day for observations of the inner satellites;

Opposite Pioneer 10: Attitude control and propulsion

but the bulk of the commands went to the imaging photo-polarimeter to obtain views of the satellites and of Jupiter itself. The immense volume of data returned took nearly a year to correlate. The description of Jupiter and its moon in the Solar System section at the back of this book is based upon the discoveries of Pioneer 10 and its successor, Pioneer 11, 1 yr later. Though scarred by the intense Jovian radiation belts—some of the instruments, particularly the asteroid-meteoroid detector, were degraded—the spacecraft survived so well that it was decided that a closer approach would be possible by Pioneer 11. Still working $2\frac{1}{2}$ yr later, in Feb 1976, then 1384M km from Earth, Pioneer 10 crossed the orbit of Saturn, and sent back data showing that Jupiter's enormous magnetic tail, almost 800M km long, spanned the whole distance between the 2 planets. This spacecraft will reach its absolute communications limit, $7\frac{1}{2}$ yr after launch, near the orbit of Uranus about 2900M km from the Sun. Any data sent back from there will take more than 4 hr to reach Earth. It is expected to leave the solar system in 1987, and should reach the vicinity of Aldebaran in *8 million years!*

Pioneer 11 L Apr 5 1973 by Atlas-Centaur from C Kennedy. Identical to Pioneer 10, apart from the addition of a 2nd magnetometer to measure Jupiter's high magnetic fields. Following Pioneer 10's success, it was possible to re-target Pioneer 11 so that it would fly 3 times closer to Jupiter, and then fly on to Saturn. On Apr 19 1974, from a distance of 675M km, the spacecraft's thrusters were fired for 42 min 36 sec, to add 230 kph to its velocity of 46,200 kph. The manoeuvre used 7·7 kg or 28% of its original propellant load. Now targeted to pass under Jupiter's S pole, Pioneer 11 sent back 130 pictures during its final approach; its imaging system and UV photometer studied Callisto,

Ganymede and Europa. Changes in cloud structure near the Red Spot were observed since Pioneer 10's flyby; and Callisto's S pole ice-cap was revealed. The results are summarized in the Solar System section at the back of this Directory; but Pioneer 11's data confirmed that Jupiter was primarily a liquid planet, consisting mostly of hydrogen, and that it radiated more heat than it absorbed from the Sun. At 17.22 GMT on Dec 3 1974, travelling faster than any previous man-made object, at 171,000 kph, Pioneer 11 passed 42,940 km below the Jovian S pole. Mission Controllers at NASA's Ames Research Centre in California, had an anxious 1-hr wait; but then, its speed brought it unscathed through the radiation belts behind Jupiter, and it was flung over the N Pole to double back across the solar system towards its planned Saturn encounter; since it was travelling in the opposite direction to Jupiter's rotation, scientists were given a look at the magnetic field, radiation belt and surface during a complete revolution of the planet. Pioneer 11 confirmed the conclusion, drawn from its predecessor's data, that Jupiter's magnetic fields consist of 2 very different regions—a weak outer field with the solar wind impinging on it, and a much stronger inner field. And it found that cloud tops on Jupiter's polar regions (which cannot be seen from Earth) are much lower than at the equator, and covered by a thicker transparent atmosphere, with some 'blue sky' similar to Earth's.

On Nov 20 1975, Pioneer 11 made the first observations of Saturn from a spacecraft—a prelude to photo-polarimeter scanning planned for Jul 1976. And on Dec 18 1975, by which time Pioneer 11 had covered a 3rd of the distance between Jupiter and Saturn—640M km—and was travelling at 64,000 kph, controllers successfully fired the thrusters to increase velocity by 108 kph. This should enable it to

Jupiter: Pioneer 11 established that it was mostly liquid hydrogen

fly either between Saturn's rings and the planet, or to come in under the rings and pass upwards outside of them. The concentration of debris in the rings is far higher than in the asteroid belt, and the risk of destruction is much greater; but if the spacecraft is still working it is hoped that at worst, pictures of the approach to Titan as well as Saturn will be sent back before that happens in Sep 1979.

Pioneer Venus 1978 Pioneers 12 & 13, to be launched early 1978, will consist of an orbiter and a lander. The latter will drop 4 small probes and then itself become a large probe which will radio data to the orbiter as it descends.

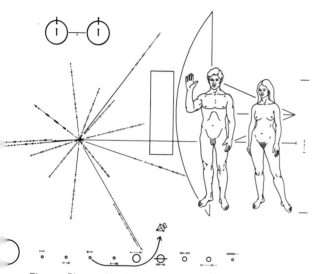

Pioneer Plaque: Key is hydrogen atom (*top left*) most common in universe. 14 lines from Sun indicate cosmic sources of radio energy. Human figures (man's hand raised in gesture of goodwill) compared with size of spacecraft. (*Bottom*) Solar system, showing Pioneer coming from Earth

RANGER Lunar Landers

History The first of 3 projects (the others being Lunar Orbiter and Surveyor) aimed at obtaining sufficient knowledge and pictures of the lunar surface to make the manned Apollo landings possible. The first 6 missions were failures, beginning with Ranger 1, launched from C Kennedy on Aug 23 1961, right through to Ranger 6, launched on Jan 30 1964. Plans to eject and hard-land a 42·6kg capsule to measure seismic activity on the Moon's surface as a mission 'bonus' were abandoned, as failures continued right through the manned Project Mercury flights, and the need for close-up pictures of the lunar surface became desperate if Project Apollo was not to be held up. For Rangers 6–9 a 170kg conical structure containing 2 wide-angle, and 4 narrow-angle TV cameras, plus video combiner, transmitters, etc, was mounted on the spacecraft to send back about 14 min of TV pictures from 2261 km above the Moon, until the spacecraft was destroyed by impact at 9330 kph. Ranger 6 looked like being successful, until it was found that the TV system had been destroyed by being inadvertently turned on early in the flight. Rangers 7, 8 and 9, however, justified the persistence of NASA's Jet Propulsion Laboratory at Pasadena, California. Ranger 7, launched on Jul 28 1964, after a 68-hr flight, returned 4308 excellent pictures—man's first close-up lunar views—before impacting in the Sea of Clouds. Ranger 8, launched on Feb 17 1965, after a 64-hr flight, returned 7137 high-quality photographs; they covered 2,331,000 sq km in the Sea of Tranquillity area. Finally Ranger 9, launched on Mar 21 1965 after a 64-hr flight covering 417,054 km, relayed 5814 excellent pictures before impacting within 4·8 km of the target point in Crater

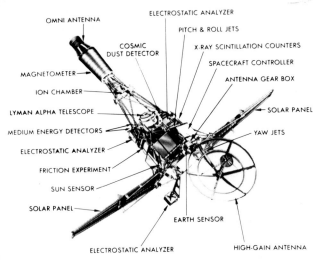

OMNI ANTENNA
ELECTROSTATIC ANALYZER
PITCH & ROLL JETS
COSMIC DUST DETECTOR
X-RAY SCINTILLATION COUNTERS
SPACECRAFT CONTROLLER
MAGNETOMETER
ANTENNA GEAR BOX
ION CHAMBER
LYMAN ALPHA TELESCOPE
SOLAR PANEL
MEDIUM ENERGY DETECTORS
YAW JETS
ELECTROSTATIC ANALYZER
FRICTION EXPERIMENT
SUN SENSOR
SOLAR PANEL
EARTH SENSOR
ELECTROSTATIC ANALYZER
HIGH-GAIN ANTENNA

Ranger 1 details

Alphonsus. Photographs covered features down to only 25 cm across; 200 of the pictures were shown 'live' on TV — probably the first live 'TV spectacular' from the Moon. The total of over 17,000 pictures, some showing craters as small as 25 cm across, revealed the lunarscape several thousand times better than the best Earth telescope had ever done, and finally justified Ranger's cost of £108M ($260M). But while they provided the evidence needed to complete the design of softlanding spacecraft, it remained for Project Surveyor to establish whether the Moon's surface was firm enough to support manned landings.

4 of Ranger's 6 TV cameras sent back overlapping views during the lunar approach; 1 took the same scene with wide-angled lens; the 6th was angled at 30°.

Spacecraft Description Rangers 6–9: Wt 366 kg; ht 3·1 m; solar panels 4·5 m. Launchers: Atlas-Agena D. Conical camera structure, mounted on hexagonal spacecraft bus. Battery of 6 TV cameras (2 wide-angle, 4 narrow-angle) covered by polished aluminium shroud, with 33cm opening near top. High-gain antenna, and 2 solar panels hinged to base. Midcourse motor provided 22·6kg thrust for up to 98 sec. Attitude control, maintained by nitrogen gas jets, Sun and Earth sensors and 3 gyros, programmed by computer and sequencer, enabled spacecraft to aim cameras at Moon while pointing high-gain antenna to Earth for transmission.

SKYLAB   Space Station

	Length	Dia	Weight
CSM	10·45 m	3·96 m	13,970 kg
MDA	5·30 m	3·05 m	6260 kg
ATM	3·35 m	2·13 m	11,180 kg
AM (STS)	1·19 m	3·05 m	22,225 kg
AM (TNL)	4·18 m	1·68 m	
IU	0·91 m	6·58 m	2065 kg
OWS	14·66 m	6·58 m	35,380 kg

Total Cluster Weight: 90,265 kg*
Total Cluster Length: 36·12 m
Total Work Volume: 361·4 m³

*Excludes Payload Shroud: 11,795 kg

The Skylab Space Station, launched on May 14 1973, and subsequently visited by 3 Apollo astronaut crews, who lived and worked in it for periods of 28, 59 and 84 days, was a remarkable example of human ingenuity overcoming what at first seemed inevitable technical disaster. By the time the 3rd crew splashed down on Feb 8 1974, America had demonstrated man's ability not only to live and work in space almost indefinitely (with all its implications for flights to the planets) but also that faulty equipment could be replaced and repairs improvised, to an extent never thought possible when spaceflight began.

Originally called the Apollo Applications Programme, and intended to make some practical use of the hardware left over from the Apollo Moonlandings, it gradually developed into a vital step in man's mastery of the space environment. Nearly 26,000 scientists and other workers were employed on preparations for the mission, which, in addition to tackling steadily increasing flights of long-duration, was aimed at studying solar-activity from outside Earth's dusty atmosphere; and looking inward towards Earth to study its potential resources, problems of pollution, and the possibilities of giving early warning of natural disasters such as floods and volcanic eruptions.

The Skylab Workshop, with a total lift-off wt of 2,822,300 kg, was launched by Saturn 5 on May 14. In a live broadcast I described the vibration as 'terrifying'; it seemed that the whole Press Stand would be brought down upon our heads. It proved a prophetic comment; although the Skylab cluster was placed in the correct 440 × 427km orbit with 50° incl, it was soon apparent that all was not well. The launch vibrations had torn away the combined meteoroid/thermal shield, which in turn ripped away one of the pair of solar array wings, and caused the other to be jammed in a partially opened position by debris. With no thermal protection, and very limited power available from the ATM batteries and the partially opened wing, internal temperatures soared. At the time, NASA officials refused to confirm space correspondents' suspicions that exact details of the damage to Skylab had been provided by the US military SPACETRACK system. More than a year later in an article in America's *Air Force* magazine disclosed that the long-range tracking radars used by Spacetrack to reconstruct the shape and size of every object in space had in fact been used to establish that the solar panel had failed to deploy.

Launch of the first crew, Charles (Pete) Conrad, overall Commander of the Skylab astronauts, and Dr Joseph Kerwin and Paul Weitz, should have taken place 24 hr after Skylab 1, but was immediately postponed for 5 days. At first it seemed very unlikely that it would ever be possible to inhabit Skylab. Various options were studied, including the possibility of a simple 'fly-around' mission to examine and report on the damage. Internal temperatures at first went as high as 190°F, but were stabilized at lower levels as Mission Control learnt to orient the cluster at the most favourable Sun angle. Meanwhile, emergency teams devised 3 different techniques for replacing Skylab's thermal shield: 1) a parasol-type sunshade, which had the advantage of being pushed from inside through one of the 2 scientific airlocks, and then opening like an umbrella; 2) a twin-pole sunshade, which would be pushed back over the Workshop by 2 astronauts from the ATM hatch; and 3) what became known as the SEVA sail, which would be deployed during a Stand-up EVA in the open hatch of the CM, with the astronauts circling the cluster and hooking the sail on with the aid of a long pole. After watertank rehearsals at the Marshall Space Centre, Huntsville, the first 2 sunshields were selected; with a variety of special tools and other equipment, they added 180 kg to the planned weight of the CSM. It was also decided to make a SEVA before docking, with 2 astronauts in the open hatch trying to pull out the jammed solar panel with a

pole, and sending back TV pictures of the damage.

By Jul 30, when an official NASA inquiry found that the launch problems were due to 'absence of sound engineering leadership' (McDonnell Douglas had prime responsibility), so much extra confidence had been created by the discovery that space repairs were not nearly so difficult as had been feared, that the report made little impact.

Note NASA's official numbers were Skylab 1 for the unmanned space station launch; and Skylabs 2, 3 and 4 for the subsequent manned flights. Newsmen referred to the manned missions as Skylabs 1, 2 and 3, to avoid confusing the public. For convenience here, the official NASA designations are used.

Launch and Orbit Procedures The unmanned Saturn Workshop (SWS) was launched by a 2-stage Saturn 5 from Kennedy Space Center's Launch Complex 39A. During the $7\frac{1}{2}$-hr lifetime of the Instrument Unit (IU), the protective Payload Shroud was jettisoned, and the SWS automatically oriented to a solar-inertial, or Sun-pointing attitude. The Apollo Telescope Mount (ATM) was rotated 90° from its folded launch position, and the 4 solar array panels, each 14·94 m long—in appearance rather like a 4-bladed windmill—were deployed into position to supply power mainly intended for the ATM. The interior of the SWS was pressurized to 0·35 kg/cm², with a mixture of 0·26 kg/cm² oxygen, and 0·09 kg/cm² nitrogen—marking a final break by NASA from the all-oxygen environment used in previous manned flights. The SWS, containing all consumables for the whole mission, including oxygen, water, food and clothing, equivalent to 1 tonne for every man-month, should then have been ready for occupation. But the pre-set automatic programmer was unable to deploy the beams

1 Ed White, making the first US spacewalk from Gemini 4, gingerly manoeuvring with an oxygen-powered spacegun. His EVA lasted 21 min, and Jim McDivitt's pictures are still used to symbolise man in space

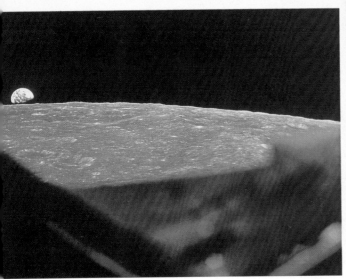

2   Above: Frank Borman's famous 'Earthrise' picture, taken from Apollo 8 during man's first flight around the Moon on Christmas Day 1968

3   Right: Ed Aldrin, Apollo 11 LM Pilot, pictured by Neil Armstrong, starts unstowing scientific equipment at Tranquillity Base during man's first visit to the lunar surface in 1969

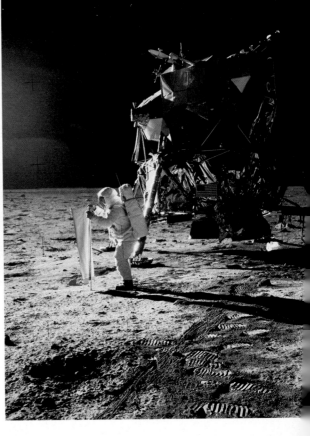

4 Left: Eugene Cernan, Apollo 17 Commander, boarding the Lunar Rover towards the end of the last Moonlanding. TV antenna and camera in foreground; South Massif in background

5 Below: Lunokhod 1 as shown at Paris Air Show. Front view shows TV 'eyes'; narrow beam antenna, top right; omni-directional antenna (cone-shaped) left

6 Landsat-1 view of Los Angeles from 915 km. Healthy crops, trees, etc, shown as red; cities and industrial areas as green or dark grey; clear water is black. Street patterns and canals visible lower right

7 Above: Both Skylab sunshades are visible in this last view taken by departing Skylab 4 crew. NASA plans to send a Shuttle, on Flight 5 in 1979, to boost Skylab into higher orbit for possible re-use

8 Right: Salyut shown at Paris Air Show. Note docking port. Future versions are planned with 2 ports for overlapping long-stay crews

9 **Left**: Pioneer 10 view of Jupiter's Red Spot—an immense hurricane 14,000 km wide and 30,000– 40,000 km long. In 1979 Voyagers 1 & 2, more advanced Pioneers, should explore Jupiter and its 13 planets in more detail

10 **Above**: Sunset on Mars' Chryse Planitia. This Viking I view is precisely what man would see standing on the Martian surface

11 Left: Soviet concept of Large Space Station, painted by Cosmonaut Leonov and A. Sokolov. Note large Salyut-type station, with a 5th Soyuz approaching multiple docking port already occupied by 4 others. Capacity would be 10 or more cosmonauts. EVAs probably required to link solar panels, making rotation possible to create artificial G

12 Overleaf: US concept of 21st Century Space Colony, as seen by resident returning from Earth holiday. Twin cylinders, 32 km long and 6400 m dia, orbit between Earth and Moon. This is the largest of 4 designs by Dr G K O'Neill, of Princeton University. Depending on interior layout, it could accommodate a population of 200,000 to several millions. Each cylinder would rotate every 114 sec to create Earthlike gravity. Lunar or asteroid materials would be used to build it in space. Teacup-shaped containers ringing the cylinders are agricultural stations; manufacturing and power stations cap the cylinders. Rectangular mirrors, hinged at far end, direct sunlight into the interior to regulate seasons and day/night cycle

containing the largest solar arrays ever flown in space; one had been torn away from the SWS, the other jammed by debris from the meteoroid shield (as described below) which should have been deployed last at the end of the first revolution.

The CSMs were launched by 2-stage Saturn 1Bs from Launch Complex 39B, into an interim orbit of $149 \cdot 5 \times 294 \cdot 5$ km. The CSMs used the 9070 kg thrust Service Propulsion System for boosting to the SWS orbit, where, in turn, they docked with the axial port of the Multiple Docking Adapter (MDA), thus completing the cluster. After the crew entered the SWS, the CSM was powered down to the maximum extent, with only essential elements of communications, instrumentation and thermal control systems operating.

Skylab's orbit and inclination enabled observations from the Workshop to cover the US, most of Europe and S America, and the whole of Africa, Australia and China. The most northerly European track took it along England's S coast, and then on, just S of Cologne and Kiev, emerging on the far side of Siberia, N of Khabarovsk. This meant that Canada, Scandinavia, and the northern half of the Soviet Union were not covered by the flight.

If there had been a complete launch failure, back-up hardware was available for a 2nd Skylab cluster, and the use of this was seriously considered when it was thought the first one would be uninhabitable. Plans to use this for a 2nd Skylab mission, to help fill the manned spaceflight gap between 1974, and the first orbital Shuttle flight in 1979, had to be dropped because money was not available, but the back-up ATM may be used on a Shuttle flight.

Rescue Capability For the first time during a US manned flight, there was a limited 'rescue capability', for use if one

Left Skylab ready for launch by Saturn 5
Right Much smaller Saturn 1B perched on 'milkstool' to launch first Skylab crew

of the crews became stranded in space, unable to return if the CM became unserviceable— for instance as a result of being unused and 'cold-soaked' for several weeks while docked with Skylab. On the first 2 flights the Saturn 1B being prepared for the subsequent mission would have been

First view of damage. Broken wires, *top left*, indicate one solar wing torn away; *bottom right*, 2nd solar wing jammed partially open

store. The snag about the system was that the stranded astronauts would have to survive for up to 45 days (depending how far into the mission the crisis occurred) before the rescue vehicle would have been ready. Nevertheless, it was thought a rescue flight would be needed during Skylab 3.

MISSION DESCRIPTIONS

Skylab 2   On what proved to be the most dramatic flight since Apollo 13, Conrad, Kerwin and Weitz were finally launched on May 25. For Conrad, it was his 4th flight; for the others, their first. Launch was by Saturn 1B (total lift-off wt 689,680 kg. CSM wt 6062 kg. Because Pad 39B had been built for Saturn 5 Moon launches, the much smaller Saturn 1B had to be launched from a 40m steel pedestal (nicknamed by Cape workers 'the milkstool'). It was the first time men were launched towards Europe, and 2 US helicopters were on standby in England in case, following an abort between lift-off and orbit, a rescue was needed in the Land's End area. Of 600 recently released war prisoners from Vietnam, only 30 accepted invitations, in personal letters from the Skylab 2 crew, to attend the launch. The 10-yr-old Saturn 1B performed faultlessly, but 7 hr later Conrad was describing how one of the 9m long solar panels had been torn completely off, while the other was jammed, partially opened, by the torn and buckled heatshield. After much conferring, the hatch was opened, and with Conrad at the controls, manoeuvring round Skylab, in imminent danger of colliding with it, Paul Weitz, with toggle cutters fixed to a long pole, tried in vain to release the panel. As the CM swept into darkness, Weitz scrambled hurriedly inside, knocking Conrad on the head with his pole as he did so. Then came a major emergency; after 4 hr and 6 attempts, Conrad failed to

used, the CM being flown by 2 pilots, with 3 couches installed below them to bring the stranded astronauts home as passengers. During the final flight, a spare Saturn 1B, and back-up CM, were prepared for use, and later returned to

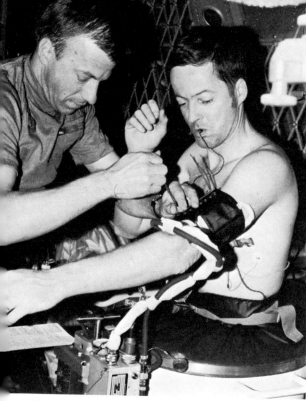

Blood pressure check for Joe Kerwin during tests in the lower body negative pressure device

dock with Skylab. An emergency flight home seemed as inevitable as the unique flow of bad language already pouring back. But spacesuits were donned for the 3rd time in a day, the docking tunnel opened, and the probe dismantled. A nut, which was to cause anxiety later, floated off into space; but the last possible docking attempt by the indomitable Conrad succeeded. With doubts which later proved groundless, as to whether the faulty probe would ever permit *undocking*, the crew rested; then Conrad remained in the CM, while Weitz and Kerwin, wearing masks and testing for possible dangerous gas, opened up the 5 hatches to gain access in turn to the Multiple Docking Adapter, the Airlock Module, and Orbital Workshop. The MDA, with undamaged thermal shield, was clean and cold; in the Workshop they could feel 'radiating heat'. After some acid exchanges with Mission Control, who, unobserved by the crew, had teleprinted up lengthy and complicated instructions that the parasol must be unpacked and refolded before being deployed, we at Mission Control were able to see, via the TV camera set up in the CM window, the orange tip of the parasol emerging from the airlock as Conrad and Weitz, taking frequent rests as they worked in the blistering heat, slowly forced it through the 20cm scientific airlock on its 7·6m handle; with difficulty, most of its wrinkles and folds were eliminated, and the handle pulled in again; section by section, so that the protective sunshade, a 6·7 × 7·3m rectangle, rested snugly just above the Workshop. Protected at last from the direct rays of the Sun, internal temperatures began falling, though more slowly than hoped. By the 4th day, the sleeping cubicles were still too hot to use; but with the crew sleeping in the MDA, and the ATM deployed, Skylab, though short of power, was operational. But while Dr Joe Kerwin was sending back the first solar flare pictures taken above Earth's atmosphere, teams on the ground continued to worry about the serviceability of the CM's docking probe, and to work out EVAs to free the jammed solar panel. On the 6th day a power failure, with the loss of 2 out of 18 remaining batteries, followed the first full day of scientific experiments. Finally, after 3 days of discussions and rehearsals, Conrad and Kerwin transformed a deteriorating situation during a hazardous spacewalk. Conrad's efforts, with the toggle cutters on a 7·6m pole, to cut away the jagged heatshield and free the jammed solar wing, were unsuccessful; without the planned, improvised handrail to support him, he made his way along the outside of the Workshop, and lay across the beam 1·2 m wide and 9·4 m long, guiding the toggle cutters at close quarters while Kerwin, remaining in the relative safety of the Workshop's 'roof', operated them by means of a lanyard. Then Conrad fastened his 9·1m rope to the beam, and, as rehearsed, stood up with it to obtain leverage over his shoulders; suddenly, the beam swung out and clicked into place. Conrad said he and Kerwin 'literally took off' as it happened. But within a few moments, electric power began flooding into the 8 Workshop batteries which had been useless since lift-off; and, as it proved, Skylab was to have plenty of power for its whole mission, despite the loss of the other solar panel. Kerwin further improved things by climbing across to the ATM and putting right 2 faulty telescope shutters; after they had climbed back inside, the 3½-hr spacewalk was marked by a congratulatory telegram from President Nixon.

By the end of the 28-day mission, triumphantly completed, the released solar wing was producing more than 5500W. Adequate electrical power enabled the astronauts to complete more than 80% of their planned activities; solar observations totalled 81 hr compared with the planned 101

hr, the major accomplishment being the monitoring of a solar flare on Jun 15. 11 out of 14 planned EREP (Earth resources) passes were completed, and 7460 out of 9000 planned photos obtained; data was obtained over 31 US states, 6 foreign countries and over the Pacific, Atlantic and other seas. All 16 medical experiments covering man's adaption to zero-G were completed. 5 student investigations were carried out. When Skylab 2 passed the Soyuz 11 record of 23 days in space, Conrad sent a message of goodwill to Russia, and wished them luck in the future after the loss of the Soyuz 11 crew during re-entry. On a final, 1½-hr spacewalk, to recover film from the ATM, Conrad even succeeded in getting one of the 2 failed ATM batteries to resume operating by thumping it with a wooden hammer. From then until the end of the 3rd mission, 23 of Skylab's 24 batteries continued to give good service. Following splashdown SW of San Diego on Jun 22, a new procedure was followed of winching the CM aboard the recovery ship with the astronauts still inside. Doctors were anxious that the astronauts should be lifted out, to provide them with maximum medical data on the effects of returning to a 1G environment. Conrad, Weitz and Kerwin disappointed them by insisting on walking out, slightly unsteadily. But they had proved there was no reason why the 2nd crew should not stay up twice as long.

Skylab 3 Skylab's Programme Director, William Schneider, told the pre-launch news conference he hoped it would be 'a dull and boring flight'. In fact the 2nd manned mission, launched on Jul 28—total lift-off wt 593,560 kg; CM 6085 kg—ran into trouble almost at once. An emergency flight home was considered, and preparations for a rescue mission hurriedly mounted. Neither proved necessary, but

View of England as Skylab 3 passes over Cornwall

sick at the end of Skylab 2. Alan Bean, Commander of Skylab 3, was making his 2nd flight; Owen Garriott, a civilian, with a doctorate in engineering, and Jack Lousma the pilot, were both making first flights. Countdown and lift-off were uneventful, but early in the flight a faulty jet thruster, not expected to cause serious trouble, had to be shut down. After 8 hr, a fly-around inspection was made of Skylab, to check the colour and condition of the parasol sunshield and establish how urgently it needed to be replaced. Bean moved in too close, and we could see on TV how, when he tried to move away, the RCS thrusters were rattling the parasol as if it was exposed, as Bean put it, 'to a 10-knot gale'. To avoid blowing it off altogether, he was able to drift away, and docking was achieved with no repetition of the Skylab 2 problems. 2 days were spent switching on lights, power systems and air conditioning, and trying to bring down the workshop temperatures of 80°F to more comfortable levels. Unstowing of the CM stores, which included 6 pocket mice, minnows and 2 spiders, named Anita and Arabella, was delayed when all 3 men developed nausea, which persisted for about a week, and made it necessary to postpone a major spacewalk to deploy the twin-pole sunshade brought up by the previous crew as a back-up to the parasol. It was noted that this crew, busy rehearsing for the EVA, had not found time for the usual pre-launch aerobatics, long ago found to be the best way of avoiding space sickness. On the 3rd day, with Lousma worst affected, and all 3 taking things very easily, it was discovered that a short-circuit in the CM had killed the pocket mice and a container of vinegar gnats. The major crisis came on the 5th day, when a leak developed in a 2nd CM thruster. An emergency flight home within 24 hr, before too much manoeuvring gas was lost, was seriously considered, with the preparation of a rescue mission (which

the crises gave the early part of the mission an Apollo 13 atmosphere. The flight plan had been extended, just before launch, from 56 to 59 days, to provide a shorter sea voyage after splashdown, because Dr Kerwin had become very sea-

Jack Lousma successfully testing the astronaut manoeuvring unit in Skylab's dome

working conditions and adequate power supplies were thus finally assured to Skylab for the remainder of its 3 missions. An EVA inspection of the CM thrusters failed to find the cause of the leaks; but here again, while preparations for a rescue flight were continued at C Canaveral, as a precaution, ground tests convinced Apollo engineers that procedures could be worked out for the CM to provide a safe passage home despite its 2 faulty RCS clusters. Following the successful EVA, everything improved at once: the astronauts, now fully recovered, asked for more work. In the big workshop dome, they successfully tested astronaut manoeuvring units (AMUs) for use during spacewalks on future flights; Garriott, who had had no previous training, quickly mastered their use. The crew described and televised the birth and development of a storm over the Gulf of Mexico. Both Anita and Arabella demonstrated they could spin webs while weightless (though Anita died before the end of the flight, and Arabella was found to be dead after splashdown). Dr Garriott's tests showed that, while all 3 astronauts experienced the expected loss of body muscle and showed signs of heart deterioration, their health levelled out from Day 39; all 3 lost about 3·1 kg in wt, and 25 mm around the calf of the legs. From Days 10–15 the crew devoted 19 man-hr per day to scientific experiments; from then on the rate increased to 27–33 man-hr per day. During 2 more EVAs, film was changed and recovered, and the CM finally prepared for re-entry packed with 75,000 pictures of the Sun, taken during over 300 hr of solar observations (which included 6 solar flares); film and photographs and tape data obtained on 39, instead of the planned 26 EREP passes, when Skylab was rolled around to study the Earth; and the results of welding and materials processing experiments, and much other material. Bean, now holding the

would require 35 days before it could be launched) as the main alternative. The latter option was selected, since it meant the astronauts, still not fully adapted to weightlessness, would have time to settle down and carry out their full programme. The recurrent crises led to 3 postponements of the EVA; but finally, on Aug 7, Garriott and Lousma spent a record 6 hr 31 min outside; it took them nearly 4 difficult hours to get the new sunshade in position, over the top of the parasol, which it had been decided not to jettison, in case the replacement should be unsuccessful. But the efforts of Major Lousma and Dr Garriott, with Bean monitoring and advising from inside, resulted in workshop temperatures quickly dropping to a comfortable average of 75°F. Good

Spider Anita demonstrates her ability to spin webs in zero-G

world record for time in space, with 69 days in 2 flights, cautiously test-fired the CM's 2 operational clusters for a 1000th of a sec—but nothing registered at Mission Control. A 2nd test of a full second, at the risk of contaminating Skylab's cameras and telescopes did register. Docking latches were released, the CM drifted clear, and re-entry made with minimum firing procedures. With a hurricane raging only 644 km from the splashdown point off San Diego, the astronauts found themselves upside down amid 2·4m waves—the roughest sea encountered during any recovery. The buoyancy balloons righted the spacecraft, but it was almost tipped over again as frogmen struggled with the flotation collar. It was 40 min before they were winched on to the recovery carrier, and a doctor, knocking politely on the hatch before opening it, leaned in to check hearts and blood pressures. All 3 had donned pressurized undergarments, to prevent blood pooling in the legs when returning to 1G after such a long period of weightlessness; they emerged, unsteadily, supported by a doctor on each side, but smiled and waved to the cameras and seamen; then took 3 steps to the forklift chairs which lifted them down to deck level. For the first time there were no speeches before 6 hr of medical tests began. Dr Fletcher, head of NASA, described it as 'one of the most significant scientific ventures of all time', and said enough material had been brought back to keep the scientists busy for 5 yr. The astronauts had travelled 38·6M km.

Skylab 4   The launch, which finally took place on Nov 16 1973, was again preceded by drama. It had to be postponed twice, for a total of 6 days, following the discovery of fatigue cracks in the Saturn 1B's 8 fins, on which it 'sits' on the launchpad. Had they not been changed, the rocket would almost certainly have broken up shortly after lift-off, and for the first time it would have been necessary for the astronauts to use the escape tower. After Gerald Carr, Dr Edward Gibson, and William Pogue—all making first flights—were safely in orbit, Walter Kapyran, Skylab Flight Director, admitted he had 'been scared to death, and sweated it out' because there had still been some cracks between the rocket's 1st and 2nd stages. Lift-off wt was 595,370 kg—CM wt 6104 kg. There had been much pre-launch argument about possible dates and length of mission. A prime object had become observation of Comet Kohoutek, discovered the previous Mar, and due to pass behind the Sun (its perihelion) at a distance of about 21M km on Dec 28; since its orbit brings it round the Sun, and past

Skylab as seen by last crew just after undocking; ATM at top

the Earth, only about once in 80,000 yr, it was expected to be the biggest astronomical event since Halley's Comet in 1910. It was decided that 2 EVAs should be made, on Christmas Day and Dec 29, for the crew to mount cameras on

the ATM truss to obtain pictures of its changing composition as it became heated by the Sun. It was also decided, following the success of Skylab 3, to aim at an 84-day mission—3 times that of Skylab 2. For this purpose, the CM was packed with spare parts for Skylab, a portable treadmill to help the crew keep their leg muscles in good condition during 3 months of weightlessness, and extra food; this included 23·1 kg of chocolate bars, plus 11·3 kg of survival food, to enable them to exist for yet another 10 days if they had to be rescued towards the end of the mission.

Carr's comment, shortly after lift-off, to Mission Control: 'You have 3 happy rookies up here', was not to hold true for the entire mission. 3 attempts were necessary before docking was achieved; the first contact was too soft, and the 2nd failed because Carr, not realizing that the capture latches had been sprung on the first attempt, had not re-set them. Soon after docking, Pogue developed nausea, and vomited. A tape recording accidentally played to the ground next day revealed the incident when all 3 crew members were heard agreeing to conceal it. Alan Shepard, Chief Astronaut, told Carr he had made 'a fairly serious error of judgment', and Carr replied: 'OK, Al, I agree with you. It was a dumb decision'. Unstowing extended into the 4th day, as a result of motion sickness and a mistake by Pogue in activating the water purification system. During their first week, the crew became tense and irritable. Gibson said they had been 'hustling around', with too much to do, especially on additional medical experiments insufficiently rehearsed following the decision to extend the mission from 56 to 84 days. 'The first 7 or 8 days were not something I would want to go through again,' he reported.

On the 8th day, however, Pogue and Gibson broke the Skylab 2 record with a spacewalk, lasting 6 hr, 34 min, 35 sec; tasks included repairing a jammed antenna, involving the tricky job of removing 6 screws with gloved hands, and installing a new electrical switching box. A misfortune, discovered later, was that they forgot to put filters on 6 of their cameras, with the result that some Earth resources data was lost. The day after the spacewalk they were woken with the news that one of Skylab's 3 attitude control gyroscopes had failed, which meant that manoeuvring would take longer and use more control gas from then on; as time went on, a 2nd gyroscope began to give trouble, causing constant anxiety, and once threatening to bring the flight to an abrupt end.

Crew morale, however, improved as they settled down to a steady routine of scientific work, including observations of Comet Kohoutek; they were able to disprove ground reports that it had broken up, though it rapidly became clear that it was not going to provide the historic astronomical spectacle which had been prophesied, and which led to the whole Skylab 4 flight plan being rewritten.

On Dec 18, when Soyuz 13 was launched on its 8-day flight, history was made because astronauts and cosmonauts were in space together for the first time; no direct communications were possible between the 2 spacecraft, nor were any sightings reported by either crew. On Christmas Day, Carr and Pogue spacewalked for 7 hr, photographing Kohoutek before it passed behind the Sun at a distance of 21M km. Because of its proximity to the Sun they were unable to see Kohoutek, but aimed the cameras on directions from Mission Control. On completion of the spacewalk, Mission Control told them where to find Christmas presents from their wives, which had been secretly stowed before the launch.

On Dec 28 Carr and Pogue carried out another EVA to observe and photograph Kohoutek moving away from the Sun and Earth on its 80,000-yr orbit, and on Dec 30 discussed their observations via TV with Dr Lubos Kohoutek, the Czech astronomer who discovered the comet.

Carr and Gibson carried out the 4th and last EVA on Feb 3, during which they retrieved the final canisters of the Sun and Comet Kohoutek, and recovered materials and collection devices placed outside the workshop for analysis of the effects of the space environment. They also placed an extra micrometeorite experiment outside for possible collection by a future visiting US or Soviet spacecraft. This brought the total EVA time for the mission to 22 hr 19 min.

On Feb 8 1974, the crew prepared a 'time capsule'—a bag containing 5 food and drink items, unused film, camera filters, clothing, flight plan pages and small electronic devices, which could be retrieved to study long-term effects if Skylab was later revisited. Though docking with Skylab should be possible, entry could only be made wearing full spacesuits.

When the crew undocked, they were on their 1213th revolution; Skylab on its 3898th. During a final, nostalgic fly-around, Gibson reported that both sunshades had faded a lot, and, as photographs later proved, one had developed a split. Before they splashed down, Skylab was a dead station, having been depressurized by Mission Control; in its 454km orbit (having been raised by an RCS burn a few days earlier), its life before re-entry was estimated at between 6 and 10 yr. The CM's re-entry was far from trouble-free, due partly to a major crew error which they admitted afterwards made them wonder whether they would survive. A helium leak developed in the No 2 RCS loop, and in their haste to complete re-entry procedures, the crew mistakenly pulled 4 circuit breakers which discon-

Skylab cluster elements

ATM Apollo telescope mount
• Solar observation unit

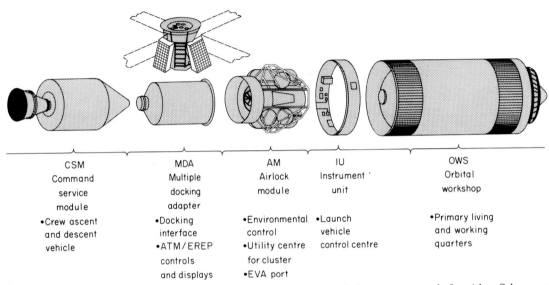

CSM	MDA	AM	IU	OWS
Command service module	Multiple docking adapter	Airlock module	Instrument unit	Orbital workshop
•Crew ascent and descent vehicle	•Docking interface •ATM/EREP controls and displays	•Environmental control •Utility centre for cluster •EVA port	•Launch vehicle control centre	•Primary living and working quarters

nected the automatic control circuit leading to the remaining No 1 thrusters. 'Our hearts fell and our eyeballs popped,' Carr confessed; but he was able to use the manual back-up system to control the spacecraft through the initial re-entry sequence, and power was later restored to the automatic circuit. Mission Control grew concerned about the possibility of the leaking fuel entering the spacecraft, and several times warned the crew to watch for either fishy or acid smells, indicating respectively that either the thruster fuel or its oxidizer was entering the cabin; and finally advised them in any case to wear oxygen masks throughout the re-entry.

Splashdown was once more 'Stable Two'—upside down; it took the crew 5 minutes to right the spacecraft, and it was

218

an hour before they had been winched aboard the recovery ship. For the first time a US manned spaceflight ended with no live TV coverage; even Mission Control was unable to see Carr and Pogue, heavily bearded (they had earlier disapproved of Skylab's onboard shaving facilities), and Gibson, clean-shaven, step on to the deck of the recovery ship. Early morale problems were more than balanced by the fact that, as a result of $1\frac{1}{2}$ hr exercise per day (treble the amount taken by the Skylab 2 crew), all 3 astronauts were in far better physical condition than either of the earlier crews. They had brought home with them about 75,000 solar pictures, 17,000 of Earth and 2500 of Comet Kohoutek, plus the results of many scientific experiments. They had become the joint holders of the record for man's longest spaceflight; and travelled about 54,804,000 km. The 3 crews had orbited the Earth 2475 times and travelled about 112,804,000 km.

Experiment Results   The long-term benefits and results of the experiments (80·5 km of magnetic tape readings and over 40,000 pictures of the Earth's surface; and 182,000 frames of solar film), will take some years to analyse. Some results emerge in the preceding descriptions of the flights themselves. So far as melting, welding and brazing of metals is concerned, 25mm pipes joined in space and brought back to Earth were described by the Principal Investigator concerned as 'the most perfect braze joint' he had ever seen. On Skylab 3, experimental manufacture of alloys and crystal was so successful that 'production runs' of several thousand kilogrammes per year are already planned during the Space Shuttle programme.

Skylab's Orbital Workshop: *Top* stowage area, showing cylindrical water tanks and food lockers; *below* grid floor, crew quarters include wardroom, partitioned off sleeping quarters, toilet facilities etc

Solar Astronomy   Skylab's 5 ATM teams were delighted at the mass of information obtained; the danger, as Dale Myers, NASA's head of Manned Spaceflight, pointed out to a Senate Committee, was that they would be overwhelmed by the flood of new information. The scientists themselves were concerned that NASA would not provide sufficient money to support the process of analysis. The head of the X-ray telescope investigators said that the 25,000 photos available to his section would keep a computer busy for 10 yr, just putting the data 'into its proper form'. Instead of the expected 1 or 2, dozens of examples had been observed of explosions in the corona, the 2M $^\circ$K outer rim of the Sun; it had been expected that these explosions would develop slowly as a 'bubble'; instead, it was found that the corona remained constant until an explosion ripped it apart and changed it permanently, with hot gases being expelled from the Sun and streaking into space towards the planets. The Sun was putting out the entire annual US energy demand 100,000 times every sec; but at the same time, the Sun itself had an energy gap, and its core was producing only one tenth of the power it was radiating away. When the gases graze the Earth, magnetic storms and communications failures could be caused, and scientists are studying what would happen if the Earth was struck 'head on' by the results of a major coronal explosion. It is now believed that the solar wind comes from the so-called coronal 'holes', darker regions on the Sun's surface, which rotate around the Sun, and are very persistent at the poles. These were first studied by the OSO spacecraft; and it seems likely that the solar wind literally blows out of the holes, and that the 'magnetic fronts' which affect Earth's weather, are rooted near the holes. Skylab has produced many clues to the mystery of the solar interior, and where and how its energy

is produced. It has enabled the scientists to move a good deal nearer to being able to predict its behaviour—which may well result in some future 'ice ages' on Earth.

Life in Skylab   6 months after the end of the record-breaking Skylab 4 flight, Col Pogue told the Society of Experimental Test Pilots what it was like living in space for such long periods—a useful guide to what men and women will have to face during long voyages to the planets. In the first day or so, due to the shift of blood and tissue fluids to the upper part of the body, the face becomes bloated, giving astronauts 'an oriental appearance'. Head stuffiness, and intermittent congestion of the inner ear continued until the 12th week. Body mechanisms assume there is a fluid surplus, resulting in a fluid-volume loss, some of which comes from the blood. This causes an abnormally high concentration of red blood cells in the remaining blood volume, and a reduction in red blood cell production. Only after 9 weeks, on the last Skylab mission, was it possible to check that the body did start producing red blood cells once more, despite zero G, when they became necessary. Physical conditions inside Skylab remained satisfactory, despite hygiene and waste management problems. Urine 'spills' were inevitable, while solid waste management was time-consuming but otherwise not much of a problem. Disposing of food packaging, dirty clothes, towels, washcloths, etc, was 'an ever-present difficulty', and made no easier as a result of a urine-bag leaking in the rubbish airlock. Grit and particles did not provide an inhalation problem, and were removed with a vacuum cleaner from fan intake screens and traps. Meal menus carefully arranged to provide the daily mineral requirements, repeated at 6-day intervals; but though satisfactory, they failed to take account of temporary

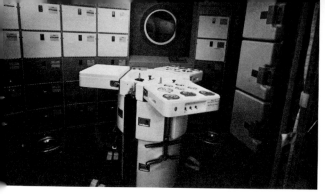

Wardroom had 58 stowage lockers; note water spigots and food preparation facilities. Earth-viewing through the window was the favourite recreation for all 3 crews

preferences for types or quantity, and they were often hungry, craving more food than permitted by the diet. At first, rest cycles were only just adequate, and they always felt tired, partly as a result of extra work caused by equipment malfunctions. But they exercised for $1\frac{1}{2}$ hr every day, and by the end of the 2nd week the work-rest cycles were 'reasonably stable'. Thanks to the treadmill device, the Skylab 3 crew suffered only half the leg-muscle loss experienced by the first 2 crews, and the heavy exercise periods enabled them to make a rapid post-flight recovery. Eating a full meal, or taking heavy exercise, temporarily relieved the head congestion. But missing a meal led to tiredness, and many of the symptoms of influenza.

Shaving in zero gravity cost more time than it was worth, and they found the best blades were only good for one shave. Washing the entire body with a washcloth and towel was done each day after exercise, and a shower was taken about once a week. Their first-aid training proved unnecessary. Small cuts and abrasions healed quickly; dryness of the skin and mucous membranes early in the mission soon disappeared. Apart from that, they only suffered from occasional minor skin rashes and infections and eye irritations. They adjusted quickly to weightlessness in a large working volume, though they had difficulty in pacing themselves during the working day.

The astronauts' strongest emotions were usually provoked by such things as equipment malfunctions, tight schedules, misunderstood instructions, poor stowage provisions, and their own errors. They found the public release of all communications, live and recorded, very irritating. The knowledge that everything they said would be transcribed, and released to the news media as part of 'Mission Commentary' resulted in a tendency to remain silent or to use euphemisms.

As time went on, rest and recreation became increasingly important. Col Pogue disclosed that 'the management had dismissed our initial unanimous suggestion' about the contents of the 57kg entertainment kit, which finally consisted of tape players and tape selections, a small library, a velcro dartboard, playing cards and rubber balls. The tapes were played daily, and particularly enjoyed during exercise on the bicycle. Each read about 4 books during the 3 months; but the most satisfying diversion during the entire flight was simply looking out of the picture window.

Summary and Conclusions The successful completion of the Skylab missions marked the end, for America, of the first era of manned spaceflight. For those who doubted whether the estimated cost of about £1000M ($2400M) was justified,

Experiments Area and Control Centre: *left*, lower body negative pressure device; trash disposal airlock at astronaut's feet; beyond him, the wardroom

it is worth recording that, although the Earth resources equipment was considered the least effective, just one potential copper deposit located by it in Nevada is estimated to have an ultimate value of billions of dollars; the value of that alone far exceeds the cost of the whole US space programme so far. As one astronaut said: 'Skylab worked better broken than anybody had hoped for if it was perfect.' The 3 crews, who kept it operational for a total of 171 days, 13 hr, 14 min, established work-patterns and techniques that should ensure the success of the even more ambitious Space Shuttle operations due to start in 1979. Man's adaptability was effectively demonstrated by the fact that the last crew, although they had flown much the longest mission, returned to Earth in the best physical condition. Discussion of the defence aspects of the flight is discouraged; but visual observations made by the astronauts of activities at Soviet missile centres, and along the Soviet/Chinese border, also established that, however good automatic cameras and sensors may be, onboard astronauts will be an essential part of future military space reconnaissance systems. The Skylab story is not yet finished. Plans are being made for the Space Shuttle to visit it during one of the 1980 missions, to boost it into a higher parking orbit, to preserve its equipment for future use.

SPACE SHUTTLE   Manned Spaceplane

History   As the Apollo programme steadily established the practicability of space exploration NASA studied methods of creating a Space Transportation System (STS) which, in the words of President Nixon, would 'transform the space frontier of the 1970s into familiar territory, easily accessible for human endeavour in the 1980s and 1990s.' The original aim was to create a spacecraft which could be used over 100 times, and a manned booster which could also be flown back to base for re-use. Financial problems proved greater than technical ones: although the technicians would sooner press on with a system in which both spacecraft and booster were fully recoverable, since this would be more efficient in the long-term, it was decided in early 1972 to develop an aircraft-like orbiter, about the size of a DC9, which would be launched vertically like a rocket, fly in orbit like a spaceship, and land like an aircraft; and an unmanned 'semi-recoverable' booster the length of a Tristar which would descend by parachute and be recovered from the sea after use. By mid 1974, after many further changes, it had been decided that the Orbiter would ride a huge External Tank, which would be non-recoverable, into orbit; and that lift-off would be assisted by twin, solid-fuelled boosters, which would be recovered by parachute (see details below). A 7-yr development plan provides for the first manned orbital flight in 1979, with the system becoming operational shortly after. Programme management has been given to Johnson's Spacecraft Centre (JSC), Houston, which is also responsible for the orbiter stage; Marshall Space Flight Centre, Huntsville, is responsible for the booster stage and shuttle engine; and Kennedy Space Centre for launch and recovery facilities. The estimated $5500M costs (about one quarter of the Apollo programme) include research, development, and 2 flight test vehicles. Each additional orbiter will cost $250M; each additional booster $50M. Another $300M will be spent on facilities. The cost of each flight is estimated at $20M. Apollo 15 cost $445M; and the hope is to reduce the cost per pound of putting a payload into space from $600–700 at present to $100. (Explorer 1, weighing 13·6 kg, cost $100,000 per pound.)

General Description   It will consist of the following main elements: 1) Orbiter; 2) External Tank (ET); 3) Solid Rocket Boosters (SRB); 4) Spacelab; 5) Space Tug. Details appear below of all except Spacelab; since that is being built by the European Space Agency, it merits a separate entry. The orbiter will have a flight crew of 3, consisting of Commander, Pilot, and Mission Specialist, who will need to be astronauts. They will occupy flight crew stations on the upper deck of the 2-level Orbiter cockpit. On the lower deck will be living quarters, systems bays, and room for up to 6 passengers. These passengers, 'Payload Specialists', usually 4 in number, are likely to be scientists or engineers rather than astronauts, since their duties will be related to operating the onboard experiments, or to releasing and recovering satellites, etc. Basic missions will consist of 7 days in orbit; but 30-day missions will be possible, though the additional consumables required will reduce the maximum payload. Acceleration limits of 3G during launch and re-entry mean that only brief training periods of 2–3 months will be necessary before scientists, doctors, artists, photographers etc, can commute to and from space stations and laboratories. As in Skylab, there will be a 2-gas atmosphere at 14·7 psi; and toilet accommodation has been designed for women as well as men. Full automatic command will be available from launch to orbit, and for re-entry to landing. The 3-man crew, however, will have the ability to override the automatic controls and take over manually; they will have full responsibility for control while in orbit, though shuttle ground controllers will be able to initiate automatic sequences for all orbital activities except docking; all docking operations will be manually controlled by the Cdr and Pilot. Because large, solid boosters are so reliable, there will be no Launch Escape System, such as was available, but never used, during Mercury and Apollo. There will be ejection seat escape systems only on Nos 1 and 2 orbiter vehicles; they will be flown by 2-man crews, on approach and landing tests from Edwards Flight Research Centre, and for vertical launch tests into low Earth orbit from C Canaveral. Assembly of the Orbiter, SRB and ET will be in the Vehicle Assembly Building which will need only minor modifications from its Apollo and Skylab days. It will be taken to Launchpads 39A and B by the Apollo crawler transporter.

By the end of 1975, about 40,000 people were working on Shuttle projects; $800M was spent on it. Orbiter 1 began

Shuttle's first captive flight test Feb 1977, the engines covered to reduce vibration. It was so trouble-free that the number of flights was reduced

to make the first horizontal test flights in Feb 1977, mounted 'piggyback' on a Boeing 747, and was to be released from heights of about 12,190 m. First manned orbital flight is due early in 1979:

Orbiter Under development by Rockwell International (formerly North American, who developed Apollo), it will have an operational 'dry' (unfuelled) wt of 68,000 kg. With 37m length and 23·7m wingspan. Its 4·6m dia and 18·3m long payload bay will carry loads ranging from 14,500 kg to 29,500 kg, depending on orbital inclination. Space Shuttle Main Engine System (SSME) consists of 3 liquid/hydrogen liquid/oxygen engines each giving 170,000 kg at sea level; this increases to 213,200 kg in space vacuum. They will burn for about 8 min at launch, and instead of being used only once, as in previous spaceflights, are designed for use 55 times before being overhauled. Orbital Manoeuvring System (OMS) consists of twin, 2700kg thrust engines mounted in pods at the base of the orbiter's vertical tail; using mono-methyl hydrazine and nitrogen tetroxide propellants, they will be used for orbital insertion, large orbital plane changes, and de-orbit. 3 RCS (reaction control system modules), 1 mounted in the nose and 2 as integral parts of the aft OMS pods, will consist of 40 408kg thrusters and 6 11kg thrust vernier motors to provide orbital attitude control around the 3 axes. The RCS system will use the same propellants as the OMS engines. The Orbiter's cargo bay doors run the full length of the bay, so they can be opened to deploy or recover free-flying spacecraft in orbit by means of the remote manipulator system. Propulsion stages will also be carried in the bay, to take satellites beyond Shuttle range. Maximum payloads will be possible only on launches due E of C Canaveral, into circular orbits of about 400 km. On polar missions of 90° incl, launched from Vandenberg, payload is reduced to 18,000 kg. One of the Shuttle's big advances is its ability to bring heavy payloads back to Earth—up to 14,500 kg. Landings will be possible on conventional, commercial runways. Postflight inspection, refurbishing and refuelling will involve a turnaround time to next launch of about 2 weeks.

External Tank This contains the propellants for the Orbiter main engines and acts as the structural member between the Solid Rocket Boosters and the Orbiter. It is 47·5 m long, 8·2 m dia, with loaded weight 0·74M kg, and dry weight of 35,400 kg. Of aluminium construction, its original conception has been simplified by removal of retromotor and electrical components. Thermal protection consists of silicone cork material for the hot surfaces, and spray-on polyurethane foam for insulating the hydrogen tank.

Solid Rocket Boosters Twin boosters, attached to the sides of the Propellant Tank, will provide total lift-off thrust of 2·31M kg. Expected to be 44·2 m long, 3·6 m dia, with loaded wt of 0·52M kg each, they will burn in parallel with the Orbiter main engines, providing thrust up to a staging altitude of 43,000 m, and a velocity of 1400 m per sec. At this point, the SRBs are separated from the ET, and parachuted into the sea. They will be towed back to port, and rebuilt for later use. Each booster will carry 454,000 kg of solid propellant at launch.

Flight Procedure The aerodynamic controls will be locked, and directional control provided by thrust vectoring of the 3 main engines and 2 solid boosters. All 5 engines will burn from lift-off until the solid boosters burn out, when they will be jettisoned and parachuted into the sea for recovery. The main engines will power the vehicle into orbit, with assistance from the OMS engines, if needed, to reach the chosen orbital inclination. The external tank will be jettisoned on a sub-orbital trajectory, impacting in the Indian Ocean, and will not be recoverable. Major control in orbit will be provided by thrust vectoring the OMS engines for a large orbital change, or by the RCS engines for smaller attitude changes. Hand controllers on small pedestals be-tween the Pilot's legs will be used for pitch-and-roll control during the aerodynamic mode, with conventional rudder pedals for yaw. Once on station in space, the Commander and Pilot will leave their side-by-side seats at the forward part of the flight deck and attach themselves to rearward-facing control stations; there the Commander will have a hand controller and instruments for manouevring, while the Pilot will have controls for opening the payload bay and operating the payload handling system. In addition to

Above Concept of Shuttle lift-off, all engines burning in parallel; *right* at 43·4 km, the 2 solid boosters are jettisoned; ET is discarded just before orbit; *below right* Shuttle Orbiter glides home to conventional runway

windows, the Pilot will have a TV monitor giving him pictures from small TV cameras on the end of each payload handling arm. The Mission Specialist, seated behind the Pilot on lift-off, will swivel his couch to a display panel, and be responsible for ensuring that the proper electrical, cooling, and other needs are provided for whatever experiments or equipment are being carried. The duties of the Payload Specialists will be to activate the experiments and monitor data collection. When the mission is completed the OMS engines will be fired to brake the orbiter into a re-entry trajectory. During re-entry a combination of RCS thrusters and aerodynamic surfaces will be used; after that, control will be entirely by aerodynamic surfaces until a conventional aircraft-type landing is completed. From

21,300 m to the 3000m final approach altitude (which will take about $3\frac{1}{2}$ min) the orbiter will be stabilized at about 300 knots. The normal landing sequence will be fully automatic; the flight crew will lower the landing gear and initiate braking and landing rollout. Touchdown speed will be 180–190 knots. Launch plans are now so advanced that they include 3 'abort profiles' to deal with an emergency: they involve flying the orbiter directly back to the launch site for an emergency landing; flying it once around the Earth on a sub-orbital track; or continuing into a low Earth orbit and making a full re-entry emergency return from there. With a 'cross range' of 2100 km after re-entry, launchings from Kennedy will be able to make emergency landings at Vandenberg; emergency landings *from* Vandenberg would be targeted for Edwards, and have the choice of several emergency landing sites across the US.

Space Tug A re-usable vehicle, carried into orbit and parked there by the Shuttle, will be needed to act as an 'upper stage' to deliver payloads into synchronous orbits or on to insert them into interplanetary trajectories. Nearly half (43%) of Shuttle payloads will need such an upper stage. Because of doubts as to whether this should be manned or unmanned, and delays for financial reasons, this will not be ready until the mid 80s. In the meantime, the US Defense Department, as a contribution to the Space Transportation System, is developing an Interim Upper Stage (IUS) to carry both NASA and Department of Defense spacecraft on missions beyond Shuttle range, from an existing expendable upper stage.

Missions America's intention is that the STS will carry into space 'virtually all of the free world's payloads, manned and unmanned'. A total of 986 payloads for 12 yr from 1980

Shuttle Orbiter places spacecraft in orbit with upper stage attached to transport it to synchronous orbit

has been identified—one third each for applications, science and the US Defense Department. The payloads would require 725 flights, at an average of 60 per yr. 4 types of mission have been identified: the delivery, retrieval and in-orbit servicing of automated satellites, and short-duration flights for carrying out experiments requiring stays of 7–30 days in orbit. One example envisages delivery by the Shuttle into 290km orbit of 1 scientific and 1 communications satellite, mounted on a Space Tug. The Tug then delivers the scientific satellite into a 925 × 2035km orbit, continues on to a synchronous equatorial orbit to drop off the communications satellite, and then collects a synchronous Earth observatory satellite due for refurbishment. When the Tug has brought the satellite back to the parked Shuttle, both the satellite and the Tug are brought back to Earth and prepared for re-use.

SPACE TELESCOPE

History   Now planned for launch in 1983, this is a much-postponed project considered of great value by world astronomers. Current project is for a 2·4m dia reflecting telescope, to be placed in orbit by the Shuttle and serviced when necessary by astronauts. It should enable international astronomers to spend 10–15 yr observing 350 times the present volume of space available to them. Stars and galaxies down to magnitude 27 should be detectable; large planets orbiting the nearest stars may be visible. Other areas of study will include things like the curvature of the universe, which could help to decide how it all started. NASA obtained $36M to start the project in 1977/78; total

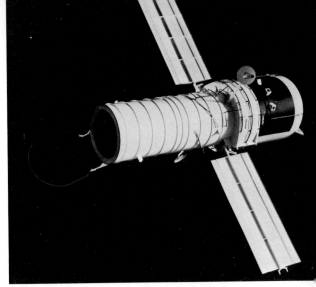

Boeing's proposal for Space Telescope project

cost was then estimated at $450M. The European Space Agency is proposing to contribute 15% of the cost in return for 15% observation time for its astronomers; ESA equipment will include a faint object camera, and the solar array.

SURVEYOR   Lunar Lander

History   The 3rd of the 3 unmanned lunar exploration

projects carried out in parallel with the 3 manned projects aimed at placing men on the Moon before 1970. Surveyors 1, 3, 5 and 6, successfully soft-landed at sites spaced across the lunar equator, achieved all Apollo objectives, enabling Surveyor 7 to be landed on the rim of Crater Tycho, to conduct digging, trenching, and bearing tests etc.

Surveyor 1   L Jun 1 1966, from C Kennedy, after a 63-hr 36-min flight, made the world's first fully controlled soft-landing in the Ocean of Storms; in the following 6 weeks, it sent back 11,150 pictures, from horizon views of mountains to close-ups of its own mirrors, etc.

Surveyor 2   L Sep 20 1966, crashed SE of Crater Copernicus when one of the vernier engines failed to fire.

Surveyor 3   L Apr 17 1967, landed safely despite a heavy bounce, in the Ocean of Storms 612 km E of Surveyor 1. In addition to returning 6315 photos, it used a scoop to make the first excavation and bearing test on an extraterrestrial body. (In Nov 1969 Apollo 12 landed almost alongside Surveyor 3, and Conrad and Bean brought back to Earth the TV camera, scoop, and other parts, for studies to be made of the effects of 31 months of lunar exposure.)

Surveyor 4   L Jul 14 1967, lost radio contact $2\frac{1}{2}$ min before touchdown, and crashed in Sinus Medii (Central Bay).

Surveyor 5   L Sep 8 1967, and major technical problems were successfully solved by tests and manoeuvres during the flight; 18,000 photos were obtained from the southern part of the Sea of Tranquillity, and the first on-site chemical soil analysis was carried out.

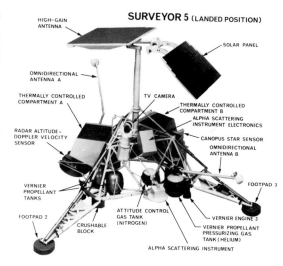

SURVEYOR 5 (LANDED POSITION)

HIGH-GAIN ANTENNA
SOLAR PANEL
OMNIDIRECTIONAL ANTENNA A
THERMALLY CONTROLLED COMPARTMENT A
TV CAMERA
THERMALLY CONTROLLED COMPARTMENT B
ALPHA SCATTERING INSTRUMENT ELECTRONICS
RADAR ALTITUDE–DOPPLER VELOCITY SENSOR
CANOPUS STAR SENSOR
OMNIDIRECTIONAL ANTENNA B
VERNIER PROPELLANT TANKS
FOOTPAD 3
FOOTPAD 2
ATTITUDE CONTROL GAS TANK (NITROGEN)
VERNIER ENGINE 3
VERNIER PROPELLANT PRESSURIZING GAS TANK (HELIUM)
CRUSHABLE BLOCK
ALPHA SCATTERING INSTRUMENT

Surveyor 6   L Nov 7 1967, landed in Sinus Medii in the centre of the Moon's front face; in addition to sending over 30,000 pictures, it performed the first take-off from the lunar surface, its 3 vernier engines, fired for $2\frac{1}{2}$ sec, lifted it to 3 m, and landed it again 2·4 m away—a test providing confidence that the surface was firm enough for a manned landing.

Surveyor 7   L Jan 7 1968, was then sent to Crater Tycho; chemical analyses suggested that the debris there had once

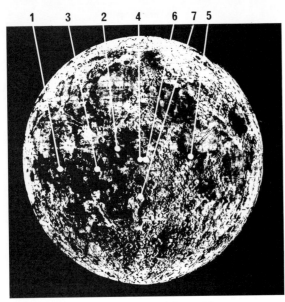

Surveyor's lunar landings prepared the way for men

been in a molten state. 21,000 photos were obtained; but Apollo astronauts' requests to visit this area were turned down because of the extra fuel needed to reach and return from such a remote area.

Spacecraft Description A 3-legged vehicle 3·05 m high, with a triangular aluminium frame, providing mounting surfaces and attachment points for landing gear, main retro-rocket engine of up to 4536 kg thrust, and 3 vernier engines of up to 47 kg thrust, etc. Launch wt of 998 kg was reduced by fuel consumption to 283 kg on landing. A central mast supported the high-gain antenna and single solar panel of 0·75 × 1 m. Aluminium honeycomb footpads were attached to each leg of the tripod landing gear. The TV camera was pointed at a mirror which could swivel 360°, and at Earth command be focused from 1·2 m to infinity with narrow or wide-angle views, returning 200 and 600-line photos. Surveyors did not go into lunar orbit before landing; in a direct approach, automatically controlled by radar, the main retro-rocket was fired at an altitude of about 96 km for 40 sec. At 40 km altitude, with speed down to 402 kph, the verniers took over, dropping the craft on to the surface, in the case of Surveyor 1, from 4·3 km at 12 kph. Launcher: Atlas-Centaur.

VANGUARD Test Satellite

History The Vanguard rocket, a US Navy development of sounding rocket technology built on Viking and Aerobee, was selected in 1955 as the most suitable launcher for America's first satellite—an unfortunate choice, as it turned out, since Russia had launched Sputnik 1 and the US Army had launched Explorer 1 before the first Vanguard satellite went into orbit. The choice had been whether to adapt sounding rockets or military rockets—in this case the US Army Redstone—for the first satellite launch. Russia chose military rockets in the same year, and got there first; in America, against a background of bitter Service rivalry, with the US Air Force advocating the use of their Atlas rocket, the deciding factor was that development of military rockets, known to be lagging behind Russia's, must not be held up.

The 1st stage of the Vanguard's rocket, using liquid oxygen and kerosene, developed 12,250 kg thrust. The 2nd stage, burning white fuming nitric acid and unsymmetrical dimethyl hydrazine, provided 3402 kg thrust. The 3rd stage, with solid propellant, added up to 1406 kg. It was 22 m long, with 1·1 m dia, and weighed 10,250 kg. The first full-scale Vanguard Test Vehicle, designated TV2, with dummy upper stages, was successfully launched from C Canaveral on Oct 23 1957, sending a 1814 kg payload on a 175km high 491km trajectory. But TV3 toppled over on the launch-pad and exploded. 2 months later, the backup vehicle veered off course and broke up at an altitude of 5·8 km. By then Sputnik 1 was in orbit. TV4 finally became Vanguard 1, and a remarkably long-lived and successful satellite, transmitting temperatures and geodetic measurements until Mar 1964. But TV5 failed; so did the first 3 operational Vanguard Satellite Launch Vehicles, designated SLV1, 2 and 3. There were 2 more failures between Vanguards 2 and 3 and the programme was over. Dr Wernher von Braun, rejoicing in the triumph of Redstone, later said Vanguard was a 'goat' through no fault of its own. For all its failures, Vanguard's 2nd and 3rd stages were later bequeathed to Thor and Atlas, and its 3rd stage also to Scout; and the swivelling motors on its 1st stage worked perfectly on Saturn.

Vanguard Test Vehicle No 3 breaking up and engulfing launchpad in flames

VIKING   Martian Landers

History   Positive, but not conclusive evidence that there may be at least some basic form of life on Mars was obtained by the Viking 1 & 2 Landers which touched down with remarkable success on Jul 20 and Sep 3 1976. Man's first pictures of the Red Planet, showing a rocky desert which was indeed red, set against a sky of pink or red, were being sent back within seconds of touchdown. At the time of writing both the Viking 1 & 2 orbiters and their landers, having survived the first solar conjunction in Nov-Dec 1976, were still operational. Originally scheduled for launch in 1973, the Viking project was postponed for financial reasons to the much less favourable 1975 launch window; a political flavour was added when it was decided to attempt the first landing, after a journey lasting nearly a year, on Jul 4 1976, the 200th anniversary of US Independence. But the pre-launch history of the project was most inauspicious; trouble both with the Titan launcher and with battery failures and helium leaks in the spacecraft, led to Vikings 1 & 2 having to be switched. Viking 1 finally took off 9 days late on Aug 20 1975, Viking 2 19 days late on Sep 9 1975. By then only 12 days remained of the 40-day launch window. The flight team of 750 scientists, engineers and technicians, with traditional NASA determination and flexibility, overcame technical problems and setbacks during the launch and the 740M km chase of Mars around the Sun; followed by the need to search for new landing sites (and abandon the Jul 4 landing) when the first orbital pictures showed the original selections at Chryse and Cydonia to be much too rough. These areas were chosen because Chryse is a valley NE of the giant 4800km long Martian Grand Canyon discovered by

VIKING SPACECRAFT
[WEIGHT-3450 KG - 7600 LB]

ORBITER INSTRUMENT SCAN PLATFORM

MARS ORBIT INSERTION ENGINE & TANKS

S-BAND LOW GAIN ANTENNA

ORBITER [WEIGHT-2360 KG - 5200 LB]

SOLAR PANELS

S & X BAND HIGH GAIN ANTENNA

LANDER CAPSULE [WEIGHT-1090 KG - 2400 LB]

Viking spacecraft. Total wt 3450 kg. Pre-launch sterilization ensured that chances of contaminating Mars with Earth micro-organisms were less than 1 in a million

Mariner 9. About 5 km lower than the mean surface, this area may once have been a drainage basin for a large portion of equatorial Mars. Cydonia is at the southernmost edge of the N Pole 'hood'—a hazy veil shrouding each polar region during the winter, and at 5·4 km below mean surface, even lower than Chryse. It was hoped (though the scientists were

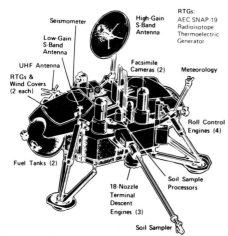

Seismometer

High-Gain
S-Band
Antenna

RTGs:
AEC SNAP-19
Radioisotope
Thermoelectric
Generator

Low-Gain
S-Band
Antenna

UHF Antenna

Facsimile
Cameras (2)

Meteorology

RTGs &
Wind Covers
(2 each)

Roll Control
Engines (4)

Fuel Tanks (2)

Soil Sample
Processors

18-Nozzle
Terminal
Descent
Engines (3)

Soil Sampler

Viking Lander. The inconclusive search for life was ended on May 30 1977, but both landers and orbiters continued to operate

larger structure. The spacecraft's total lift-off wt of 3520 kg was one factor which led to the 11-month journey to Mars, instead of 5 months for the Mariner missions.

Orbiter With a total weight of 2325 kg (propellant 1422 kg) it is octagonal, 2·4 m across, with 3·3 m ht and 9·7 m wide with solar panels extended. These panels have an area of 15 sq m and provide power from the Sun for the radio transmitter, and during flight, to the Lander. Rechargeable batteries provide power when the spacecraft is not facing the Sun, during correction manoeuvres and Mars occultation. The orbiter carries 3 instruments and 4 experiments. 2 narrow-angle TV cameras are first used to check the landing site, then provide high-resolution imaging of the Martian surface. A water detector maps the Martian atmosphere, while an infrared thermal mapper covers the surface for signs of warmth. Other instruments provide data on the planet's size, gravity, mass density and other physical characteristics. The communication system is used as a relay between the Lander and Earth via an antenna on the outer edge of a solar panel; data (a total of 1280M 'bits') can be stored aboard the Orbiter on 2 8-track digital tape recorders and transferred to Earth at the rate of 16,000 'bits' per second.

Lander With total wt of 1067 kg (fuel 491 kg), it is approx 3 m across and 2 m tall. Before launch it was encased in a bioshield to ensure that it did not contaminate Mars with Earth organisms. Inside was an aeroshell to act as heatshield during Martian entry. The main body is a hollow, 6-sided aluminium box, with alternate 109cm and 56cm sides. 3 landing legs were provided because a 3-legged structure will always rest on all 3 legs, while a 4-legged structure also usually rests on 3 legs; they are attached to the shorter

to be disappointed) that these would be smooth, calm areas, with enough moisture to carry organic life.

A summary of the final triumph of this mission, which continued the Martian exploration already started by Project Mariner, follows below. The cost of Vikings 1 & 2 was $930M, plus $132M for the 2 Titan-Centaurs.

Spacecraft Description Design was based on the successful Mariner 9, but the addition of the Lander dictated a

Viking Orbiter 1 view W of Chryse Planitia landing site shows how floodwater channels have cut through old craters. New craters are superimposed on flood channels

sides. Mounted on the outside of the body are 2 TV cameras with 360° scan, meteorology boom, surface sampler boom, seismometer, power generators, antennas, 3 main descent engines (each containing 18 tiny nozzles to spread exhaust gases and minimize contamination on touchdown), field tanks, inertial reference unit, various control boxes and the soil inlets for the organic, inorganic and biology instruments. Inside the body is an environmentally controlled compartment for the biology instrument, gas chromatograph/mass spectrometer, computer, tape recorder, data storage memory, batteries, radios, data acquis-

ition and processing unit, and command control and support units. Basic power is provided by 2 SNAP nuclear generators. It is hoped they will enable the Landers to continue operating for 2 yr.

Viking 1 L Aug 20 1975 by Titan 3E-Centaur from Cape Canaveral. Wt 3399 kg. Trajectory adjustments compensated for the launch 9 days late, and the Viking team maintained that the proposed landing on the US Bicentennial day would still be possible; privately, however, scientists made it clear they were opposed to the plan. With a high probability of the spacecraft crashing on touchdown, a hurried landing for political purposes could well increase that chance. Failure on Jul 3 or 5, it was argued, would be quickly forgotten; failure on Jul 4 would be remembered (and politically resented) for ever. During the 310-day flight there were indications that one of the 3 'ovens' to be used for analysing soil samples in the search for Martian life, was not working; a slight helium gas leak 12 days before insertion into Mars orbit also caused anxiety. A course correction on Jun 10 1976 which slowed the spacecraft by 180 kph instead of the planned 14 kph resulted in arrival at Mars 7 hr late, and although the jerk was not entirely successful in clearing the regulator and stopping the leak, other ways were found of overcoming the problem. Although Earth and Mars were 380M km apart—almost their maximum distance on opposite sides of the Sun—astonishingly this, as it turned out, in no way reduced the success of the mission. On Jun 19 Viking 1's main engine was fired for 38 min, consuming 1800 kg of the total fuel load of 2260 kg, and slowing it by 4296 kph to the required orbital speed of 16,500 kph. This placed it in a considerably higher 42·6-hr elliptical orbit than scheduled, and a trim manoeuvre was required after one orbit to put it

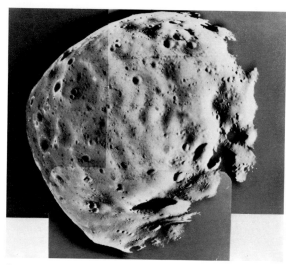

Orbiter 1 passed within 480 km to get this view of Phobos. Viking scientists now believe Phobos and Deimos were captured from the asteroid belt

radar data from Earth and high resolution photographs from orbit, a smooth area in the western reaches of Chryse Planitia (Gold Plain) could be found. On Jul 20 the Lander separation command was transmitted, initiating a 3-hr 21-min automatic separation and entry sequence, during which pyrotechnic devices and springs pushed the Lander away from the orbiter. The lander deorbit sequence started with a 23-min propulsion burn using 8 of the 12 hydrazine monopropellant RCS engines; the remaining 4, each producing 3·6 kg of thrust, were used for attitude control. During a 3-hr coast period, the Lander was rolled 180° to sterilize the base cover of the aeroshell with solar ultraviolet radiation. At 244,000 m the Lander dipped into the upper levels of the Martian atmosphere; at 6000 m the base of the aeroshell was jettisoned and a 16m dia parachute deployed. The initial entry speed of 16,500 kph had been reduced to about 1090 kph, and the 1 min parachute descent further reduced this to 220 kph. By that time the altitude was 1430 m. Finally, 3 hydrazine engines slowed the Lander to a vertical touchdown speed of about 8·7 kph. More than 19 min after the actual touchdown, signals confirming a successful landing reached Earth. By a coincidence it was 7 yr to the day after Armstrong and Aldrin had landed on the Moon. Already dramatic cliff falls into valleys, and meteorite impacts into watery mud, completely different from the Moon, had been revealed by orbital photographs. Now man's first view of the Martian surface, transmission of which began 25 sec after touchdown, showed a red jumble of boulders and rocks interspersed with what was probably sand. Later panoramic pictures showed a creamy pink sky. Scientists were relieved that only about 1·5% of the atmosphere was argon gas, instead of the 30% suggested by Russia's Mars 6 lander, since this gas could have had adverse effects on the

in the originally planned 24·66-hr orbit, inclined at 38° to the Martian equator, with a 1500km periapsis and 32,597km apoapsis. The orbiter was soon transmitting spectacular pictures of the proposed Chryse landing area, revealed to be much rougher and more cratered than expected, with islands and channels cut by massive flows of water – which of course had long since disappeared. The Jul 4 landing was postponed for a fortnight, so that, with a combination of

Left Lander 2 pushing aside rock to look for life in soil beneath; *right* trench after soil removal

equipment used to analyse soil samples. The only major disappointment was that the seismometer package, which should have dropped on to the surface to detect Mars-quakes, remained stuck in its protective cage because the locking pin failed to eject. The search for Martian life by testing soil samples proved to be dramatic. The locking pin on the sampler arm, stuck at first, was finally shaken out 5

days after touchdown, by means of a technique worked out in the Viking simulator. 8 days after landing the first soil sample was scooped up and transferred to the internal laboratories. Confirmation was provided by a clear picture of the straight trench which had been dug on the Martian surface. On Jul 31 the amount of oxygen released in the 'gas exchange experiment' was announced to be 15 times greater than expected. This could have been caused by biological activity; but the biology team cautiously decided it could equally have been caused by an inorganic chemical reaction in the soil which was not understood. The temptation to announce that there WAS life on Mars was resisted. The Orbiter's pictures of the surface commanded less attention, but had their own fascination. The formation of early morning ground fog was shown in low lying areas such as craters and channels; pre-dawn pictures, when compared with those of the same area taken 30 min later, showed white patches where slight warming of the sub-zero surface had caused small amounts of water vapour to condense in the colder air just above the surface. Pictures of the Cydonia region showed strange geometric markings, like an aerial view of ploughed fields.

Viking 2 L Sep 9 1975 by Titan 3E-Centaur from C Canaveral. Wt 3399 kg. Viking scientists, fully occupied with the Viking Orbiter and Lander, had to start running down activities on those as Viking 2 approached Mars. On Aug 7 1976 this was successfully placed in an orbit with 1500 km periapsis with a 27·6-hr period and 55° incl. With the Lander still attached, the Orbiter was soon transmitting pictures of 'huge striped patterns' in northern regions which resembled sand dunes created by wind in the Sahara desert. In the light of Viking 1 experience, a fresh search was begun

Afternoon, Sep 5 1976, as seen by Lander 2. Some rocks are porous and sponge-like, others dense and fine-grained; largest are about $\frac{1}{2}$ m across. Slope of horizon due to the spacecraft's 8° tilt.

for a suitable landing area. Sites in the Cydonia and Alba Patera regions were rejected because they had too many craters, ridges and channels; 12 days after Viking 2's arrival a new site, Utopia Planitia, 7420 km NE of Viking 1 was chosen. Final confirmation on Sep 3 that Lander 2 was safely down, with one footpad perched on a rock, and tilted at an angle of 8·2°, was not obtained until 8 hr later. The shock of separation from the Orbiter caused a temporary electrical fault which deprived mission controllers of their primary communications link through the Orbiter. However, the Lander's onboard computer guided it down to a point later established as 225·86°W Long and 47·97°N Lat. Because the radar was apparently misled by a rock or highly reflective surface, an unscheduled extra firing of the retrorockets 0·4 sec before touchdown cracked the surface crust and scattered dust. It also gave an extremely soft touchdown. This time the seismometer was successfully freed, to start its watch for Marsquakes; and first pictures showed a bleak landscape covered with a variety of pitted rocks and small sand drifts among them. Scientists were somewhat disap-

pointed that the area did not differ so much as expected from the Lander 1 site. The weather at the new site was found to be less complex than at Chryse Planitia. Winds followed a daily pattern, with average velocity of 15 kph and maximum gust speed of 26 kph. With the night 2·5 hr shorter at the Utopia site, temperatures were slightly warmer; minimum at 04.08 Martian time was −81°C (−114°F); maximum, at 15.50 was −30·5°C (−23°F). Surface pressure was 0·6 of a millibar higher than at Chryse, suggesting that the Utopia site was 1·3 km lower. 9 days after touchdown the first soil sample was scooped up and transferred to the 3 biology experiments. But as weeks passed with repeated experiments, Lander 1's first dramatic results were not repeated and no positive evidence was found of even the most primitive life forms. By the end of Oct, with Lander 1 having been virtually shut down for 50 Sols (Martian days) so that work could be concentrated on Lander 2, a small rock had been carefully pushed aside by the surface sampler arm, and soil which had been protected beneath it transferred and tested. When this was completed, just before the end of the primary mission on Nov 10 1976, it was admitted that, if there was life on Mars, the Vikings were unlikely to detect it. With no communications possible for a month while Mars was behind the Sun, all 4 spacecraft—2 Orbiters and 2 Landers—were shut down. But Viking's mission controllers' confidence that it would be possible to reactivate them, and continue working with them for up to 2 yr appears justified. During the period of conjunction, plans were made to manoeuvre Orbiter 1 to within 50 km of Phobos, to add to the detailed pictures already obtained of both Martian moons.

Future Missions A mobile version of Viking, able to operate for 2 yr, and travel up to 500 km, is under development. The 3 landing pads would be replaced with barrel-shaped, 'continuous elastic-loop tracks'. They would combine the suspension and drive systems and enable the spacecraft to climb slopes of 30–40°.

WESTAR   Domestic Communications

History The first US domestic communications satellite system, owned by Western Union. Westars 1 & 2 (L Apr 13 1974 and Oct 10 1974; wt 572 kg) were placed in synchronous orbits at 90°W (W of the Galapagos Isles) and at 0·4° over the equator S of Los Angeles. Each can provide 12 colour TV channels or 14,400 one-way phone circuits through 5 Earth stations located near New York, Atlanta, Chicago, Dallas and Los Angeles. Originally 3 satellites were due to be orbited in 1974, but trouble with the Delta launch vehicles caused postponements. In mid 1977 the 3rd was still to be launched. Western Union was repaying NASA the full launch costs. The satellites had a 7-yr operational life, and 1M yr orbital life.

Spacecraft Description Main body is a spin-stabilized cylinder with 1·8m dia and 1·6m ht. On top is a spoon-shaped mesh antenna, despun in orbit so that it points continuously towards Earth. More than 20,000 solar cells cover the exterior of the cylinder to provide 300W of power.

LAUNCH VEHICLES

History The wide range of US launch rockets has enabled NASA and the Department of Defense to place in orbit payloads as light as 9 kg right up to Saturn 5's 136,000 kg. In his 1976 comparison of US and Soviet launchers, Dr Charles Sheldon, Chief of the US Science Policy Division, points out that the manned Mercury spacecraft launched by Atlas rockets, weighed about 1360 kg—the same orbital lift capacity as Russia's original ICBM, which she began using for her space programme in 1957. Gemini, weighing about 3630 kg was launched by Titan 2. The first manned Apollo flight, launched by Saturn 1B, was about 20,410 kg; the lunar Apollos, launched by Saturn 5, carried about 52,600 kg to the Moon. It must be remembered that the further the desired orbit is from Earth, the greater the velocity required to send it there; when longer flights are required, either a smaller payload or a larger rocket must be used. With the phasing out of the Saturn rockets in 1975, America faced a 5-yr period until the Space Shuttle became operational, during which its largest launch vehicle was the Titan 3E-Centaur, with a maximum payload into Earth-orbit of 13,600 kg.

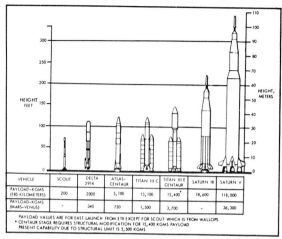

VEHICLE	SCOUT	DELTA 2914	ATLAS-CENTAUR	TITAN III C	TITAN III E CENTAUR	SATURN IB	SATURN V
PAYLOAD-KGMS (185 KILOMETERS)	200	2000	5,100	13,100	15,400*	18,600	118,000
PAYLOAD-KGMS (MARS-VENUS)	-	340	730	1,500	3,700	-	36,300

PAYLOAD VALUES ARE FOR EAST LAUNCH FROM ETR EXCEPT FOR SCOUT WHICH IS FROM WALLOPS
* CENTAUR STAGE REQUIRES STRUCTURAL MODIFICATION FOR 15,400 KGMS PAYLOAD
PRESENT CAPABILITY DUE TO STRUCTURAL LIMIT IS 5,500 KGMS

NASA Launch Vehicles

AGENA The versatile upper stage Agena rockets have been used for both classified and unclassified missions since 1959. They have been placed on top of Thor, Atlas, Thrust Augmented Thor (TAD), Long-Tank Thor, Thorad and Titan 3B. Typical length 7·09 m. Thrust 7257 kg.

The Agena-D version, launched as the 2nd stage of an Atlas, was used as a docking target on 5 of the Gemini flights. The first, on Gemini 6, was lost during launch; on the 3rd launch, for Gemini 9, docking was impossible because the protective shroud around the docking ring failed to open properly, giving the Agena the famous 'angry alligator' look. On Gemini 8, however, history's first space-docking was achieved with the Agena target by Armstrong and Scott. Agena was selected as target vehicle because of its record of achieving precise predetermined orbits, and the fact that, once in orbit, it could be successfully stabilized and then maintain constant orientation to the Earth. On Agena 8 it became the first spacecraft to be commanded and controlled by another space vehicle; the astronauts could control it, whether docked or merely in its immediate vicinity. Its main, gimballing engine, was restarted 8 times by ground command after Gemini 8 had returned to Earth. Gemini 10 saw the first dual space rendezvous, when the docked spacecraft and Agena were manoeuvred to the still orbiting Gemini 8 Agena target, so that Astronaut Collins could spacewalk across to it and retrieve a device fitted to record the effect of micro-meteoroid bombardment.

ATLAS

Lift-Off Weight:	117,934 kg
Lift-Off Thrust:	166,460 kg
Height (with Mercury):	28·35 m
Height (as Missile):	24·08 m
Diameter:	3·05 m
Speed at Burn-Out:	28,166 kph

History It was on top of an Atlas D that John Glenn, on Feb 20 1962, became the first American in orbit. There were 10 Atlas D flights in Project Mercury, the pioneering US manned programme; 5 unmanned test flights were followed by MA 5 with a chimpanzee, and 4 manned flights starting with Glenn. Technical problems with the rocket delayed Glenn's flight; many believed it would never be reliable enough for manned flight. In fact all 4 manned flights were successful, and the rocket has now been used in more than

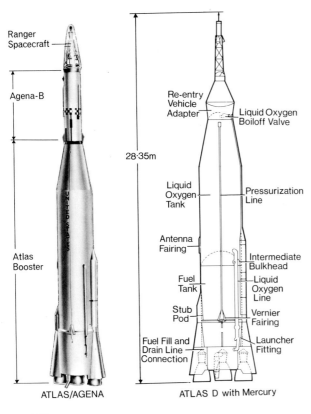

Ranger
Spacecraft

Agena-B

28·35m

Atlas
Booster

ATLAS/AGENA

Re-entry
Vehicle
Adapter

Liquid Oxygen
Boiloff Valve

Liquid
Oxygen
Tank

Pressurization
Line

Antenna
Fairing

Intermediate
Bulkhead

Fuel
Tank

Liquid
Oxygen
Line

Stub
Pod

Vernier
Fairing

Fuel Fill and
Drain Line
Connection

Launcher
Fitting

ATLAS D with Mercury

25 space programmes. It has been fired over 400 times. It has sent Ranger spacecraft to the Moon, and Mariners to Mars and Venus in combination with Agena as an upper stage. Originally developed as America's first ICBM (Intercontinental Ballistic Missile), it became operational in Sep 1959; a total of 159, with a range of over 8050 km, were at one time deployed at sites across the US. Later versions had a range of 12,875 km.

General Description Atlas D is 20 m from base to Mercury adapter section, and 3·05 m diameter at tank section. Construction is of such thin-gauge metal that it must be pressurized during ground transport, and as its propellants are consumed during flight, to maintain structural rigidity. All 5 engines, which burn highly refined RP-1 kerosene and liquid oxygen, are fired for take-off: the central sustainer engine 27,215 kg thrust, the 2 outer booster engines 68,040 kg thrust each, and the 2 small vernier engines used for mid-course corrections during powered flight. System components, including command receivers, telemetry, guidance, antennas etc., are in pods on the side of the fuel tank located immediately above the main engines. Launch-vehicle guidance is provided by a combination of onboard and radio ground guidance equipment. At T + 130 sec, ground command shuts off the booster engines, and ground guidance controls the sustainer engine; the boosters are jettisoned at shut-off, taking with them the large flared 'skirt' around the tail, which provides stability during the initial flight stages. At about T + 300 sec, when insertion parameters are attained, ground control shuts off sustainer and vernier engines. Spacecraft separation is at an altitude of about 64·5 km, range of about 72·5 km from the launchpad, and velocity of 28,232 kph.

First Delta to be fitted with 6 solid fuel strap-on rockets to give additional lift-off thrust

Atlas-Agena General purpose launcher developed in various versions from original ICBM, used by both USAF and NASA. Lift-off thrust about 158,760 kg. Overall ht 31·7 m. By Sep 1975 total firings totalled 26, but only 20 were successful.

Atlas-Centaur High-energy, 2-stage rocket, able to lift medium-wt spacecraft of 4500 kg, into 555 km orbits, or 1810 kg into synchronous transfer orbits. Atlas provides 195,500kg lift-off thrust; overall ht 35·8 m; max dia 3·04 m. From first test flight in May 1962 to the end of 1976, 36 Atlas-Centaur launches included planetary missions to Mars, Venus, Mercury and Jupiter; there were 6 failures.

Centaur Used as upper stage on Atlas and Titan launchers. First US space vehicle to use liquid hydrogen as a propellant; capable of putting 4570kg payloads into orbit, and sending large payloads to planets. Thrust 13,610 kg. Ht 9·14 m.

Delta Versatile launcher based on Thor 1st stage, used for wide variety of medium-sized satellites and small space probes. Can be used as 2 or 3 stages, augmented with 3, 6 or 9 solid-fuel strap-on 1st stage motors able to lift 900 kg to synchronous orbits. Thor 1st stage provides 216,800 kg thrust, including 6 strap-ons. Overall ht 35 m; max dia 2·4 m. By end 1975, 114 Deltas had been launched with 91% success rate.

Juno *See* Explorer entry

Jupiter *See* Explorer entry

LTTAD Long Tank Thrust Augmented Delta. Modified Thor 1st stage, with 7800 kg thrust, augmented by 3, 6 or 9 strap-on solid propellant rockets giving combined thrust of

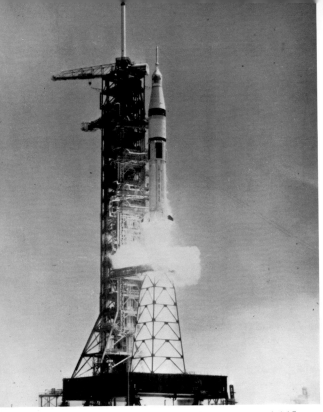

Saturn 1B lifting off with Apollo-Soyuz crew on Jul 15 1975—probably the last ever Saturn launch

152,200 kg. Ht 31·9 m approx. Able to launch 158 kg to stationary orbits.

SATURN 1B

Lift-Off Weight (with CSM):	587,675 kg
Overall Height:	68·28 m
Lift-Off Thrust:	0·73 million kg
Payload:	18,145 kg
Diameter—1st stage:	6·53 m
2nd stage:	6·60 m

History The Wernher von Braun organization, then working with the Army Ballistic Missile Agency, first proposed the need for a launcher of 680,400 kg thrust, able to place between 10 and 20 tonnes in Earth orbit, or send 3 to 10 tonnes on escape missions, in Apr 1957. By 1959 the project had been named Saturn. This was because Saturn was the next planet after Jupiter in the solar system, and the Saturn rocket was the next von Braun project following completion of Jupiter missile development. Saturn 1, with engines and tanks in clusters in order to make use of equipment already developed, was a 2-stage vehicle. The first stage had 8 H-1 engines, burning RP-1 kerosene and liquid-oxygen, each generating 82,250 kg thrust. The 2nd stage (designated S4), had 6 liquid-oxygen, liquid hydrogen RL-10 A-3 engines, each generating 6800 kg thrust. 10 Saturn 1s were fired between Oct 1961 and Jul 1965, with an unprecedented record of 100% success. The 5th placed a 17,100 kg payload into Earth orbit, and the 6th and 7th each placed unmanned 'boilerplate' models of Apollo spacecraft in Earth orbit. The 9th orbited a Pegasus meteoroid technology satellite. Meanwhile, it had been decided that elements of Saturn 1 and of the planned, much larger Saturn 5, should be combined to form a new mid-range vehicle, Saturn 1B. This would have a 50% greater capability than Saturn 1, and enable complete Apollo spacecraft to be tested in manned Earth orbital flights 1 yr earlier than would be possible with Saturn 5.

General Description Saturn 1B's 1st stage retains the same size and diameter, and is 24·49 m high, with 6·53 m diameter; but its weight was reduced, and payload correspondingly increased by 9070 kg, by a new fin design, and by resizing and redesigning tail section and spider beam etc. The 8 H-1 engines have been uprated to 90,700 kg. In 150 sec from lift-off, they burn 158,980 litres of RP-1 fuel, and 253,615 litres of liquid oxygen, to reach an altitude of 67·6 km at burn-out.

The 2nd stage, an enlarged S4, designated S4B, 17·81 m high, and 6·60 m in diameter, has a single liquid-hydrogen, liquid-oxygen J2 engine, giving 90,700 kg thrust. It burns 242,250 litres of liquid-hydrogen, and 75,700 litres of liquid-oxygen, in 450 sec of operation, to achieve orbital speed and altitude. The Instrument Unit (IU), only 0·91 m high, with 6·60 m diameter, is known as 'the wedding ring', since it joins launcher and spacecraft. Unpressurized, it houses the instrumentation concerned with vehicle performance from lift-off to insertion of payload into orbit; it contains tracking command, measuring and telemetry systems, an electrical power supply and thermal conditioning system, and in the Skylab mission was designed for a 7½-hr life after lift-off. Throughout the launch the IU systems measure the vehicle's rate of acceleration and attitude, calculate what corrections are needed to keep it on course, and issue commands to the engines, shortening or lengthening

their burn-time, so that the vehicle achieves the exact height and speed needed for the mission.

9 Saturn 1B firings have maintained Saturn 1's record of success. The first launch, which was also the first flight test of a powered (unmanned) Apollo spacecraft, was on Feb 26 1966; after 3 more test flights, the first manned flight of the series, Apollo 7 was successfully launched on Oct 11 1968. 3 Saturn 1Bs were used to send the 3 rotating crews of astronauts to the Skylab space station in 1973, followed by the equally successful ASTP launch in 1975. Of the 12 originally built, 3 were unused. The Saturn 1 programme cost £319,580,000 ($767M); Saturn 1B an additional £469,587,000 ($1127M).

SATURN 5

Lift-Off Weight (Apollo 11):	2,938,312 kg
1st stage Thrust (Apollo 11):	3,425,500 kg
Height—(with Apollo):	111 m
(with Skylab):	109 m
Diameter—	
1st and 2nd stages:	10·06 m
3rd stage:	6·60 m

History Saturn 5, designed and developed at NASA's Marshall Space Flight Center under Dr Wernher von Braun, for the Apollo Project was also used to launch the Skylab orbiting workshop. 15 were built, and 2 were unused.

Saturn 5 had an impressive record of success since development began in Jan 1962. Its first flight was on Nov 9 1967, the unmanned Apollo 4 mission, which successfully tested both the rocket and the spacecraft. It was first used for a manned flight on the Apollo 8 mission, on Dec 21 1968,

Saturn 5 on Launch Complex 39A during fuelled countdown rehearsal for Apollo 16. Note size of men on transporter platform

when Frank Borman commanded man's first historic flight around the Moon. Its flexibility was demonstrated when it survived a lightning strike during the Apollo 12 launch; and on Apollo 13, when the centre engine of stage 2 shut down 2 min early. The other 4 engines automatically burned to depletion to make up the lost thrust, and the 3rd stage automatically burned for an extra 10 sec to complete the task. There was still ample fuel left to complete the translunar injection. Structural weight reductions, with improved engine performance and operational techniques, enabled the Saturn 5s used in the final Apollo missions to place payloads of 152 tonnes into Earth orbit and to send 53 tonnes to the Moon.

General Description *First Stage* (S1C) Built by the Boeing Co at NASA's Michoud Assembly Facility, New Orleans. Its 5 F-1 engines consume kerosene and liquid oxygen at 13,319 kg per sec, and boost the vehicle to approx 8530 kph and a ht of 61 km in $2\frac{1}{2}$ min. Major components are the forward skirt, oxidizer tank, intertank structure, fuel tank and thrust structure. One engine is rigidly mounted on the stage's centreline; the other 4, mounted on a ring at $90°$ angles around the centre engine, are gimballed to control the vehicle's attitude during flight.

Second Stage (S2) Built by the Space Division of North American Rockwell corporation at Seal Beach, California. Its 5 J-2 engines ignite as the first stage separates and falls away, and develop a total thrust of 526,165 kg, burning a mixture of liquid hydrogen and liquid oxygen. They must raise the vehicle's speed to approx 24,625 kph, and a ht of 183·5 km in 6 min. Major components are the forward skirt, liquid hydrogen and liquid oxygen tanks (separated by an in-

The 3000tonne Crawler Transporter, with a crew of 15, formerly used to transfer Apollo/Saturns from VAB to launchpad, will in future be used to transfer Space Shuttle assemblies

sulated common bulkhead), a thrust structure, and an interstage section connecting it with the first stage. As on the first stage, the centre engine is rigid, and the outer 4 can be gimballed.

Third Stage (S4B) Built by the McDonnell Douglas Astronautics Co at Huntingdon Beach, California. The function of this stage is quite different. Its single, J-2, gimballed engine, powered by liquid hydrogen and liquid oxygen, has a maximum thrust of 104,325 kg and can be shut off and restarted. Its first job is to take over when stage 2 falls away, and burn for about $2\frac{1}{2}$ min (the exact time is controlled by computer) to increase the speed to the necessary orbital rate of 28,000 kph. Major components are the aft interstage and skirt, thrust structure, 2 propellant tanks with common

bulkhead and forward skirt. On a Moonflight, the S4B shuts down after placing the vehicle into a parking orbit, and remains attached to the spacecraft. Usually on the 2nd orbit it is fired a 2nd time for approximately 5 min to accelerate the vehicle to over 39,270 kph, thus injecting it into a translunar orbit. Shortly after (about 3 hr after lift-off), the Apollo Command Module separates from the nose of the S4B, turns around, docks and withdraws the Lunar Module, protectively housed for take-off just below the Command Module, and inside the top of the S4B. This done, the S4B is either sent off into solar orbit to ensure that its path is well clear of Apollo, or it is sent on to impact at a selected point on the Moon's surface, for seismometer and other scientific tests.

Instrument Stage As described under Saturn 1B (*qv*).

Total Saturn 5 costs were £2,519,166,000 ($6046M); including Saturns 1 and 1B, expenditure on Saturn programmes totals £3,308,300,000 ($7,940M).

REDSTONE

Lift-Off Weight:	29,935 kg
Lift-Off Thrust:	35,375 kg
Height (with Mercury):	25·30 m
Height (as Missile):	21·13 m
Diameter:	1·77 m
Speed at Burn-out:	7080 kph
Range:	320 km

History Redstone was used to launch the 2 suborbital flights which inaugurated America's manned space programme. It was first developed by Dr Wernher von Braun for the US Army from his original V2 missile; with a 965km

Test firing of Mercury spacecraft escape tower from Redstone rocket. In practice, it never had to be used

range, it was deployed in Europe in 1958, and became known as 'Old Reliable'. Some 800 engineering changes were made to transform it into a booster for the first experimental manned flights. In addition to major engine improvements, 20 sec were added to the burn-time by lengthening the tank section by 1·83 m. This, with the spacecraft and escape tower, added 2265 kg to the original 27,670 kg missile weight. Telemetry provided 65 measurements covering attitude, vibration, acceleration, temperature, thrust level etc. Redstone also became established as a space launcher in 1958 when used as the first stage of the Jupiter C rocket which put Explorer 1, America's first satellite, into Earth orbit.

6 Mercury-Redstone flights were made, starting in Nov 1960. The first 4 were test-firings; then came the historic 15-min 'lob' which made Alan Shepard America's first man in space, on May 5 1961, followed by Virgil Grissom on Jul 21 1961. Details of these flights appear under Mercury (qv).

Scout America's smallest and only solid-fuel launcher, used for large variety of small scientific payloads, and high-speed re-entry experiments. 4 stages; with lift-off thrust of 48,580 kg. Ht 22 m, max dia 1·12 m, can place 190 kg in 480km orbit due E from Wallops Island, Va. By end 1975, 64 Scouts had been launched with 91% success rate.

Thor Originally America's first IRBM, now provides in various forms her most frequently used launcher. Fuelled by liquid oxygen and kerosene. Starting in 1957 with a 1st stage thrust of 68,000 kg, the latest Long Tank Thor now provides 149,700 kg. Ht 21·5 m. Constant dia 2·4 m. The US Air Force developed a 2nd stage, Thor-Able, in 1957, capable of sending small probes to the Moon; but it was abandoned

Thor-Agena D being prepared at Vandenberg to launch OGO satellite

when 2 out of 5 launches failed; a 3-stage version, Thor-Delta, was then developed with a successful first launch in May 1960. The augmented Thor followed, with 3 strap-on solid propellant rockets on the 1st stage. Several other versions have Agena or Delta upper stages. Famous Thor payloads, which totalled about 500 by end 1975, include the first communications satellites, Telstar and Early bird.

Thorad-Agena D Uprated version of TAD (Thrust Augmented Delta) with Agena 2nd stage. 1st stage, with 3 strap-on solid motors gives lift-off thrust of 147,870 kg. Ht 33 m.

TITAN 2

Lift-Off Thrust:	195,000 kg
Second Stage Thrust:	45,360 kg
Lift-Off Weight:	148,325 kg
Height with Spacecraft:	33·22 m
Height as Missile:	31·40 m
1st stage Diameter:	3·05 m
Length of 1st stage:	21·34 m

History Titan 2 was successfully used to launch all 10 of the 2-man Gemini spacecraft which followed the Mercury-Atlas flights. 15 Titan 2s, adapted as space launchers from the original Titan series of ICBMs, were ordered by NASA, and on Apr 8, 1964 an unmanned Gemini was put into orbit at the first attempt. From the start Titan proved to be one of the most dependable US missiles; Titan 1 had only 4 failures in 47 launches: Titan 2, first launched in 1962, had a range of over 9655 km, and its storable propellants gave it a capability of being launched from underground silos in less than a minute. Titan's reliability enabled Project Gemini—completed in only 20 months—to mark the start of sophisti-

Nightview of Titan 2, carrying Gemini spacecraft, being prepared for launch at Cape Canaveral

cated manned spaceflight; this was demonstrated by Gemini 7 and 6 being launched from the same pad within 11 days; the craft were regularly placed in such precise orbits that it was possible in less than 2 yr to master the technique of rendezvous and docking.

General Description A 2-stage vehicle, with a rigid structure of high-strength aluminium, Titan has the advantage over Atlas of not needing to be pressurized to maintain its rigidity on the launchpad; the fuel-tank walls double as the outer skin of the vehicle. Both stages are liquid-fuelled; they burn unsymmetrical dimethyl-hydrazine (UDMH), with nitrogen tetroxide as oxidizer. The fuels are storable and hypergolic (that is, they ignite on contact, so that an ignition system is unnecessary). 'Fire-in-the-hole staging' is employed—that is, 2nd-stage engine ignites before separation from 1st stage is complete at Lift-off + 2 min 30 sec. 2nd-stage engine shuts down at $5\frac{1}{2}$ min, at 161 km altitude, and 854·5 km downrange; spacecraft separation follows at 5 min 50 sec. Most important of the modifications, first installed on the man-rated Titans, is the Malfunction Detection System (MDS). This continually monitors performance of the sub-systems, and signals reports to the astronauts, thus enabling them to decide whether the mission should be continued or aborted.

Titan 3 Designed in 3 versions, originally for military launches, such as Big Bird reconnaissance satellites, but with Saturns no longer operational, is now America's heaviest available launch vehicle. Titan 3A is a modified Titan 2 core with new upper stage and control module. Titan 3C, with 2 strap-on solid fuel rockets provides 1,043,300 kg lift-off thrust, followed by 220,000 kg thrust from main core booster. This can place 10,570 kg into 555km orbits, or 3600 kg into synchronous orbits, or send 1200 kg to Mars or Venus. Titan 3D consists of the core vehicle minus some

Titan—Centaur launching Viking 1 on its way to Mars on Aug 20 1975

systems required for manned flight and the transtage which can be replaced by Agena or Centaur. Titan 3E-Centaur (until the Shuttle becomes available) provides America's maximum launch capability of 13,600 kg into Earth orbit, 3500 kg into synchronous orbit, or 4000 kg to the planets. Its first test flight was in 1974; 2nd flight in Dec 1974 successfully launched Helios, Germany's solar observatory; 3rd and 4th flights in Aug and Sep 1975 launched the Viking/Mars craft. Such launches cost $29·3M. Ht of 4-stage Titan 3C: 38·7 m. Max dia 3 m. With Centaur 4th stage, overall ht 48·7 m.

Vanguard *See* Vanguard Satellites.

USSR SPACE PROJECTS

AUREOLE   Soviet-French Satellites

Aureole 1   L Dec 27 1971, by C1 from Plesetsk. Wt ?300 kg. Orbit 410 × 2500 km. Incl 74°. The first Soviet-French satellite, launched under a new international project, 'Arcade'. It studied the aurora borealis, or Polar lights, believed to be huge 'plasma explosions' far from Earth, which change the ion composition of Earth's upper atmosphere and affect radio conditions. Instruments were designed and supplied by Russia's Space Research Institute and France's Toulouse Centre of Study of Space Radiation. Orbital life approx 70 yr.

Aureole 2   L Dec 26 1973 by C1 from Plesetsk. Wt ?400 kg. Orbit 400 × 1975 km. Incl 74°. Continued the previous satellite's work; little was heard of the results. Life approx 30 yr.

COSMOS

Launch Rate

1962:	1–12 = 12	**1970**:	318–389 =	72
1963:	13–24 = 12	**1971**:	390–470 =	81
1964:	25–51 = 27	**1972**:	471–542 =	72
1965:	52–103 = 52	**1973**:	543–627 =	84
1966:	104–137 = 34	**1974**:	628–701 =	74
1967:	138–198 = 61	**1975**:	702–786 =	85
1968:	199–262 = 64	**1976**:	787–887 =	101
1969:	263–317 = 55	**1977**:	888–	

Left Cosmos 2, for ionospheric research   *Right* Cosmos 3, for investigation of low energy particles

History   On Mar 29 1977 the Soviet Union launched the 900th in the Cosmos series. Since they began with Cosmos 1 on Mar 16 1962, satellites have averaged 59 per year. Between 1970 and 1976 satellites have averaged 81 per year, with a record 101 in 1976. (It should be noted, however, that up to 8 satellites are frequently placed in orbit by 1 rocket; so the number of rocket firings is much lower than the number of satellites.)

Cosmos numbers have always been used by Russia as a convenient way of concealing the inevitable failures that still accompany any space programme; interplanetary probes are only given designations in the Mars, Venus etc series when they are on course and working well. If they are unsuccessful they are merely given Cosmos numbers with the usual routine announcement about their launch. Cosmos 359 is an example. It should have been Venus 8, the 2nd of a pair, but failed to achieve escape velocity after being launched from its Earth-parking orbit.

The Cosmos series is launched from all 3 Soviet launch sites, at Tyuratam (Baikonur), Kapustin Yar and Plesetsk. The last seems to be principally a military site; its existence was first made public in 1966, as a result of the tracking activities of Kettering Grammar School in England. When Cosmos 500 was launched about 100 of the series were still in orbit, though not all were active. Some of the series at this stage were undoubtedly testing improved hatches for Soyuz spacecraft, following the catastrophic depressurization that caused the death of the Soyuz 11 crew.

Soviet scientists pointed out, when the 700th Cosmos was launched that the tremendous role which space was beginning to play in the Soviet economy was becoming increasingly obvious. Most Cosmos satellites were standardized both in the design and in the composition of the onboard systems: this made 'flow line production of these spaceworkers possible', and together with the use of standard rocket carriers had enabled considerable savings to be made in the cost of space exploration.

At least half of the series are military satellites, though this has never been admitted; in 1970, for instance, 57 of the 72 Cosmos launches were believed to be for military purposes. In 1975, military spacecraft peaked at 62 out of 85; in 1976 about 90 out of 101. Their activities, ranging from the development of the ability to intercept and destroy other satellites, to the regular launching of pairs of overlapping 'spy' satellites, are described later. The variety of activity is reflected in the range of payloads; their wt varies from 129 kg to about 7484 kg. The non-military

Cosmos satellites continue to range over a wide area of research, and have achieved remarkable results; these results are usually announced months, or even years, after launch. According to the Soviet *Encyclopaedia of Spaceflight* (1969) the research programme 'includes study of the concentration of charged particles, corpuscular fluxes, radio-wave propagation, distribution of the Earth's magnetic field, solar radiation, meteoric matter, cloud formations in the Earth's atmosphere, solution of technological problems of spaceflight (including docking, atmospheric entry, effect of space factors, means of attitude control, life support, radiation protection etc), and flight testing of many structural elements and spaceborne systems'. The review issued after the 700th launch broke the series down more simply into 3: 1) research into physical phenomena in space, and the upper Earth's atmosphere; 2) tackling technical problems in connection with spaceflights, development work on structural members and onboard systems of future spacecraft; and 3) experiments of an applied character in the interests of terrestrial sciences and the national economy.

In addition to studying the ionosphere, etc, the first group also had the task of probing the radiological situation in space before manned craft were launched. Development work carried out by the 2nd group included automatic link-ups, 'very important for cosmonautics of both today and of the nearest future'. The 2nd group also included biosatellites, such as Cosmos 690, to study the effects of spaceflight on living organisms. Clearly Group 2 is connected with the plans seen to be nearing maturity in 1975, to start building up large space stations by automatically joining a series of Salyut-type vehicles, and then manning them for periods even longer than America's 3-month stay in Skylab. Group 3, said briefly to include meteorological, radio and TV

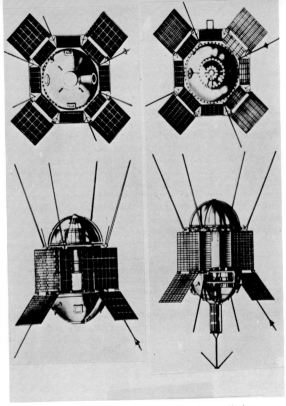

Left Cosmos 137, for investigation of Earth's radiation belt   *Right* Cosmos 97, carrying a molecular generator

satellites, no doubt also includes the still unmentioned military craft. (But it should be noted that there is evidence that some recent Soyuz/Salyut flights also appear to be used in part for military experiments and surveillance.) It is interesting to note that, so far as I can trace, no public mention was made of the fact that Cosmos 637 (L Mar 26 1974) was Russia's first-ever synchronous satellite—probably because by then America had been using synchronous satellites for 12 yr.

Military Cosmos   These are usually launched at the rate of 2 a month, often in overlapping pairs. The launch rate rises during periods of tension, as was observed during clashes between Soviet and Chinese forces on 2 occasions in 1969. Reconnaissance satellites are mostly launched from Plesetsk and Tyuratam; the latter mostly have inclinations of 72°; others 52° and 65°. Until the summer of 1968 Russia's recoverable reconnaissance satellites remained in orbit for 8 days or less before being brought back for their film and electronic recordings to be processed. Starting with Cosmos 228, on Jun 21 1968, they began to stay in orbit for 12 days or longer before ejecting their film packages; during 1971 the staytime of this type was frequently 14 days; presumably larger film packs made it possible to reduce the number of launches. Cosmos 251 in Oct 1968, and some later satellites such as 264 and 280, displayed limited manoeuvring capability to enable more precise coverage of the targets allotted to them.

Interceptor satellites, able to approach, inspect, and possibly destroy other satellites, began to appear in Oct 1968, with Cosmos 249 and 252. Western observers were able to study the new technique during Oct 20–30 1970, by observing Cosmos 373, 374 and 375. As can be seen from the following list, 373 was probably sent up to play the part of an 'enemy satellite'; about 3·9 m long, with 2·1 m dia, it was probably able to report back on 'miss' distances. Cosmos 374 went up 3 days later, and passed very close to 373 on its 2nd orbit, and then exploded into over 16 pieces. Cosmos 375 was launched on Oct 30, and also passed very close to 373 on its 2nd orbit, 230 min after launch; it then exploded into 30 pieces. By the end of 1971 12 interceptor satellites had been identified. By then Russia had succeeded in developing the ability to intercept and destroy satellites at relatively low level—much more difficult than at high level, because as the altitude decreases the target satellite moves faster in relation to a ground location. As can be seen later, Cosmos 462 exploded on Dec 3 1971, when approached by 459. During earlier interception tests target satellites had been destroyed at altitudes ranging between 579 and 885 km. On this occasion the target was destroyed below 257 km.

A possibly more aggressive use of space techniques began to be developed with 2 unnumbered launches (designation Cosmos U1 and U2) in Sep and Nov 1966. Later it was designated FOBS, as Western monitoring devices watched the tests with growing concern. The principle of this space bomb is that it can be fired into an orbit of 160 km, but is slowed down by retro-rockets so that it re-enters and causes its nuclear warhead to fall on the target before completion of the first orbit. This provides a capability of attacking Western targets via the 'back door'—i.e. by travelling three-quarters of the way round the world via the S Pole, instead of by the shorter, more obvious N Pole route, which is monitored by BMEWS. In 1967 there were 9 FOBS tests, with Cosmos 139, 160, 169, 170, 171, 178, 179, 183 and 187. In the following 4 yr there were 9 more tests, at less frequent intervals, with none in the first half of 1972.

'Scarp' rockets, 34·5 m long with 3 m dia, are used for FOB launches; the payload, which has been optically sighted from RAE, Farnborough, is estimated at 2 m long and 1·2 m dia. Undoubtedly some of the navigational satellites will be for use by nuclear submarines in targeting their ballistic missiles.

Russia is also launching astonishing numbers of military communications satellites to provide direct communications between ships, planes and bases. By the end of 1975 octuple launches—8 satellites orbited by one C1 (Skean + restart stage) rocket, were being sent up 4 times a year. The first experiment in this technique, Cosmos 336–43, was made in Apr 1970, and steadily developed. A minimum of 24 of these 100cm spheres, weighing 41 kg with a 2-hr orbit, is needed to provide global coverage; Soviet military chiefs apparently prefer 36–48 to ensure plenty of redundancy and to ensure that Western jamming of them all would be impossible. Each has an operational life of 2–3 yr; but since their orbital life is 7–10,000 yr, with a 20,000-yr life for the rocket stages, their orbits become increasingly cluttered with dead ironmongery—already more than 200 dead satellites and rocket stages. Collecting this debris will one day become an important task for a Soviet space shuttle.

Finally, 1975–76, as can be seen from this list below, marked a sudden rise in Soviet military activity in space, matching increased military production for surface forces. After a 3-yr gap, tests were resumed (Cosmos 803/4) with 'killer' satellites able to intercept and destroy enemy satellites; early-warning synchronous satellites were introduced; military navigation satellites were increased in density; and Salyut space stations were tested for strategic reconnaissance.

General Description   Satellites placed in 49–56° orbits are usually cylindrical, about 1·8 m long by 1·0 m dia, and 360 kg in wt. Military or reconnaissance satellites are also spheres, with an approx wt of 3175 kg. The series as a whole, however, is so varied that they range from small uninstrumented spheres to large vehicles like Cosmos 110, which carried 2 dogs with sufficient supplies to enable it to be recovered after remaining in orbit for 22 days. The orbits range from 145 km to 60,000 km; the lifetime of the satellites varies from less than 1 orbit in the case of the military FOBS tests, to a possible 50,000 yr. A complete, detailed list of Cosmos satellites would be too long for a book of this size, but typical examples, giving launch dates, orbits and inclinations, followed by a summary of the spacecraft's purpose and achievements, are listed below:

Cosmos 1   L Mar 16 1962, by B1 from Kapustin Yar. Wt ? 200 kg. Orbit 217 × 980 km. Incl 49°. At first classified as Sputnik 11; used radio methods to study structure of the ionosphere; dec. after 70 days.

Cosmos 2   L Apr 6 1962, by B1 from Kapustin Yar. Wt ? 400 kg. Orbit 212 × 1560 km. Incl 49°. At first classified as Sputnik 12. Returned data on radiation belts and cosmic rays and re-entered after 499 days.

Cosmos 3   L Apr 24 1962, by B1 from Kapustin Yar. Wt ? 400 kg. Orbit 228 × 719 km. Incl 49°. Returned radiation belt and cosmic ray data. Dec. after 176 days.

Cosmos 4   L Apr 26 1962 by A1 from Tyuratam. Wt ? 4000 kg. Orbit 298 × 330 km. Incl 65°. The first spacecraft to be recovered; it was about 5 m long, and 2 m in dia, and re-entered after 3 days. The first military satellite, since its task was to measure radiation before and after US nuclear tests.

Cosmos 7   L Jul 28 1962, by A1 from Tyuratam. Wt ? 4000 kg. Orbit 209 × 368 km. Incl 65°. Its job was to watch for solar flares during the manned Vostok 3 & 4 flights. Re-entered or dec. after 4 days.

Cosmos 97   L Nov 26 1965, by B1 from Kapustin Yar. Wt ? 400 kg. Orbit 220 × 2098 km. Incl 49°. First experiment in measuring mazers; tested a molecular quantum generator, which makes it possible to communicate with, and control other spacecraft, and to send information great distances. Also checked aspects of the theory of relativity. Dec. 492 days.

Cosmos 110   L Feb 22 1966, by A2 from Tyuratam. Wt ? 4000 kg. Orbit 186 × 904 km. Incl 52°. Biological satellite carrying dogs Veterok and Ugolyok, who were successfully recovered after 330 orbits in 22 days.

Cosmos 122   L Jun 25 1966, by A1/2 from Tyuratam. Wt ?. Orbit 625 × 625 km. Incl 65°. Meteorological satellite. Launch witnessed by Gen de Gaulle. Expected life 50 yr.

Cosmos 144, 156, 184, 206   L between Feb 1967 & Mar 1968 into circular 628km orbits at 81° incl with lifetime of 50–60 yr. Part of Meteor system. C 144 & 156 carry equipment for TV and infrared photography, providing pictures of cloud layers, and of snow and ice-fields for about 8% of Earth's surface: also measures radiation streams reflected and emitted by Earth and its atmosphere over about 20% of Earth's surface on each orbit. The area scanned by one is reviewed by the 2nd 6 hr later. C 206 is about 20 min behind 184, so that forecasters can check weather received from the first.

Cosmos 166 & 215   L Jun 16 1967 & Apr 19 1968 by B1 from Kapustin Yar. Wt ? 400 kg, into orbits between

Top left Cosmos 166, for solar radiation investigation, exhibited at Moscow Exhibition of Economic Achievement, 1967. Others, at the same exhibition, were described as 'A satellite of the Cosmos series'

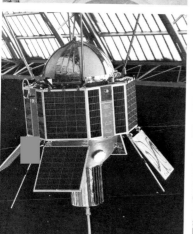

260 × 577 km. Incl 48°. Studied solar radiation. The 2nd had 8 mirror telescopes, an X-ray telescope and 2 photometers to observe the radiation of hot stars in various wavebands—a first step, according to Pravda, towards placing a big telescope beyond the confines of Earth's atmosphere. Their orbits decayed after 130 and 72 days respectively.

Cosmos 186 & 188 L Oct 27 & 30 1967 by A2 from Tyuratam. Wt ? 6000 kg. Orbits between 209 × 276 km. Incl 51°. Carried out world's first automatic docking, and Russia's first docking of any kind. C 186, the 'active' craft, automatically manoeuvred to rendezvous and dock with C 188. After 3½ hr they were commanded to undock and manoeuvred into different orbits. The mission, in Soyuz orbits, was clearly a rehearsal for manned flight. Both craft were recovered during the 4th day.

Cosmos 212 & 213 Cosmos 212 & 213 achieved a 2nd automatic docking on Apr 15 1968, 5 months before the joint Soyuz 2 & 3 flights, but the latter did not succeed in docking.

Cosmos 187 L Oct 28 1967 by F1r from Tyuratam; orbit 145 × 209 km. Incl 50°. Believed to be the first test of FOBS (Fractional Orbital Bombardment System), which enables a nuclear warhead to be sent round the world, avoiding existing missile early warning systems. The warhead re-enters and descends on its target without completing 1 orbit. C 187's weight was unknown; it was cylindrical, approx 2 m long × 1 m dia.

Cosmos 243 L Sep 23 1968 by A1/2 from Tyuratam. Wt ? 4000 kg. Orbit 209 × 319 km. Incl 71°. Regarded by Soviet scientists as landmark in the series; first satellite to study heat ray emissions from Earth and its atmosphere. It enabled

Cosmos 149, Earth-oriented meteorological satellite, was in orbit 17 days

an Antarctic ice map to be made, showing temperature distributions around the world. Registering heat radiation in this way enabled scientists to determine moisture content in the atmosphere, and to discover focal points of intensive precipitation concealed by thick clouds. Oceans were huge accumulators of solar energy, said Soviet scientists, which they emitted as evaporation heat; this heat 'fed' the cyclones which made the Earth's weather. Cross-sections of water surface temperatures in the Pacific from the Bering Sea to the Antarctic were 'mapped in tens of minutes'. C 243 was probably recovered after 11 days.

Cosmos 248, 249, 252 L Oct 19 Oct 20 & Nov 1 1968, by F1m from Tyuratam into orbits ranging from 547 × 2175 km. Incl 62°. Military reconnaissance satellites with ability to manoeuvre and lives of 10, 100 & 200 yr.

Cosmos 251, 264, 280 L Oct 31 1968, Jan 23 & Apr 23 1969 from Tyuratam. Wt ? 4000 kg. Orbits ranging from

198 × 402 km. Incl 65°, 70° & 51°. Military reconnaissance satellites, with manoeuvring capability for more precise target coverage. Cosmos 251 ejected a capsule about half its own weight after 12 days in orbit, and was recovered after 18 days. 264 and 280 ejected capsules after about 11 days and were themselves recovered on the 12th day. The capsules each remained in orbit for several days before decaying; their purpose is unclear though the recovered portions undoubtedly brought back film packages. There were 4 capsule ejections in 1969; 9 in 1970; and 24 in 1971.

Cosmos 261   L Dec 20 1968 by B1 from Plesetsk. Wt ? 400 kg. orbit 217 × 669 km. Incl 71°. This paved the way for the Intercosmos programme. Bulgaria, Czechoslovakia, GDR, Hungary, Poland and Romania collaborated in experiments exploring air density in the upper atmosphere, and the nature of the Polar auroras. Dec. Feb 12 1969.

Cosmos 262   L Dec 26 1968 by B1 from Kapustin Yar. Wt ? 400 kg. Orbit 262 × 965 km. Incl 48·5°. First satellite to study vacuum ultraviolet (VUV) and soft X-ray radiation (SX) from the stars, Sun and Earth's upper atmosphere. Carried 3 16-channel photometers. Results announced Oct 1969. Dec. after 4 months' operation, Jul 18 1969.

Cosmos 336–343   L Apr 25 1970 by C1 from Plesetsk. 40 kg. Orbits from 1313 × 1554 km. Incl 74°. First octuple, or 8-satellite launch, each believed to be spheroid, about 1 m long, 0·8 m dia.

Cosmos 359   L Aug 22 1970 by A2e from Tyuratam. Wt 1180 kg. Orbit 889 × 208 km. Incl 51°. Almost certainly intended to be Venus 8, but failed to achieve escape velocity. (Venus 7 was launched 5 days earlier.)

Cosmos 373   L Oct 20 1970, by F1m from Tyuratam. Wt ?. Orbit 543 × 472 km. Incl 62°. Target for satellite intercept system; probably about 4 m long, 2 m dia, containing devices to measure 'miss' distances. 10-yr life; see below.

Cosmos 374   L Oct 23 1970, by F1m from Tyuratam. Wt ?. Orbit 2142 × 521 km. Incl 62°. Satellite intercept test; passed close to C 373 on 2nd orbit and exploded into 16 pieces.

Cosmos 375   L Oct 30 1970, by F1m from Tyuratam. Wt ?. Orbit 2100 × 524 km. Incl 62°. Continued satellite intercept test; passed close to C 373 on 2nd orbit and exploded into 30 pieces.

Cosmos 381   L Dec 2 1970, by C1 from Plesetsk. Wt ?. Orbit 1013 × 971 km. Incl 74°. Navigational satellite; provided information on physical characteristics of ionosphere layers covering almost entire global surface. Expected life 1200 yr.

Cosmos 382   L Dec 2 1970, by D1 from Tyuratam. Orbit 5045 × 305 km. Incl 51°. Possibly test vehicle for future interplanetary probes; life approx 1000 yr.

Cosmos 394   L Feb 9 1971 by F1m from Plesetsk. Orbit 619 × 574 km. Incl 65·9°. Orbital intercept target for 397; life 40 yr.

Cosmos 397   L Feb 25 1971 by F1m from Tyuratam. Orbit 2317 × 593 km. Incl 65·8°. Orbital intercept test; passed close to 394. Life 150 yr.

Cosmos 400   L Mar 18 1971, by F1m from Plesetsk. Orbit 1006 × 903 km. Incl 65·8°. Orbital intercept target for 404. Life 1200 yr.

Cosmos 404   L Apr 3 1971, by F1r from Tyuratam. Orbit 1009 × 811 km. Incl 65·1°. Orbital intercept attacker; passed

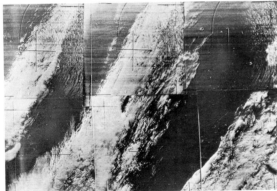

Above Cosmos 144, another meteorological satellite
Below Montage of its TV pictures, taken on Mar 4 1967
over the S Atlantic, showing SW Africa coastline and cloud

close to 400; did not explode and deorbited after about 5 orbits.

Cosmos 411–418   L May 7 1971, by C1 from Plesetsk. 2nd octuple launch; orbits 1539 × 1482 km approx. Wt 41 kg. Incl 74°.

Cosmos 419   L May 10 1971 by Proton from Tyuratam. Orbit 145 × 159 km. Incl 51°. Attempted Mars probe; failed to leave Earth orbit. Dec. 2 days.

Cosmos 433   L Aug 8 1971, by F1r from Tyuratam. Orbit 259 × 159 km. Incl 49°. FOBS test; possibly recovered just short of 1 orbit.

Cosmos 444–451   L Oct 13 1971, by C1 from Plesetsk. Wt 40 kg. Orbits 1574 × 1492 km approx. Incl 74°. 3rd octuple launch; said to be 'carrying radio systems for measuring elements of the orbit, and radio telemetric systems for relaying to Earth data about the functioning of instruments and scientific equipment'. Life 6–10,000 yr.

Cosmos 459   L Nov 29 1971, by F1m from Plesetsk. Orbit 277 × 266 km. Incl 65°. Orbital intercept test; passive target for 462; probably measured 'miss' distance. Life 28 days.

Cosmos 462   L Dec 3 1971, by F1m from Tyuratam. Orbit 1840 × 237 km. Incl 65°. Caught up with 459 on 2nd orbit near Plesetsk and exploded into more than 16 pieces.

Cosmos 480   L Mar 25 1972, by C1 from Plesetsk. Orbit 1174 × 1203 km. Wt ?. First Cosmos using new 83° incl. Believed to be cylinder with paddles 1·8 m long, 0·9 m dia. Life 3000 yr.

Cosmos 482   L Mar 31 1972 by A2e from Tyuratam. Wt ?. Orbit 204 × 9800 km. Incl 52°. Intended to be Venus 9, to accompany Venus 8, launched 4 days earlier; escape stage fired only partially; life 6 yr.

Cosmos 496 L Jun 26 1972, by A2 from Tyuratam. Wt ?. Orbit 195 × 343 km. Incl 51°. Probably test of equipment for manned spaceflight; possibly redesigned Salyut/Soyuz hatch. Life 10 days.

Cosmos 500 L Jul 10 1972 by C1 from Plesetsk. Wt ?. Orbit 505 × 549 km. Incl 74°. Believed to be cylinder with paddles, about 2 m long, 1 m dia. 'To continue the exploration of the upper layers of the atmosphere'. Life 7 yr.

Cosmos 504–511 L Jul 20 1972, by C1 from Plesetsk. Wt 40 kg approx. Orbits 1446 × 1497 km approx. Incl 74°. 4th octuple launch. Lifetimes 5–10,000 yr; rocket 20,000 yr.

Cosmos 520 L Sep 19 1972 by F1r from Tyuratam. Orbit 227 × 669 km. Incl 62°. There was US speculation that this was intended as a high-altitude intercept, due to meet 516 head-on over Atlantic. If so, approach was not close enough to be successful.

Cosmos 605 & 690 L Oct 31 1973 & Oct 22 1974 by A2 from Plesetsk. Wt 5500 kg, orbits approx 213 × 405 km. Incl of both 62°. Biological missions, using the original Vostok spacecraft. Cosmos 605 contained several dozen rats, 6 boxes of tortoises, a mushroom bed, 4 beetles, and living bacteriological spores. Recovered after 3 weeks, 605 provided data on the reaction of mammal, reptile, insect, fungal and bacterial forms to prolonged weightlessness. Cosmos 690 contained specially trained albino rats. A large body of biologists, physicians, physicists and engineers from Russia, Czechoslovakia and Romania took part in individually subjecting the rats, on ground commands, to daily radiation doses from a gamma-ray source carried on board. When they

Soviet artist's rendering of Cosmos 186 and 188 about to achieve world's first automatic docking

were recovered, 20½ days later, many rats had developed lung trouble (subpleural petechiae), and their blood and bone marrow showed that radiation in space had a much greater effect than in ground conditions. Presumably these experiments, which would be likely to arouse some controversy in non-Communist countries, were part of Russia's long-term planning for sending men to the planets. (*See* Cosmos 782.)

Cosmos 637 L Mar 26 1974 by D1e from Tyuratam. Wt ?. Orbit 35,407 × 35,760 km. Incl 25°. Russia's first ever synchronous satellite, and a rehearsal for the new Molniya 1S series. Orbital life 1M yr.

1975 COSMOS LAUNCHES

To provide a complete picture of the current Cosmos pattern, a list of the 84 satellites launched in 1975 is given overleaf. It was only one less than the 1973 record. Only 12 appear to have no military application. Launch sites are abbreviated to 'T' for Tyuratam and 'P' for Plesetsk.

Cosmos	Date	Site	Orbit	Incl	Details
702	Jan 17	T	204–314	71	Like 701, a 6tonne military recce satellite, recovered after 12 days, from rare Polar Orbit.
703	Jan 21	P	198–1517	82	400 kg science/?military sat. 1-yr life.
704	Jan 23	P	170–304	73	Follow-up to 701/2. Recovered in 12 days.
705	Jan 28	P	272–502	71	400kg science/?military. 9-month life.
706	Jan 30	P	623–39,774	63	$1\frac{1}{4}$tonne military communications or Early Warning, similar Molniya. 10-yr life.
707	Feb 5	P	504–547	74	Navigation or electronic eavesdropping. 10-yr life.
708	Feb 12	P	1369–1414	69	$\frac{1}{2}$tonne military applications. Unusual incl. 6000-yr life.
709	Feb 12	P	182–310	63	$6\frac{1}{2}$tonne manoeuvrable recce. Recovered in 13 days.
710	Feb 26	T	175–335	65	6tonne recce. Recovered in 12 days.
711–718	Feb 28	P	1482–1538	74	Military communication satellites. Spheres, each 40kg. Life 7–10,000 yr.
719	Mar 12	T	175–307	65	$6\frac{1}{2}$tonne manoeuvrable recce. Recovered in 13 days.
720	Mar 21	P	212–273	63	6tonne recce. Recovered in 12 days.
721	Mar 26	P	208–228	81	6tonne recce. Recovered in 12 days.
722	Mar 27	T	172–336	71	$6\frac{1}{2}$tonne manoeuvrable recce. Recovered in 13 days.
723	Apr 2	T	249–265	65	Pair of Ocean survey sats in low orbit; after 6-week operation, moved to
724	Apr 7	T	248–265	65	900km, 600-yr orbit to allow for radioactive decay.
725	Apr 8	P	270–481	71	400kg science/?military. 9-month life.
726	Apr 11	P	956–996	83	?Military applications. 1200-yr life.
727	Apr 16	T	172–333	65	6tonne recce. Recovered after 12 days.
728	Apr 18	P	204–323	73	6tonne recce. Recovered after 11 days.
729	Apr 22	P	980–1010	83	Military applications, probably similar to 726.
730	Apr 24	P	169–269	81	6tonne reece. Recovered in 12 days.

Cosmos	Date	Site	Orbit	Incl	Details
731	May 21	T	203–296	65	6tonne recce. Recovered in 12 days.
732–739	May 28	P	1471–1556	74	Another 8-satellite launch replenishing military comsats. Life 7–10,000 yr.
740	May 28	T	172–327	65	6tonne recce. Recovered in 13 days.
741	30	P	211–323	81	6tonne recce. Recovered in 12 days.
742	Jun 3	P	148–328	63	$6\frac{1}{2}$tonne very low manoeuvrable recce. Recovered in 13 days.
743	Jun 12	P	169–298	63	$6\frac{1}{2}$tonne manoeuvrable recce. Recovered in 13 days.
744	Jun 20	P	602–636	81	Military applications in orbit similar to Meteor. 60-yr life.
745	Jun 24	P	264–513	71	400kg science/?military. 6-month life.
746	Jun 23	P	180–325	63	6tonne recce. Recovered in 13 days.
747	Jun 27	P	193–291	63	6tonne recce. Recovered in 12 days.
748	Jul 3	P	178–317	63	$6\frac{1}{2}$tonne manoeuvrable recce. Recovered in 13 days.
749	Jul 4	P	509–550	74	Probably electronic eavesdropper. Life 10 yr.
750	Jul 17	P	272–803	63	400kg science/?military. 2-yr life.
751	Jul 23	P	197–313	63	6tonne recce sat. Recovered in 12 days.
752	Jul 24	P	481–515	66	Probably military research or applications in unusual orbit. Life 10 yr.
753	Jul 31	P	181–330	63	$6\frac{1}{2}$tonne manoeuvrable recce. Recovered in 13 days.
754	Aug 13	T	204–326	71	$6\frac{1}{2}$tonne manoeuvrable recce. Recovered in 13 days.
755	Aug 14	P	974–1013	83	Navigation Satellite with re-start stage. Life 1200 yr.
756	Aug 22	P	622–634	81	Possibly electronic eavesdropper in Meteor-type orbit. Life 60 yr.
757	Aug 27	P	182–316	63	$6\frac{1}{2}$tonne manoeuvrable recce. Recovered in 13 days.
758	Sep 5	P	174–326	67	Recce satellite at unusual incl. exploded into 76 pieces after 1 day — believed intentionally to prevent hardware falling into US hands.
759	Sep 12	P	231–276	63	6tonne recce. Recovered in 12 days.
760	Sep 16	T	174–335	65	$6\frac{1}{2}$tonne manoeuvrable recce. Recovered in 14 days.

Cosmos	Date	Site	Orbit	Incl	Details
761–768	Sep 17	P	1402–1484	74	8-satellite launch replenishing military comsats. Life 7–20,000 yr.
769	Sep 23	P	203–307	73	6tonne recce. Recovered in 12 days.
770	Sep 24	P	1169–1210	83	Probably military applications. Life 3000 yr.
771	Sep 25	P	203–219	81	6½tonne manoeuvrable recce. Believed one of regular spring/autumn ice survey missions, to navigate ships through ice-fields.
772	Sep 29	T	195–299	52	Apparently unmanned Soyuz test. Recovered 3 days.
773	Sep 30	P	791–808	74	Possibly electronic eavesdropper or military communications. Life 120 yr.
774	Oct 1	T	204–315	71	6½tonne manoeuvrable recce. Recovered in 14 days.
775	Oct 8	T	35,900–35,900	0·1	Russia's 1st early warning sat, apparently to monitor US missile subs in Atlantic. Life 1 M yr.
776	Oct 17	P	200–288	63	6tonne recce. Recovered in 12 days.
777	Oct 29	T	123–412	65	Possibly in orbit-test of ion engines as Cosmos 699. Apparently deliberately destroyed after 93 days.
778	Nov 4	P	978–1004	83	Navigation satellite. Start of new policy placing them 30°, not 60° apart. 1200-yr life.
779	Nov 4	P	182–341	63	6½tonne manoeuvrable recce. Recovered in 14 days.
780	Nov 21	T	201–278	65	6tonne recce. Recovered in 12 days.
781	Nov 21	P	505–551	74	Probably electronic eavesdropper. Life 10 yr.
782	Nov 25	P	218–384	63	Biological flight, also carrying US livestock. Recovered early on 20th day due to snowstorms.
783	Nov 28	P	795–815	74	Possibly electronic eavesdropper or military comsat. Life 120 yr.
784	Dec 3	P	215–232	81	6tonne recce. Recovered in 12 days.
785	Dec 12	T	251–261		Ocean survey, probably failed prematurely in low orbit. Moved Dec 13 into 900km, 600-yr orbit to allow for radio-active decay.
786	Dec 16	T	174–326	65	6½tonne manoeuvrable recce. Recovered in 13 days.

After a 5-yr gap, development of the ability to intercept and destroy satellites was resumed, with 3 target satellites and 4 interceptors in 1976. Cosmos 909/910 in May 1977 were further tests. C 905, L Apr 26 1977 from Plesetsk into a rare 67° incl, was apparently a 2nd-generation spy satellite, with 1-month photo lifetime before being exploded; Russia's 1000th launch compared with 680 US launches at that date.

ELEKTRON   Dual Launch Satellites

Elektron 1 and 2   L Jan 30 1964 by A1 from Tyuratam. Wt 350 and 445 kg. Orbits 394 × 7126 km, and 441 × 67,988 km. Incl 61°. Russia's first dual launch, the satellites being separated from the launch vehicle while the last stage was still firing. They studied the inner and outer zones of the Van Allen radiation belt and Earth's magnetic field, gathering information for radiation protection of manned spacecraft. Orbital life 200 and 30 yr.

Elektron 3 and 4   L Jul 11 1964 by A1/2 from Tyuratam. Similar wt, orbits and missions. Orbital life 200 and 23 yr.

LUNA   Unmanned Moon Explorers

History   Nearly 20 yr of lunar exploration is covered in this entry. The most notable triumphs achieved were by Lunas 16 and 20, which brought samples of lunar soil back to Earth; and by Lunas 17 and 21, which landed the Lunokhod robots on the surface. Their surface explorations covered a total of about 48 km. But, especially in recent

Elektron-2 in Soviet assembly shop. Gaps between solar cells are temperature control louvres; *top left* magnetometer probe folded down

263

years, it is a programme which has encountered many setbacks. Luna 23 was the 3rd failure in 5 attempts during the 1969–75 period to bring back soil samples.

Soviet scientists must have been planning lunar flights and exploration several years before their first satellite, Sputnik 1, was orbited in Oct 1957. Luna 1, 15 months later, was only Russia's 4th space launch. Details given below, and also under Lunokhod, show the Russian technique of tackling and solving each problem with a group of spacecraft, and then pausing for a year or two of research and development before going on to the next phase. Thus the 'flyby' technique was mastered in 3 flights in 1959, culminating in the transmission by Luna 3 of the first historic pictures of the Moon's farside. The next phase, preceded by a failure (Sputnik 25, on Jan 4 1963) began the development of soft-landing techniques; though none appears to have been completely successful, much was learned from Lunas 4–8. Cosmos 60, on Mar 12 1965, and Cosmos 111, Mar 1 1966, were almost certainly 2 of only 5 complete failures in this long programme. The group of 5 Lunas launched in 1966 marked the biggest leap forward, achieving both soft-landing and orbiting techniques. The following entries reveal an interesting example of the Soviet policy of strict secrecy about long-term planning: it was only during Lunokhod 1's first activities on the lunar surface that we were told that Luna 12, 4 years earlier, had space-tested the electric motor for the robot's wheels. There were follow-up trials of the gears on Luna 14. There seemed to be a touch of desperation about Luna 15, which in Jul 1969 crashed on the lunar surface during an attempt to make an automatic recovery of lunar samples and get them back to Earth a few hours before the Apollo 11 crew. This was the first time that Russia had tried a soft-landing from parking orbit; previous attempts had always been by direct flight. The more sophisticated parking-orbit technique, which gives time for orbital corrections and makes pin-point touchdown possible, has always been preferred by the Americans, and was finally employed by Russia from Luna 16 onwards. (Luna 16 was apparently preceded by 2 launch failures, Cosmos 300 and 305, in Sep and Oct 1969.) When Lunokhod exploration began, Academician Boris Petrov claimed that robots could carry out such missions for less than one-tenth the cost of a similar manned flight. Such a claim is difficult to sustain against the background of many Luna failures. And certainly the fact that all 5 attempts at robot recovery of soil samples were made in the Sea of Crises, a relatively easy area to reach, and that 3 of those were failures, showed some lack of confidence and technique. However, Luna 22, which was successfully manoeuvred in a whole variety of lunar orbits over a period of 15 months, provided the data and experience for Lunokhod and soil recovery operations in the near future on the Moon's polar regions and farside under the command of at least one orbiting satellite.

Luna 1 L Jan 2 1959, by A1 from Tyuratam. Wt 361 kg. Intended to impact on the Moon, this was the world's first spacecraft to reach 'second cosmic velocity', or 40,234 kph. Launch believed to be by 3-stage vehicle with total thrust of 263,000 kg. The spherical craft, equipped with instruments for measuring radiation, etc, and its separated 3rd stage, both passed within 5955 km of the Moon, and went into solar orbit. Also named 'Mechta', it was only Russia's 4th space launch.

Luna 2 L Sep 12 1959, by A2 from Tyuratam. Wt 390 kg. The first spacecraft to reach another celestial body; impacted E of Sea of Serenity, carrying USSR pennants.

Luna-2, first spacecraft to reach the Moon

Luna 3 L Oct 4 1959 by A1 from Tyuratam. Wt 278·5 kg. First spacecraft to pass behind the Moon and send back pictures of farside; placed on an elliptical Earth orbit with apogee of 480,000 km, so that without midcourse corrections, lunar gravity pulled it around the Moon at a distance of about 6200 km. Equipped with a TV, processing and transmission system, it took an unannounced number of farside pictures which were sent back as Lunar 3 moved back towards Earth; 3 were published, including a composite full view of the farside, and 2 large seas were named Mare Moscovrae (Moscow Sea) and Mare Desiderii (Dream Sea). Dec. after 11 orbits totalling 177 days.

Luna 4 L Apr 2 1963, by A2e from Tyuratam. Wt 1422 kg. First of 5 spacecraft aimed at solving problems of soft-landing instrument containers. Contact lost as it missed Moon by 8529 km, leaving it in 89,782 × 692,300 km Earth orbit.

Luna 5 L May 9 1965, by A2e from Tyuratam. Wt 1476 kg. First soft-landing attempt. Flight lasted 82 hr, and at 60 hr Soviet scientists for first time made advance announcement of their plans. Retro-rocket, due to be fired about 64 km, from surface, malfunctioned, and spacecraft impacted in Sea of Clouds 5 min earlier than planned.

Luna 6 L Jun 8 1965, by A2e from Tyuratam. Wt 1442 kg. During midcourse correction, manoeuvre engine failed to switch off. Spacecraft missed the Moon by nearly 161,000 km, and passed into solar orbit.

Luna 7 L Oct 4 1965, by A2e from Tyuratam. Wt 1506 kg. After 86-hr flight, retro-rockets fired early; crashed in Ocean of Storms.

Luna 8 L Dec 3 1965, by A2e from Tyuratam. Wt 1552 kg.

Luna-3, seen before launch, gave man his first view of the Moon's farside

After 83-hr flight, retro-rockets fired late; crashed in Ocean of Storms.

Luna 9   L Jan 31 1966, by A2e from Tyuratam. Wt 1583 kg. First successful soft-landing, followed by first TV transmission from surface. After 79-hr flight, the spacecraft ejected an egg-shaped 0·6m dia instrument capsule, weighing 100 kg, as a probe touched the surface and switched off the retro-rocket. The capsule, weighted so that it rolled into an upright position, was stabilized on Earth command by 4 spring-ejected 'Petals' to serve as legs. 3 panoramas of the lunar landscape, with different Sun angles, on the eastern edge of the Ocean of Storms, were transmitted over a 3-day period.

Luna 10   L Mar 31 1966 by A2e from Tyuratam. Wt 1600 kg. First lunar satellite; wt was 245 kg when it was fired into 349×1017km lunar orbit, with $71°$ incl. In addition to broadcasting the 'Internationale' several times, it studied lunar surface radiation, magnetic field intensity, etc. Communications were maintained for 2 months and 460 orbits, providing opportunities for tracking the strength and variation of lunar gravitation.

Luna 11   L Aug 24 1966, by A2e from Tyuratam. Wt 1640 kg. 2nd, heavier lunar satellite. Placed in 159×1200km lunar orbit with $27°$ incl. Possibly carried TV system which failed to operate. Data was received during 277 orbits until Oct 1 1966.

Luna 12   L Oct 22 1966 by A2e from Tyuratam. Wt 1640 kg. 3rd lunar satellite; placed in 100×1739km orbit. A TV system transmitted large-scale pictures of the Sea of Rains surface and Crater Aristarchus areas, showing craters as small as 15 m across. Tested electric motor for Lunokhod's

Luna-9 mockup in Moscow Space Pavilion. Black ball at top is ejected at touchdown, and rolls clear; 4 petal-like panels open to release antenna and expose TV camera

wheels. Communications continued for 602 orbits until Jan 19 1967.

Luna 13 L Dec 21 1966, by A2e from Tyuratam. Wt ?1583 kg. 2nd successful soft-landing. Capsule was again bounced on to the Ocean of Storms. In addition to sending back panoramic views, 2 Meccano-like arms, 1·5 m long, were extended, and measured soil density and surface radioactivity; communications lasted for 6 days from landing.

Luna 14 L Apr 7 1968, by A2e from Tyuratam. Wt not known. 4th Soviet lunar satellite, placed in 160×870km orbit, with 42° incl. Studied Moon's gravitational field, and 'stability of radio signals sent to spacecraft at different locations in respect to the Moon'; made further tests of geared electric motor for Lunokhod's wheels.

Luna 15 L Jul 13 1969, by D1e from Tyuratam. Wt ? 1814 kg. A bold but unsuccessful attempt to obtain lunar samples and return them to Earth a few hours before America's first men on the Moon (Apollo 11) could do so. Launched $3\frac{1}{2}$ days before Apollo 11, Luna 15 was placed in lunar orbit, and remained there during the Apollo flight and first manned landing. 2 orbital changes were made, however, and American scientists, concerned about possible conflict of radio frequencies, asked Frank Borman, who had recently visited Moscow, to seek assurances from Russia that there would be no radio conflict. Such assurances were quickly given. 2 hr before Apollo 11 was due to lift-off from the Sea of Tranquillity, Luna 15, on its 52nd revolution, began descent manoeuvres, apparently aimed at the Sea of Crises. But Russia's first attempt at an automatic landing from lunar orbit went wrong, and crashed at the end of a 4-min descent. It was more than a year later before the Luna 16 flight revealed what had been intended.

Luna 9: the first picture of the Moon's surface

Luna 16 L Sep 12 1970, by D1e from Tyuratam. Wt ? 1814 kg. First recovery of lunar soil by automatic spacecraft. Luna 16 was placed in a 110km orbit with 70° incl, corrected prior to landing to 110 × 15 km with 71° incl. On Sep 20 Russia's Long Range Space Communications Centre gave the command to fire the descent engine. At 20 m above the surface this was switched off, and final touchdown on the Sea of Fertility, 1·5 km from target point, was controlled by 2 vernier engines. Luna 16 consisted of landing and ascent stages; as in the case of America's Lunar Module, landing or descent stage served as a launch platform for the ascent stage. On Earth command, an automatic drilling rig was deployed; with a 0·9m reach, it was capable of penetrating

the surface just over 30 cm. About 100 gr was lifted by the drill into the loading hatch of a spherical capsule at the top of the ascent stage, which was then hermetically sealed. After $26\frac{1}{2}$ hr on the surface, the ascent stage was launched on the ballistic trajectory back to Earth; it consisted of the lunar soil container, an instrument compartment, and a parachute compartment, containing braking and main parachutes and 2 gas-filled balloons, presumably in case of descent on water. No midcourse corrections were made on the return flight; 3 hr before re-entry, the instrument compartment was jettisoned. The sample capsule's transmitters enabled it to be located and recovered in Kazakhstan on Sep 24.

Luna 17 L Nov 10 1970 by D1e from Tyuratam. Wt ? 1814 kg. Carrying the first Moon robot, Lunokhod 1, was placed in an initial circular 84km orbit with 141° incl; the following day, the perilune was lowered to 19 km. On Nov 17 a successful soft-landing was made in a shallow crater in the north-western Sea of Rains (Mare Imbrium). After checks by TV cameras that there were no boulder obstructions, one of 2 alternative ramps was lowered by commands from Russia's Deep Space Communications Centre, and Lunokhod 1 rolled down on to the surface. It proved to be the Soviet Union's greatest technical space success. Full details are given under Lunokhod.

Luna 18 L Sep 2 1971, by D1e from Tyuratam. Wt ? 1814 kg. Intended as a soft-lander, most probably with an ascent stage for a 2nd soil recovery, this was placed in a 100km circular lunar orbit with 35° incl. After 54 orbits taking $4\frac{1}{2}$ days, (twice as long as Luna 16 & 17), an attempt was made to land in highland terrain in the Sea of Fertility. Communications ceased shortly after Earth command had started the descent engine, probably as a result of impact. Novosti reported that the landing 'in difficult topographic conditions had been unlucky'; the first time, within the author's knowledge, that Russia had publicly admitted a space failure.

Luna 19   L Sep 28 1971, by D1e from Tyuratam. Wt ? 1814 kg. 5th Soviet lunar satellite, initially placed in circular 140km orbit with 40° incl. Communications continued for over a year; on Oct 3 1972 it was stated the experiment was nearing completion. By then Luna 19 had made more than 4000 lunar orbits, and over 1000 communication sessions had been held. A systematic study was made of the Moon's gravitational field and the effect of its mascons, coupled with TV pictures of the surface. At least 10 powerful solar flares were observed. Lunar radiation was compared with similar measurements made by Mars 2 & 3 in Martian orbit and by Venus 7 & 8, and Prognoz 1 & 2.

Luna 20   L Feb 14 1972 by D1e from Tyuratam. Wt ? 1814 kg. 2nd successful soil recovery. Placed in initial circular lunar orbit of 100 km at 65° incl. This was lowered 1 day later to 100 × 21 km. After site selection photography, the descent engine was fired for 4 min 27 sec, and 7½ days after launch, a safe landing was made on a mountainous isthmus pockmarked with large craters, S of the Sea of Crises, and on the Sea of Fertility's extreme NE; it was 120 km N of Luna 16's sampling point. A 'photo-telemetric device' relayed to Earth pictures of the surface, and from these a site with 'a grey cloudy structure' was chosen from which to take samples. A rotary-percussion drill, of improved design as a result of Luna 16, drilled into the rock; it sank quickly to a depth of 100–150 mm; then because of the rock's hardness, drilling had to be done in stages, with intervals, so that the drill did not overheat. The samples were then lifted into the

Luna 13 depicted on Moon, stabilized by petal-like panels. TV cameras at top; *left* extending arm examines soil; *right* arm measures radiation

return capsule on Luna 20's ascent stage, which was fired back to Earth on a ballistic trajectory 1 day after the touchdown. It was recovered with some difficulty on an island in the River Karakingir, 40 km NW of Dzhezkazgan, in Kazakhstan, after landing in a blizzard on Feb 25. Newsmen were present 2 days later, when the contents of the hollow drill brought back from the Moon were poured out. The ash was light-grey compared with the dark, 'black slate colour with a metallic glitter', of the Luna 16 samples. Scientists expected these samples to be about 1000M yr older than the Luna 16 samples, which were estimated at 3 to 5000M yr old.

Luna 21   L Jan 8 1973 by D1e from Tyuratam. Wt ? over 1814 kg. Carried Lunokhod 2 to the Moon; during trans-

Luna 13 view of Ocean of Storms

lunar flight when false telemetry signals almost aborted the mission, there was also, apparently, a problem with the Lunokhod's power supply, and the rover's solar batteries were exposed to sunlight during most of the translunar coast. Descent was made on the 41st orbit, from a height of 16 km, and Lunokhod 2, wt 840 kg (84 kg heavier than Lunokhod 1), was placed on the eastern edge of the Sea of Serenity. Touchdown was only 180 km from the Apollo 17 landing point. (*See* Lunokhod 2.)

Luna 22   L May 29 1974 by D1e from Tyuratam. Placed in 220km circular lunar orbit, at $19 \cdot 35°$ incl, with 2 hr 10 min revolution period, on Jun 2 1974. Its object was said to be to continue the lunar orbit observations begun by Luna 19; for 4 days from Jun 9, the orbit was lowered to 244×25 km, to obtain TV panoramas of high quality and good resolution; simultaneously, altimeter readings were taken of the relief of the area, and chemical rock composition determined by gamma radiation. Luna 22 was then returned to a 299 × 181km orbit to continue studying the gravitational field. It was still operational, and had been used for over 1000 communications sessions, when 5 months later it was joined in orbit by Luna 23; but there is no evidence that Luna 22 was used experimentally, or in any other way, to help control the Luna 23 operations on the surface. Completion of its programme, after 15 months of operation and over 4000 lunar revolutions, was announced in Oct 1975. Final manoeuvres included lowering the orbit to within 30 km of the surface for TV pictures; it was finally abandoned in lunar orbit of 100×1286 km with $21°$ incl to the lunar equator. During the flight, 30,000 radio commands were used to control the spacecraft. The official report said that Luna 22 did 'more than planned in its extensive programme', and 'many photographs' of lunar landscapes had been returned, but did not specify the number or quality.

Luna 23   L Oct 28 1974 by D1e from Tyuratam. Wt 1814 kg. Intended 'to continue scientific exploration of the Moon and nearby', it was placed in a lunar orbit of 104×94 km with $130°$ incl on Nov 2. By Nov 5 the orbit had been reduced to 105×17 km, and a landing was attempted the following day on the southern part of the Sea of Crises. Moscow radio reported that touchdown was in a sector with unfavourable relief, and the device for taking lunar rock samples to a depth of 2·5 m was damaged. No drilling or soil collection was possible; a revised and reduced programme had to be abandoned after 3 days. It was the 5th attempt at soil recovery with only 2 successes. It was also noteworthy that all these attempts had been in the same area. This suggested that scientists felt their techniques were not yet well enough developed to attempt sending their robots to remote areas, such as the lunar Poles, which had been beyond the reach of America's manned Apollo landings.

Luna 24 L Aug 9 1976 by D1e from Tyuratam. Wt ? 4000 kg. Russia's 3rd successful sample recovery operation; with a much bigger soil carrier, 2 m in length, a much larger sample was obtained than the 100 g and 150 g now believed by US scientists to have been obtained by Lunas 10 & 16. The initial circular lunar orbit of 115 km with $120°$, achieved on Aug 14, was lowered to 120×12 km on Aug 16–17; touchdown, after about 53 lunar orbits, was in the SE of the Sea of Crises ($12·75°$N $62·2°$E). Luna 23 had attempted to land in the same area. The site was 17 km from a large crater 10 km in dia and 2 km deep, and the hope was to obtain ejecta samples. Core samples were drilled to a depth of 2 m, and the return rocket was launched from the lunar surface 23 hr after touchdown. After recovery, 200 km SE of Surgut, W Siberia, the sample was taken to the Moscow Institute of Geochemistry and Analytical Chemistry, where it was said to be 'silvery in colour with a brown tint'. The material was

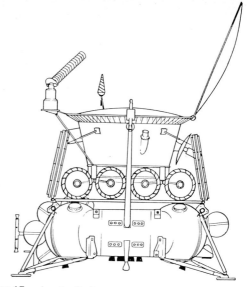

Luna 17, using the 2nd generation lunar orbiter and lander, with Lunokhod 1 roving vehicle mounted on it before deployment of the folded landing ramps (*Copyright drawing by D. R. Woods*)

lighter with small grains up to 8 mm across, and included more large grains than the Luna 16 sample. US scientists were once again given samples totalling 3 g probably because US techniques for high quality age-dating and chemical analysis are superior to Soviet techniques. Divided into 6 soil samples and one rock fragment, they were allotted to 30 investigators.

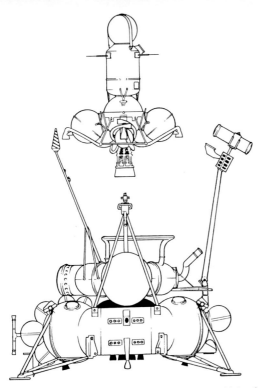

Luna 16, illustrating the separated Earth-return vehicle after transfer of Luna material from the drill unit into the recoverable capsule (*Copyright drawing D. R. Woods*)

LUNOKHOD   Soviet 'Moonwalker'

Weight:	756 kg
Wheelbase:	160 cm
Diameter of wheel:	51 cm
Lid:	215 cm
Overall length of wheels:	222 cm

General Description   Lunokhod consists of a circular instrument compartment, 2·1 m dia, mounted on an 8-wheel chassis. The 8-spoked, wire-mesh wheels are in 4 pairs; each has an independent electric motor; if any wheel seizes up, or gets stuck, a powder charge can be fired to break the drive shaft, enabling the wheel to become a passive roller; movement should still be possible with only 2 wheels on each side operational. Sensors provide automatic braking, overriding Earth commands, if slope angles become so steep that there is a danger of overturning. Disc brakes hold the craft at rest on slopes and during lunar nights. There are 2 forward speeds, and possibly 2 in reverse.

The instrument compartment, made of magnesium alloy, is designed so that it warms up easily and releases heat slowly, thus enabling it to survive lunar nights which are equivalent to 14 Earth-nights. When not in use, the lid is kept tightly shut. During the lunar nights the instruments, cameras, etc are kept warm and operational despite outside temperatures of −150°C by the circulation of heated gas. The instrument compartment contains radio transmitters and receivers, remote control, electric power and heat control systems, electronic-transformer units, and 2 TV systems. One of these, transmitting a frame every 3 to 20 sec, enables the operators to monitor its progress. The 2nd TV

Lunokhod 1 at 1973 Paris Air Show; 'Moon's eye view' of panoramic TV camera at rear

system, to obtain panoramic pictures of the locality, the horizon, Sun and Earth, consists of 4 identical telephoto cameras; 3 look to the sides and rear; the fourth, mounted alongside the low-rate camera, looks through forward portholes. The fore and aft cameras have a combined range of 30° horizontally and 360° vertically, the side cameras 180° horizontally and 30° vertically.

Lunokhod 1 The first 'Moonwalker' was landed on the Sea of Rains (Mare Imbrium) by Luna 17 on Nov 17 1970. Its expected life was 90 days; but it continued to operate for 11 months. During that time, it travelled 10,540 m photographing an area of more than 80,000 sq m. Over 200 panoramic pictures and 20,000 separate photographs were returned; physical and mechanical soil analyses were carried out at 500 locations and chemical analyses at 25. When at last Lunokhod's equipment froze during the 11th lunar night following exhaustion of its isotopic fuel, the vehicle had been placed so that its French laser-reflector could continue to be used indefinitely for Earth–Moon measurements.

Academician Blagonravov said its performance had surpassed all expectations, and that Lunokhod could, in principle, make a 'round-the-Moon' journey—though this would presumably involve the use of a lunar satellite for communications and control on the farside. Soviet scientists pointed out that Lunokhod was specifically designed for the Moon; planetary robots would necessitate different designs, involving walking, crawling or hopping, according to conditions. In the case of Mars, with its atmosphere saturated with carbon dioxide, frictional surfaces would probably be unsuitable, because they would wear off much too fast; and the great distances from Earth would mean revision of the remote-control methods employed for Lunokhod, which provided great difficulties themselves. Radio signals from Earth to Moon take 1·3 sec; from Earth to Mars they take 20 min. A 'Marsokhod', therefore, would have to be self-controlled by means of an onboard computer.

Operating Technique Lunokhod 1 was driven by a 5-man team—commander, driver, navigator, systems engineer and radio operator—working in the Deep Space Communications Centre, believed to be near Moscow. The need for co-ordinating their efforts in the early days was said to put great psychological and physical strain upon them; the technique was reported to be so difficult that it beat many highly experienced drivers and pilots. This was mainly because of the time-delays resulting from sending commands, and waiting to observe the response, over a distance of 386,240 km. They had to remember that the robot was already several metres ahead of the slow-scan TV picture they were watching; and that if there were rocks or boulders to be avoided, it would have moved on still further before the signals to take avoiding action would reach it. However, the robot's automatic 'stop' system worked so

Lunokhod 2: Cone, *left*, is omni-directional antenna; *top right*, narrow beam directional antenna. The TV cameras returned 80,000 pictures

well that throughout its 11-month life it never did overturn. The first task at the beginning of each lunar day was to open the lid to enable the solar cells on the lid's inside to start recharging the depleted batteries. When being parked at the end of the previous lunar day, it had been carefully turned towards the E, to obtain the maximum intensity from the Sun's rays as soon as the lid was opened. Soon afterwards it became necessary to keep the upper part of the hull cool; a mirror-like surface reflected into space as much heat in an hour as a man engaged in heavy manual labour generated. As soon as movement began, telemetry poured in separately from each of the 8 wheels, so that action could be taken immediately if any one of them showed signs of over-heating or freezing. The revolutions of each wheel were also counted separately to check whether it was skidding or not. Initially they were the main source of anxiety; it was thought they would build up charges of static electricity because of the vacuum conditions on the lunar surface; without air it was feared that it would be impossible to run off the electricity, with the result that particles of soil would cling to the wheels, and ultimately clog them and make the robot inoperable. The fear was unjustified; in practice the soil fell off like volcanic sand—a factor which should make it possible to simplify future Lunokhods. Orientation proved another problem at first; on the 3rd lunar day, scientists became excited by a big rock, estimated to be 150 m away. They asked the crew to move up close to the boulder; but when they did so it was nowhere to be seen. It turned out to be only a small stone, which on the screen appeared to be a huge rock. Attempts to steer Lunokhod closer to it, resulted in its being passed and left further behind. It was soon found that TV pictures of the wheel ruts provided a useful distance scale for navigators and drivers, as well as information for scientists about the physical and mechanical properties of the lunar soil. By the end of the 2nd lunar day, the robot was climbing slopes with gradients of $23°$, weaving, turning and zig-zagging to avoid steep craters and large boulders, and being sent with confidence through shallow craters 150 m across. Probably its most successful instrument was RIFMA (Roentgen Isotopic Fluorescent Method of Analysis), able to analyse the chemical composition of lunar rocks by sending out a stream of electrons, or X-rays. Under the action of the X-rays, the atoms of the rocks rearrange their electronic shells, and send back an X-radiation of their own. Measurements of this response show how much aluminium, silicon, magnesium, potassium, calcium, iron and titanium is con-

274

Lunokhod 2 detail shows wheel mechanism with, *right*, separate wheel for measuring distance.

tained in the rocks. On Dec 12 1970 RIFMA reported a solar flare, which, according to Pravda, would have created hazardous radioactivity for men on the Moon.

Lunokhod 2 Landed on the eastern edge of the Sea of Serenity in the Le Monnier crater on Jan 16 1973, Lunokhod 2 appears to have ceased operating abruptly early in the 5th lunar day, which began on May 8. But though it operated for only half as long as its predecessor, it was much more active; its 9th, pedometer, wheel recorded that it had covered 37 km, $3\frac{1}{2}$ times the distance; 86 panoramic pictures and 80,000

TV pictures of the lunar surface were transmitted. The Le Monnier landing site was chosen to provide information about a transitional 'sea/continental' area, and because there is a 16km tectonic fault nearby. Improved equipment and additional instruments added nearly 100 kg to Lunokhod 2's weight. Control and communication systems had been improved, and one of the panoramic TV cameras, sited in the vehicle's centre, had been raised to improve picture quality and reduce time taken to relay signals back to Earth, thus in turn giving the crew better control.

Equipment carried included a 'corner reflector' provided by French engineers, to continue the Franco-Soviet collaboration programme with laser-reflectors begun with Luna 17. Measurements made from Earth and circumlunar orbit had already established that the visible side of the Moon averages 2 km lower than the middle-radius, and that the farside is $9\frac{1}{2}$ km higher. Additional measurements by means of the French corner reflector were aimed at filling in lunar contours with more precision, checking the theory that its continents, like those on Earth, are subject to 'drift' etc. French laser-ranging staff were present at Mt Lock on Jan 26 when the University of Texas McDonald Observatory succeeded in locating the reflector with a powerful laser attached to a 2·7m telescope. The reflected signal strength was comparable with that from the Apollo 15 array, the largest of the NASA arrays deposited by Apollos 11, 14 and 15.

First TV pictures showed that Lunokhod 2 was standing on an even plain between 2 small craters, with the low peaks of the Taurus mountains about 6 km away. Owing to the high position of the Sun, the crew at the Space Communications Centre at first had difficulty in maintaining control; during the 3rd session there was a near collision

Luna 21 landing stage seen from Lunokhod 2. Landing ramps used for Lunokhod's descent visible on left

with the Luna 21 landing stage; the 2 vehicles were less than 4 m apart when Lunokhod 2 stopped. Its speed was more than double that of Lunokhod 1, and operating it over the rising, terraced ground towards the Taurus mountains was described as 'very taxing' for the crew. They had to undergo medical checks, and eyesight and hearing reaction tests before each session, and wore sensors just as if they were themselves operating in space. First chemical tests of the regolith made by the Rifma instrument suggested that its composition was much the same as had been found on the Sea of Rains; the landing area generally, however, was more rugged, with more craters and boulders. During the first lunar night, the temperature at the end of the probe fell to a record −183°C, though the wheel temperatures never fell below 128°C.

During its 2nd lunar day, occasionally sinking to the hubs of its wheels in loose rock, Lunokhod 2 moved steadily towards the Taurus mountains, taking magnetic measurements on the way. (Unlike Earth, the Moon has no general magnetic field, probably because it has no liquid core; but the lunar seas and mainlands have local magnetic fields.) The robot zig-zagged up the mountain slopes, wheel-slip sometimes registering 80%. From a 400m peak, views were obtained of the opposite shore of the Le Monnier bay, and of the main mountain peaks 55–60 km away. It also sent back an unexpected view of Earth, appearing as a 'thin sickle' in the local sky.

During the 3rd day, and throughout the 4th, Lunokhod 2 was worked along the narrow precipice, or tectonal abyss 30–50 m deep, and 300–400 m wide. Moving cautiously along the westward edge, and then in the reverse direction along the eastern slope, it established that it was solid basalt rock which had been split by tectonic forces. On May 9 it was reported that movement had begun, at the start of the 5th day, away from the abyss in a north-easterly direction. There were no further reports until, on Jun 3, it was announced that the robot's research programme was completed. No indication was given as to whether there had been a mechanical failure, or whether it had been lost or damaged by falling into a crater or fissure.

Perhaps Lunokhod 2's most interesting discovery was that while the Moon almost certainly has admirable conditions for astronomic observations during the lunar night, conditions are not nearly so good as had always been imagined during the day. An astrophotometer (an electron telescope without lens), registered glow in the lunar sky, both in the visible and ultraviolet bands of the spectrum, to establish whether the lunar 'atmosphere' contained cosmic dust which could affect observations from the surface. The after-sunset glow proved much higher than expected—10–15 times brighter than on Earth, suggesting that the lunar sky is surrounded with a swarm of dust particles, with the effect of scattering solar light. This high luminosity would make daytime astronomy 'ineffective'.

MARS   Soviet Spacecraft

History   Details of Soviet probes and spacecraft to Mars, given below, show that, as in the case of the Moon, Russian scientists were well advanced with plans for an interplanetary expedition long before the launch of Sputnik 1. First official Martian launches were Sputniks 22 & 24 in late 1962, preceded by Sputniks 19–21 aimed at Venus, and followed by Sputnik 25 aimed at the Moon. As the launch table shows, it was Soviet practice from the start to make 3 launches on each occasion; in 1962 none was successful, although much was learned from the only one successfully placed on course. Then in May 1971 the 2 planets were so favourably placed—factors such as the relative inclinations of the orbital planes providing good radio communications are important, as well as distance—that it was possible to put spacecraft on suitable trajectories with a velocity only a few hundred metres above the minimum escape speed. This time 2 out of 3 spacecraft reached the planet. The flights of Mars 2 & 3 occupied respectively 192 and 188 days. Such favourable conditions will not occur again until 1986, which explains why Russia prepares triple launches, and the Americans usually double launches. It was no coincidence that Mars 2 and 3 were in Martian orbit at the same time as America's Mariner 9.

The Soviet technique, at least in these early landing attempts, is to separate the landing craft for a direct approach before the parent craft is placed in Martian orbit. The US preference is first to place the whole vehicle in parking orbit, thus providing time for adjustments before the descent attempt is made. In the case of Mars 2 and 3 it meant that the landings had to go ahead despite the dust-

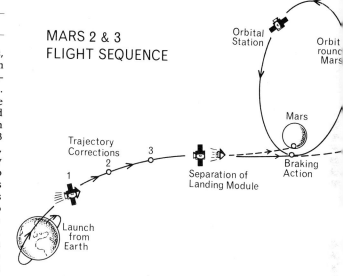

MARS 2 & 3
FLIGHT SEQUENCE

storm raging on the surface; there was no possibility of holding the mission in Martian orbit, though presumably it was possible to adjust the targeted landing site during the final, pre-separation manoeuvre. It should also be noted that a Martian landing presents technical and design problems which are the exact opposite of those encountered on Venusian landings. Atmospheric pressure on the Venusian surface is about 100 times greater than on Earth; on the Martian surface pressure is about 100 times less. Thus, while the dense Venusian atmosphere will quickly decelerate an approaching spacecraft, on Mars, if the approach angle is

Mars 3: Lander is mushroom-shaped capsule at top; piped section inside right solar panel is radiator for temperature control; antenna for French 'stereo' experiment is mounted on right solar panel

not exactly right, it will either pass through the thin atmosphere into space without slowing down, or enter too fast to enable the parachute to operate. No doubt these factors contributed to the bitter disappointment of 1973–4, after Russia had successfully despatched a fleet of 4 spacecraft. This occasion again emphasized the shortcomings of the 'slingshot' approach. One lander missed the planet completely, the other crashed just before touchdown; the orbiters too were unsuccessful. It seems likely that the decision not to use the 1975 launch window was because Soviet techniques were being reviewed in the light of the Mariner 9 success, and the more sophisticated methods being tried by America's Viking programme. Despite their

disappointments, however, their Martian experience has convinced Soviet scientists that radio communications are possible over distances of 1000M km; eventually they should be possible throughout the entire solar system, even though the signals reaching Earth from its edge will be so weak as to be measured in thousandths of a microvolt. Experience with Mars 2 & 3, combined with Lunokhods 1 & 2, is being used to develop 'Planetokhods' for robot exploration of the Martian surface. These, however, would have to be operated mostly from onboard sensors rather than by Earth commands, because of the time required for signals to be transmitted and obeyed.

Spacecraft Description Photographs released shortly after Mars 2 and 3 had been placed in Martian orbit showed that they were about 3·6 m in ht, and that more than half the launch wt consisted of fuel. The main hypergolic fuel tanks provided the central cylinder, around which the spacecraft was built. The S-band, directional antenna dish, and 2 deployable solar panels were attached. A doughnut-shaped compartment, or toroid, at the base of the structure, containing electronics and instruments, completely enclosed the engine. The landing craft, protected by a heatshield shaped like an inverted saucer, consisted of a sphere, resting in a toroid, with its engine nozzle projecting through it. The latter contained the parachute system; the sphere was described as the 'automatic Mars station'. The heatshield provided maximum aerodynamic braking as the spacecraft entered the thin Martian atmosphere; it was discarded shortly before the drogue and main parachutes were explosively deployed; the parachutes were jettisoned at about 30 m to reduce final landing wt just before a solid-fuel braking engine was fired by a radar altimeter. Separation of the lander, followed by the placing of the orbiter into an elliptical path around Mars, was carried out by an onboard navigation/computer complex, updated by Earth commands. The Mars 3 lander carried mass spectrometers for chemical analysis of the atmosphere, and equipment to analyse the chemical and physical content of the surface soil, wind anemometer, thermometer and pressure gauge and at least 1 TV camera. Data was to have been relayed via the orbiters, but only 20 sec of signals were obtained. Each orbiter carried 2 cameras, 1 wide-angled, 1 with a telescopic 4° narrow-angle lens. The photographs were automatically developed and the image transmitted directly to Earth by the 200cm S-band aerial, but the system was apparently less successful than the US Mariner system, which breaks down the pictures into coded dots which are then reconstructed on Earth by computer. Other orbiter instruments provided ultraviolet, infrared and optical examination of Mars and its atmosphere; by using infrared sensors to measure the thickness of the carbon-dioxide atmosphere, variations in surface ht could be found, and a relief map built up. Surface colouring could also be defined. Mars 3, also 6 & 7 carried French equipment to collect solar emission data for comparison with similar measurements on Earth. With both US and Soviet spacecraft in Martian orbit at the same time, a teleprinter link was established between NASA's Jet Propulsion Laboratory at Pasadena, California, and the Soviet Co-ordinating and Computing Centre, for the exchange of information; Soviet information supplied over this link was not allowed to be published, but, apparently, merely duplicated Russian official statements.

Sputnik 22   L Oct 24 1962 by A2e from Tyuratam. Wt 893·5 kg. From an initial Earth orbit of 180 × 405 km, with 64° incl, the spacecraft and final rocket stage blew up when being accelerated to escape velocity. The debris passed within range of the West's BMEWS radars, which indicated a possible ICBM attack—especially alarming at that moment, occurring as it did during the Cuban missile crisis amid maximum East–West tension. BMEWS computers, which assess trajectory and impact points, proved the alarm to be false within a few seconds.

Mars 1   L Nov 1 1962 by A2e from Tyuratam. Wt 893·5 kg. World's first Mars probe, successfully launched from Earth orbit similar to Sputnik 22. Mars 1 was 3·3 m long; with solar panels and radiators deployed following trans-Martian insertion, width was 4 m. It consisted of 2 hermetically sealed compartments, 'orbital and planetary'. The orbital section contained midcourse rocket engine, solar panels, high- and low-gain antennas, etc; the planetary section contained a photo-TV system and other instruments for studying the Martian surface during the period of closest approach. The Soviet Deep Space Tracking Station at Yevpatoriya in the Crimea held 61 communication sessions, first at 2-day, then at 5-day intervals until Mar 21 1963; contact was then lost, because the craft's antenna could no longer be pointed towards Earth, following a fault in the orientation system. At that point the craft had travelled 106M km from Earth. It should have passed Mars 3 months later at a distance of between 998 and 10,783 km.

Sputnik 24   L Nov 4 1962 by A2e from Tyuratam. Orbit 196 × 590 km. Incl 64°. This too disintegrated during an attempt to place it on course for Mars from Earth parking orbit; 5 major pieces of debris were tracked by BMEWS.

Cosmos 419   L May 10 1971 by D1e from Tyuratam. Wt ? 10,000 kg. First use of Proton launcher for a planetary mission. First of 3 craft intended for Mars; placed in 145×159km Earth orbit, but failed to separate from 4th stage and decayed in 2 days.

Mars 2 L May 19 1971 by D1e from Tyuratam. Wt 4650 kg. Reached Mars on Nov 27 after 192-day flight, during which 3 mid-course corrections were made. Following the final manoeuvre, a landing capsule was separated from the orbiter and made a first, unsuccessful attempt to soft-land. Few details were given, except that it had delivered a pennant with the USSR hammer-and-sickle emblem to the Martian surface. It was presumably destroyed on impact. After the lander was released, the orbiter's retro-rockets were fired; it became the 2nd artificial Martian satellite, with an 18-hr, $1380 \times 25,000$km orbit, at 48° incl. Within a few days Mars 2 established that at least nine-tenths of the Martian atmosphere consisted of carbon-dioxide; there was almost no nitrogen, and water vapour was very scarce.

Mars 3 L May 28 1971 by D1e from Tyuratam. Wt 4650 kg. Reached Mars 5 days after its predecessor, on Dec 2. About $4\frac{1}{2}$ hr before atmospheric entry the Martian lander was automatically separated from the parent craft, and placed on a shallow approach angle for a 3-min descent. Aerodynamic braking on the saucer-shaped heatshield began at a speed of 21,600 kph; the speed was still supersonic when the drogue parachute was explosively deployed. After main parachute deployment, the heatshield was jettisoned and radio antennas deployed. The soft-landing engine was switched on at about 30 m and the parachute fired to one side by another jet engine to prevent its canopy covering the landing capsule. 90 sec after touchdown, a timing device switched on the video-transmitter. For 20 sec 'a small part of a panoramic view' was transmitted by its TV camera, but in any case surface details were probably obscured by the dust storm then in progress. It was probably a combination of a heavy-landing and the dust-storm that overwhelmed the capsule. It had touched down in a pale-coloured region in the southern hemisphere, between Electris and Phaetontis (45°S, 158°W), in a rounded hollow about 1500 km across. A few days after the landing Soviet scientists said they assumed the surface was covered by 'something resembling the sands of the Earth'; the reason for its changing colour during the Martian seasons was still unclear. After releasing the lander, the orbiter stage of Mars 3 fired its retro-rockets and went into an 11-day, $1500 \times 200,000$km orbit, from which it sent back data for about 3 months. US estimates are that, because it seldom approached Mars closely, its total observing time was about 2 hr; and because it could not be reprogrammed, it automatically exposed all its photographic film towards a planet completely hidden by the great dust storm.

Summary of Results Results of data and pictures received from Mars 2 and 3 were reported in Aug 1972, when it was stated that their work had been completed. Mars 2 had made 362 revolutions, and Mars 3, in its much more elliptical orbit, 20. The differing orbits had enabled comparative studies to be made of Mars's magnetic field, essential 'to understand not only the nature of Mars, but the origin and evolution of the solar system'. The spacecraft had travelled more than 1000M km; 687 communication sessions had taken place, 448 of them while in Martian orbit. Like America's Mariner 9, Mars 2 and 3 were able to study the 3-month dust-storm which began in Oct 1971, and Soviet scientists concluded that winds were not constant, but most likely occurred only in the initial phase, and that the particles, mostly of silicate composition, took months to fall 'in the absence of supporting vertical currents in the atmosphere'. The dust-clouds were at least 8–10 km high; during the storm the surface temperature dropped by 20–30°C, while the atmosphere got warmer as it absorbed solar radiation. When the storm ended surface temperatures taken from the southern hemisphere, where summer was ending, across the Equator to the northern hemisphere, ranged from 13°C to −93°C; at the N Pole cap they dropped to −110°C. The surface cooled very quickly after sunset, indicating a low conductivity, such as dry sand or dust in a rarified atmosphere. The Martian 'seas', or dark areas, were warmer than the 'continents', or light areas, and in some cases the 'seas' took longer to cool, suggesting a rock surface with greater heat conductivity. Atmospheric pressure varied with altitude but at average surface level was about $\frac{1}{200}$th of Earth's. Mars 3 measured heights of 3 km and depressions of 1 km below average level. During the dust-storm the dust-clouds rose as high as 10 km. Water-vapour content of Martian atmosphere did not exceed 5 microns of precipitated water—2000 times less than Earth's. Immediately above the surface the atmosphere was mainly carbon-dioxide, but at 100 km altitude was broken up by ultraviolet solar radiation into a carbon-monoxide molecule and an oxygen atom. Traces of oxygen were recorded up to altitudes of 700–800 km, where its concentration was 100 atoms per cu cm. Photographs had revealed interesting twilight phenomena, including an airglow 200 km beyond the terminator, or border line between day and night, and changes in surface colour close to the terminator. The severe climatic conditions did not rule out the possibility of some form of life, though at best it was likely to be no more than micro-organism or plants.

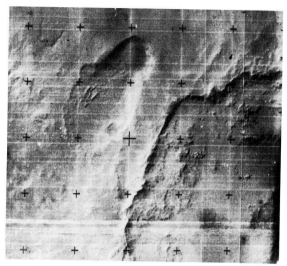

Mars 5 picture, showing western half of 150km dia Martian crater, 3 km deep, in area where it was hoped to land.

Mars 4 & 5   L Jul 21 & 25 1973 by D1e from Tyuratam. Wt 4650 kg. The start of Russia's most ambitious onslaught on Mars, with 2 pairs of spacecraft (see below) being launched within 3 weeks—the first time a complete series had been successfully launched. But although all 4 reached the vicinity of Mars, and returned some data, the results were disappointing. Mars 4's retrorockets failed to fire, and it passed the planet on Feb 10 1974 at a distance of 2200 km,

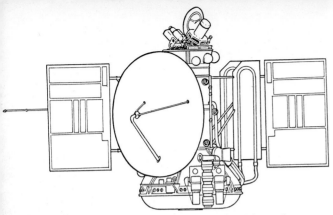

Drawing by D. R. Woods of Mars 4 & 5 orbiters, after release of landers. TV cameras exposed at top; optical sensors for astro-orientation system *bottom right*

returning one swath of pictures and some radio occultation data as it flew by. Mars 5 which, like Mars 4 had travelled about 460M km since launch, was successfully placed in a 32,500 × 1760km orbit on Feb 12 1974, with 35° orbit and 25-hr period of revolution. However, it continued to operate for only a few days; during that time it returned a few photographs of excellent quality showing a small portion of the southern hemisphere in the region where a landing was intended.

Mars 6 & 7 L Aug 5 & 9 1973 by D1e from Tyuratam. Wt 4650 kg. The 2nd pair was launched only a few hours before the 2-yearly launch window ended. The object of the 4-craft fleet was stated to be 'comprehensive exploration of the planet'. Both contained French instruments for investigat-

ing radio emissions from the Sun in the 1m waveband, as well as solar plasma and cosmic rays. Mars 6 was said to differ from the previous pair, and would carry out part of its scientific exploration with the use of equipment on Mars 4. Since it was also stated that Mars 6 & 7 were 'analogous', presumably 7 was intended to work with 5. 6 reached the planet on Mar 12 1974, and the final trajectory correction was performed automatically 'by the onboard system of astro-navigation'. The lander module separated, its descent engine fired as planned, and, following aerodynamic deceleration, the parachute deployed. For 148 sec of parachute descent, man's first direct measurements of the Martian atmosphere were transmitted back to Earth via the parent craft. Its telemetry ceased abruptly when the landing rockets were fired, and it presumably crashed at approx 24°S and 25°W. But its descent through the thin atmosphere included measurements of the electric current drawn by the mass spectrometer pump. The pump current increased in a completely unexpected manner, and Soviet scientists interpreted this as indicating the presence of argon, a rare gas. Since large quantities of this gas would also affect both the landing sequence and the operation of the US Viking atmospheric gas spectrometer, NASA scientists made changes as a direct result of Mars 6. Mars 7 was an almost complete failure. It reached the planet 3 days before Mars 6; the official report merely said that 'the descent module was separated from the station because of a hitch in the operation of one of the onboard systems, and passed by the planet 1300 km from its surface'. However, Soviet scientists were able to claim that, overall, new information about the Martian surface and atmosphere had been obtained, including the fact that the atmosphere over some areas contained several times more water vapour than had been thought.

METEOR   Weather Satellites

History   This is the basic satellite in the Soviet weather programme. By the end of 1976 Meteor 26 had been launched, and the first of a new series, Meteor 2-01 had been orbited on Jun 11 1975. Meteor 2-02 followed on Jan 6 1977. Reports by Meteor satellites in the USSR, and warnings of tropical storms, have saved many lives. Irrigation planning has been improved as a result of data on snow cover in the Tien Shan and Himalaya Mountains, and 10% of sailing time for all Soviet ships is estimated to have been saved because Meteors can provide the ships with the best courses to avoid storms, sea states, winds and ice conditions. This maritime contribution alone is officially estimated to be worth 'over a million roubles in profit'. The 12th International Botany Congress, held in Leningrad in Jul 1975 was also told that TV pictures from Meteors had enabled shepherds in Soviet Central Asia to begin using new pastureland in a vast area between the Caspian and the Pamirs. This, too, was worth 'millions of roubles' because it helped both to conserve plant resources and to use them rationally.

Meteor photographs, by now numbered in hundreds of thousands, have revolutionized Soviet weather-forecasting techniques. In 1 hr the 2 TV cameras aboard each satellite cover 30,000 sq km, automatically adjusting their exposure times to match lighting conditions on the daylight side, storing and then transmitting the pictures. On Earth's dark side heat emissions and cloud formations are measured by infrared sensors. Radiometers measure the radiation balance of the Earth-atmosphere system; clouds, ice and snowfields reflect about 80% of solar radiation, and measurements of incoming energy are regarded as essential for reliable

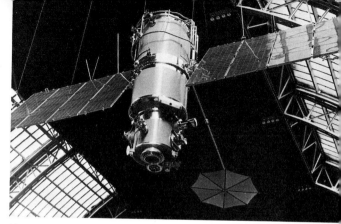

Meteor weather satellite, with cameras and sensors pointing Earthward, as shown at Moscow Space Pavilion

forecasts. 2 diagrammatic maps are compiled every 24 hr.

Basically the system provides a daily weather review from more than two-thirds of the globe. Soviet sources say it started to function on Apr 27 1967, with the launch of Cosmos 154 and 156. Western observers noted 14 Cosmos meteorological test satellites, beginning with 44 and 45 in Sep 1964. The 2nd of these returned a film capsule. Subsequent Cosmos, announced as 'metsats', included 144, 156, 184, 206 and 226; their orbits, varying by only 12 sec, were phased to provide almost continuous coverage of Earth's surface. Space 'panoramas' are transmitted to meteorological organizations in Siberia and the Soviet Far East, as well as to several other countries. The satellites are also studying clear air turbulence—strong currents in the upper layers of the atmosphere used by intercontinental jets.

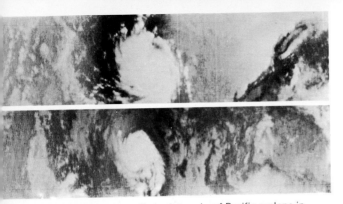

Meteor photos, at 7-day intervals, of Pacific cyclone in 1970. Dia is 250–300 km, with 120 kph windspeeds. *Bottom*, cloud mass down to 200–250 km dia, with max wind 90 kph

Spacecraft Description No weight has been given; the satellites are believed to be 5 m long, 1·5 m wide. A large cylindrical body carries meteorological instruments at the Earth-facing end; a pair of large solar panels, rotated by a drive mechanism so that they remain Sun-oriented, charge the electrical storage batteries. The instruments are kept pointing towards Earth by a 3-axis attitude control system, which includes reaction wheels.

Meteor 1 itself was launched from Plesetsk on Mar 26 1969 into a 644×713 km orbit with $81°$ incl. Launches continue at a rate of 4 a yr, so that 2 or 3 are always operational, providing a continuous survey of atmospheric conditions from Pole to Pole in a band up to 1500 km wide. Information must be 'dumped', in passes lasting only a few minutes, into the 3 reception centres of the USSR Hydrometeorological Service in Moscow, Novosibirsk (Siberia) and Khabarovsk (Pacific Coast); processing the data takes $1\frac{1}{2}$ hr. Orbital life of Meteors 1–10 is about 60 yr; later Meteors, usually in very precise orbits of approx 870×910 km and $81·3°$ incl, have 500-yr lives before they decay.

Meteor 2 Although described by Novosti as 'another Meteor 2' satellite, the first of this series was apparently launched on Jul 11 1975. It is believed to have been much the same size and shape as earlier Meteors. Orbit 877×906 km. Incl $81·2°$. Stated to have 'perfected equipment' for obtaining meteorological information' as well as data on streams of radiation penetrating near-Earth space, and capable of sending back TV pictures of cloud cover. It was undoubtedly the start of a new '3-stage system' of long-lasting satellites announced 5 months earlier. Their mission was 'to watch high-speed natural phenomena; the paths of cyclones, tidal action, dust storms and tsunami waves'.

Meteor 2-02 L Jan 6 1977 by A1 from Plesetsk. Wt ? 2750 kg. Incl $81°$. Orbit 890×906 km.

MOLNIYA   Communications Satellites

History Molniya is part of the Soviet Union's communication system, providing TV, telephone and telegraph links to 33 Orbita ground stations in the USSR, plus 1 in Cuba and 1 in Mongolia. It is regularly used to transmit facsimiles of newspaper pages to remote areas. The 24-hr TV service provided to the Far North and Far East territories, said Moscow radio in Feb 1972, had 'saved the national economy hundreds of millions of roubles' by eliminating the need for a network of ground relay stations. A radio-isotope power supply had already been tested in a Cosmos flight; it was intended to replace Molniya's solar batteries with atomic reactors powerful enough to transmit directly to TV aerials without being boosted by Orbita stations. By the end of 1976 launch totals were Molniya 1-38, Molniya 2-20 and Molniya 3-6. Including Cosmos 637 (L Mar 26 1974), which astonishingly was Russia's first-ever synchronous satellite, brought the series total to 58.

There were 4 sets of Molniyas circling the Earth in orbital planes spaced at 90° intervals, a distance later reduced to 45°. Each set contained 1 of each of the 3 Molniya versions—a total of 12 operational satellites. Launches were still continuing, despite the fact that Statsionar 1, the first of a more advanced series of synchronous satellites, had been successfully launched on Dec 22 1975.

While the US had moved steadily from low-altitude, tape-storage satellites, and passive, balloon-type reflectors, to synchronous satellites, like Intelsat, Russia followed a different path with the Molniya 1, 2 & 3 series, which began in 1965. Molniya 1s, much heavier than corresponding US comsats, had about 10 times the power output of the original Early Bird. The principle was that they flew in an orbit inclined at 65° to the equator, with a low point of about 500 km over the Southern Hemisphere, but a long time at the peak of the orbit (about 40,000 km) in the Northern Hemisphere, so that each was in view of major Soviet ground terminals for about 8 hr per day, including those at near Polar latitudes, which are more difficult for the US equatorial systems to reach. The satellites' high power enabled low-cost ground terminals to distribute Moscow's TV programmes to remote areas, as well as carrying telephone and computer data traffic. By spacing 3 satellites in 12-hr orbits,

Molniya 2 communications satellite shown at Paris Air Show 1973. *Bottom,* tri-lens camera used for Earth photography

it was possible to give full daily coverage, with each satellite appearing at the same points in the sky relative to the Earth at a certain hour each day. In addition to using Molniyas 1-20 and 1-30 to orbit French satellites, the system has also been used for experimental colour TV between Moscow and Paris.

Under a US–USSR agreement of Sep 30 1971 Molniya 2 satellites are used as 'hotline stations', with a complementary service being provided by US Intelsat satellites. A ground installation at Fort Derrick, Maryland, about 72 km from the White House, links the US President, via Molniya 2s, to the Kremlin. It has 2 receive/transmit antenna stations with parabolic reflectors of over 12 m dia. An Intelsat ground station is being located near Moscow. This 'hotline' system to remove the danger of accidental war and allow direct conversation between the heads of the Super Powers in times of crisis, will replace the 'terrestrial' cable hotline, routed Washington-London-Copenhagen-Stockholm-Helsinki-Moscow, operated by ITT with backup route Washington-Tangier-Moscow, operated by RCA Globcom.

Molniya 1   These are placed in highly elliptical orbits from Plesetsk by A2e with approx 40,000km apogee in the N Hemisphere, and 482km perigee in the S Hemisphere, to provide 8–10 hr continuous communications through the USSR and associated countries. Orbital changes are made by onboard rocket engines, to synchronize available hours. By the end of 1976, 38 Molniya 1s had been launched. The 1st, L Apr 23 1965 from Tyuratam, provided 'many months' of TV and phone links between Moscow and Vladivostok. The 2nd, L Oct 14 1965 from Tyuratam, provided experimental TV, phone and telegraph communications for 5 months before decaying. 3rd, L Apr 25 1966 from Tyuratam was used for an exchange of colour TV with France, and for Earth pictures showing cloud-patterns on a global scale. Colour photos of Earth started in 1967. Molniyas also provide communication links during manned Soyuz flights. Performance has been steadily upgraded to improve radiation resistance, decrease noise levels and increase power output. Some of this series also carry a steerable camera to add wide area weather views to the Meteor data. The 20th Molniya 1, L Apr 4 1972 from Plesetsk, also carried France's SRET 1 (environmental research satellite) into orbit, under a 1970 collaborative agreement. SRET 1 was released 2 sec later than Molniya. France's SRET 2 was given a similar piggyback launch by Molniya 1-30 on Jun 5 1975. Latest versions of Molniya 1 weigh about 816 kg. Cylindrical in shape with conical top, they are 3·4 m long and 1·6 m dia, with 6 paddle-wheel solar wings, providing 500–700W of power. 2 0·9m dish antennas extend from the base, which also contains an orbital correction engine. 3 transceivers have estimated life of over 40,000 hr; TV transmissions at 625 lines per frame.

Molniya 2   More advanced and slightly heavier, approx 1250 kg, and using higher frequencies. 20 launches to end of 1975 by A2e from Plesetsk, in Nov 1971, and May, Sep and Dec 1972, into approx 483 × 40,230km orbits. Like recent Molniya 1s, they are believed to be windmill-shaped, with 6 solar wings and 5-yr life.

Molniya 3   First launch of this even heavier series (approx 1500 kg) was on Nov 21 1974 by A2e from Plesetsk. Orbit 628 × 40,685 km. Believed to be windmill-shaped with 6 vanes, with 4·2m length and 1·6m dia. Molniya 3-2 followed on Apr 14 1975. Molniya 3-4 on Dec 27 was Russia's last launch of 1975. They have an orbital life of about 11 yr.

Molniya 1S   L Jul 29 1974, by D1e from Tyuratam. Orbit 35,787 × 35,790 km. Incl 0·4°. Russia's first synchronous comsat, no reliable details of wt and size available. The official announcement made no reference to the fact that this was Russia's first operational synchronous satellite. It stated that 1S was launched 'under the programme of further perfecting communication systems with the use of satellites'; the period of revolution in its circular orbit was given as 23 hr 59 min. Orbital life over 1M yr.

POLYOT   Manoeuvrable Satellites

Polyots 1 and 2   L Nov 1 1963 & Apr 12 1964, 'the first manoeuvrable Earth satellites'. With estimated wt of 600 kg, Polyot 1, placed in an initial orbit of 339 × 592 km, was successfully manoeuvred several times, ending in a 343 × 1437km orbit with a 25-yr life. Polyot 2, which had an 18-day life, made several similar manoeuvres, and also changed its inclination from 58° to 60°. America's first orbital manoeuvres were made by Gemini 3 on Mar 23 1965.

PROGNOZ   Solar Radiation Satellites

History   Satellite designed specifically to study solar radiation, the physical processes taking place on the Sun, and the interference they can cause with space communications.

Prognoz ('Forecast') 1   L Apr 14 1972 by A2e from Tyuratam. Sphere-shaped, with 4 cruciform solar panels.

Prognoz 2: picture shows instrument array for studying solar radiation

Wt 857 kg. Orbit 965 × 200,000 km, gives it a 10-yr life. Incl 65°.

Prognoz 2   L Jun 29 1972 into 550 × 200,000km orbit, with 65° incl. In apogee when Prognoz 1 is in perigee; this provides simultaneous studies of solar wind from different points of near-Earth space, during their 4-day orbits. Prognoz 2 carries French equipment to study solar wind, outer regions of magnetosphere, gamma rays and solar neutrons.

Prognoz 3   L Feb 15 1973. Wt 845 kg; orbit 590 × 200,000 km. Incl 65°. This was equipped to study solar gamma-ray and X-ray emissions. Its readings are synchronized with measurements from Earth-based observatories. The highly eccentric orbit gives comparative measurements from both near-Earth and from outside the upper atmosphere and magnetosphere. The practical benefits of maintaining a continuous watch on the Sun were mentioned in a report on Prognoz 3's successful operations. Solar radiation, it was pointed out, could hinder communications with spacecraft, and a thorough understanding of the problem could lead to a solution.

Prognoz 4   L Dec 22 1975 by A2e from Tyuratam. Wt 905 kg. Orbit 634 × 199,000 km. Incl 65°. Apparently an identical spacecraft.

Prognoz 5   L Nov 25 1976 by A2e from Tyuratam. Wt 930 kg. Orbit 510 × 199,000 km. Incl 65°. French and Czechoslovak equipment were added to Soviet instruments, possibly accounting for its slight increase in weight. Official reports emphasized that the continuing watch on solar radiation was important for space flights (presumably manned). Solar activity increasingly attracted the attention of doctors, because 'when the Sun is active, a person's concentration is reduced, heart ailments worsen, and radio communications are affected'.

PROTON   Cosmic Ray Satellites

History   A series of 4 satellites to study the nature of cosmic rays of high and super-high energy, and their interaction with the nuclei of atoms. Intercosmos 6, nearly 4 yr later, was described as a logical development of the Proton stations.

Proton 1–3   L Jul 16 1965, Nov 2 1965 and Jul 6 1966, by D1 from Tyuratam. Wt 12,200 kg, they were of record size at that time. Their orbits were 190 × 630 km, and they had a 3-month life.

Proton 4   L Nov 16 1968, by D1 from Tyuratam. Wt set a new record of 17,000 kg. Orbit of 493 × 255 km gave an 8-month life. Experiments included study of the energy spectrum and chemical composition of primary cosmic particles, and the intensity and energy spectrum of gamma rays and electrons of galactic origin.

SALYUT   Space Station

History   By the end of 1976 at least 6 Salyut space stations had been launched, although only 5 had been announced. They are the basis of Russia's long-term manned programme, which explains the persistence with which it is being pursued despite many setbacks. By the time Soyuz 23 had failed to dock with Salyut 5, resulting in a dramatic emergency return to Earth by its crew, there had been 7 failures in 11 attempts to complete space station missions. But the 17-day flight of the Soyuz 24 crew in Feb 1977 went well. The programme began in 1971 with Salyut 1, followed by 2 failures, Salyut 2 and Cosmos 557. With the launch of Salyut 3, it became clear that 2 quite different types were to be used for parallel civil and military programmes. Only then was it possible to conclude that Salyut 2 and Cosmos 557 should have been respectively military and civilian space stations. Nos 3 & 5 turned out to be military with Nos 4 and the expected 6 for civil activities. Speculation that, when more powerful launchers become

Salyut 4 control centre, as shown at Paris Air Show 1975; note console with rotating globe at left showing exact position, and shade for covering window when opposite Sun

until the Paris Air Show 1975, when a full-scale mockup was displayed in the Soviet Pavilion, and the author was the first Western journalist to gain admittance. Probably the mock-up was not fully equipped and instrumented; but even allowing for this, its controls and equipment seem sparse and unsophisticated compared with the US Skylab. But Soviet persistence with Salyut was at last rewarded. After 5 of the first 8 attempts to complete space station missions had failed for different technical reasons, the splendid Soyuz 18 flight, when the crew completed 63 days in space, more than doubled Russia's previous duration record. It should be noted that much of this mission's success was due to running repairs made by both the Soyuz 17 and 18 crews—no doubt inspired by the example of the Skylab crews—to the solar telescope and cosmic ray equipment, as well as to more fundamental things like air conditioning. Soviet cosmonauts revealed that Salyut 6, expected at the time of writing to be launched in mid 1977, would have 2 docking ports, and would be used for extensive space-walking activities—an area strangely neglected by Russia after Leonov's 'first'.

Space Station Description   A docked Soyuz appears to be an integral part of the complete Salyut system, and the combined system is always described as 'over 25 metric tonnes' with an internal area of 100 cu m. Since Soyuz is about 6575 kg, Salyut would be over 18,425 kg (usually estimated at 18,600–18,900 kg). Length of Salyut 3 was given as 21 m, and Salyut 4 as 23 m, probably (according to Dr Charles Sheldon, of the US Library of Congress) because external attachments like radio transponders are sometimes included and sometimes not. The 1st compartment, serving as transfer tunnel from Soyuz to Salyut, is 3 m long and 2 m dia. The main habitable area consists of 3 sections: the small

available, a Salyut may be placed in lunar orbit, and others developed for planetary flight, may yet prove well-founded. The fact that Salyut 3 remained operational far longer than the planned 6 months encouraged the Russians to try doubling the life of Salyut 4. In fact it survived in orbit for over 2 yr. Whether it was intended to be occupied by 3 rotating crews, for periods which would have matched America's Skylab flights, is doubtful; it is more likely the Russians decided to move instead to a station which could be utilized equally well for a long period in either a manned or an unmanned role. Few details about Salyut were revealed

cylinder 3·8 m long and 2·9 m dia; the large cylinder 4·1 m × 4·15 m; and a cone connecting them which is 1·2 m long. An unpressurized service module 2·17 m long and 2·2 m dia, completes the station. TV views show that it contains 8 chairs, 7 of them at work stations; there are 20 portholes. Externally there are 2 double sets of winglike solar cell panels, placed at opposite ends on the smaller compartments. The heat regulation system's radiators, orientation and control devices are also external. Scientific instrumentation is mounted both internally and externally.

Salyut 1   L Apr 19 1971, into an initial orbit of 200 × 222 km, it remained in near-Earth orbits for nearly 6 months until a final re-entry manoeuvre caused it to burn up on Oct 11, after about 2800 revolutions. Moscow described the launch as an important step towards the creation of long-term scientific stations in near-terrestrial space, each to be used for up to a year. But Salyut 1's orbit was so low (possibly because the weight of the 3-man Soyuz 10 and 11 craft was near the limit of Soviet rocket launch capability) that repeated posigrade firings were necessary to avoid re-entry. Following the death of the Soyuz 11 crew on Jun 29, temperature and atmospheric pressure were maintained at life-supporting levels. It seemed likely that at least one more Soyuz docking was intended, until the inquiry into the Soyuz failure showed that major modifications would be needed. Salyut 1's engine was then retro-fired; destruction by natural decay was not permitted because of the risk that parts of the station would penetrate the atmosphere and fall on populated areas. The dockings of Soyuz 10 and 11 are described under Soyuz. Cosmonaut Dr Feoktistov said Salyut 1 contained 8 working positions. The docking tunnel, first entered by boarding cosmonauts, contained part of the astrophysical equipment, with several control panels. A hatch led into the main compartment, containing '2 working seats', facing control and instrument panels in front, and Soyuz-type command and signal equipment at the side. Orientation and guidance is carried out either by ground control, or manually by the crew; a small lever, when turned right, activates small thrusters which rotate the station to the right. When pulled back or pushed, the forward part, to which Soyuz is docked, is raised or lowered. Personal comforts include a well-stocked refrigerator, and the ability to heat both food and water. Drinking water, in rubber containers, is kept potable for 90 days with the addition of a small quantity of ionic silver. A future possibility is that fresh water may be 'pumped' up to such space stations in the form of steam passed along a laser beam.

Salyut 2   L Apr 3 1973. Orbit 260 × 215 km. Incl 51°. This launch was the start of a further series of setbacks in Soviet spaceflight. The object was said to be 'to check the improved design of onboard systems and equipment and for conducting scientific and technical experiments in space'. A week later, after 130 orbits, it was said to be stable, and the orbit had been raised to 261 × 296 km. No indication was ever given as to whether it had been intended for manned occupation, except that on Cosmonauts' Day, Apr 10, Gen Shatalov said in an interview that Salyut 2 was similar 'in design and purpose' to Skylab, due to be launched the following month. Shortly after, Western observers reported it had become unstable; on Apr 14 it apparently broke up into 25 fragments, and finally decayed on May 28.

Cosmos 557   L May 11 1973. Orbit 214 × 243 km. Almost certainly a back-up Salyut 2, possibly launched in haste, 3 days before America's Skylab. (A rocket, thought to have been carrying yet another Salyut, had blown up during an attempted launch between Salyut 2 and Cosmos 557). This too was a failure, and re-entered 11 days later.

Salyut 3   L Jun 25 1974. Initial orbit 270 × 219 km. This launch took place 2 days before President Nixon arrived in Moscow; but he had already left when, 8 days later, Soyuz 14 was launched to dock with it. Improvements and changes included 3 solar arrays, automatically rotating through 180° so that they can always face the Sun, instead of 4 arrays on Salyut 1. Total wt with Soyuz was said to be 'over 25 tonnes'; total length 'over 21 m'. Its 2 cylinders were divided into 4 functional sections: equipment, working, rest and transitional. The living area included a bedroom and kitchen with a small table for 2, equipped with clamps for the crockery. A 'psychological trick' was used to give the illusion of a floor and ceiling, by painting the 'floor' dark and the 'ceiling' light. Its use for military reconnaissance is discussed under Soyuz 14. A completely successful mission was at least partially based on the parallel use of a ground-based Salyut in which problems were examined and solutions checked before being passed to the crew in orbit. Before Popovich and Artyukhin left after their 15-day stay, they prepared Salyut 3 for continued automatic use, and for a possible further manned visit.

The failure of Soyuz 15 to dock with Salyut 3 is described under the Soyuz entry. The planned 90-day programme was stated to have been 'completely fulfilled' when a 're-entry module' was ejected on revolution 1451 from a 275 × 253km orbit, and recovered in Russia. There had been American speculation that one of 2 spherical airlocks on Salyut 3, officially said to be for ejecting body wastes during manned

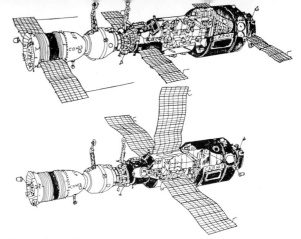

Comparison of Soyuz 11/Salyut 1 with Soyuz 17/Salyut 4 drawn by C. P. Vick. Solar panels on Soyuz ferry craft have been removed, and 3 large steerable panels placed on Salyut

missions, could be used for returning ground reconnaissance film. (The code words used during Soyuz 14 may have been associated with this.) For the first time, Salyut 3 was still in orbit when its successor was launched; the Salyut 4 launch announcement also stated that the flight of S3, when 'a number of concluding operations had been carried out by command from the ground', would be terminated. Its 6 months of controlled flight, involving by Dec 25 2950 orbits, had been double that of the initial programme. Over 400 scientific and technical experiments had been conducted on board, by means of over 800 control commands; 70 TV and 2500 telemetric communication sessions had been held; the stabilizer engines had been fired about 500,000 times, and the power system had generated 5000KWh of electricity. * The parameters of heat-exchange and energy supply were studied, and engineering and technical tasks included perfection of the onboard computer complex control system and of the energy supply system using revolving panels of solar batteries. Throughout the flight conditions in its compartments were similar to those on Earth, with atmospheric pressure of 835–850 mm of mercury and temperatures of 21–22°C. Orbit at that time was 270 × 235 km.

Salyut 4   L Dec 26 1974 by D1 from Tyuratam. Wt 18,900 kg. Initial orbit 219 × 270 km; incl 51·6°. Salyut 3 was still in orbit when its successor was launched; the purpose of Salyut 4 was stated to be further tests of its design, systems and equipment; 11 days later its orbit was raised to 355 × 343 km. The Soyuz 17 crew docked, apparently without difficulty, 17 days later—the longest gap between launch and boarding of a space station. Their activities are described under Soyuz 17; when they left Salyut 4 on Feb 9 1975, it was in a 361 × 334km orbit, and remained under automatic control. It was to have been occupied a 2nd time 2 months later, by the Soyuz 00 crew, whose launch was aborted. It therefore had to continue under automatic control until what was presumably the back-up Soyuz 18

*Salyut 3, it was claimed, was the first orbital station to have been permanently pointed towards Earth throughout its operation; stabilization was ensured by a system of liquid and air motors, and also by a special electro-mechanical stabilization system in combination with other systems. The flight ended on Jan 24 1975, when Salyut's retro-rockets were fired by ground command. It was placed on a descent trajectory to re-enter over the western Pacific, so that debris which came through the atmosphere fell harmlessly into the sea. Salyut 3 had in fact functioned for 7 months, more than twice as long as originally planned.

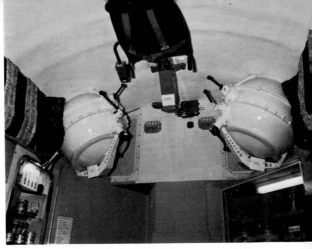

Salyut interior: food stowage and living accommodation— probably sleeping bags at sides

crew boarded it for their record 63 days' stay from May 24–Jul 26. Their final orbital change left the space station at 349 × 369 km, with about a year's life before natural decay. At the end of Oct 1975, when it had completed 4878 revolutions, and was in a 338 × 358km orbit, its equipment was still being used to study the constellation of Monoceros.

4 months after the departure of the 2nd crew, Soyuz 20, unmanned, was automatically docked with Salyut 4 on Nov 19 1975. It separated, and returned to Earth, 91 days later. Cosmonaut Feoktistov said when the docking took place that it demonstrated the ability to resupply a space station (though later it was denied that on this occasion fresh

supplies of propellant had been delivered), and to use an unmanned Soyuz as an ambulance to collect sick crew members. It was not clear whether he was envisaging the use of a 2nd docking port by the automatic Soyuz, so that 2 craft could be docked at the same time.

On Feb 3 1977, $1\frac{1}{2}$ days before natural decay would have resulted in re-entry, Salyut 4 was manoeuvred into a re-entry trajectory so that sections which were not burnt in the atmosphere would fall safely into the Pacific. The longest-lived Salyut to date, it had completed more than 12,000 revolutions; 2 crews had occupied it for a total of 93 days, and the unmanned Soyuz 20 had remained docked with it for 3 months.

Salyut 5 L Jun 22 1976 by D1 from Tyuratam. Wt ? 19,000 kg. Orbit 219 × 260 km. Incl 51·6°. Salyut 4 was still in orbit when its successor was launched. Amid speculation that Salyut 5 had 2 or more docking ports for unmanned resupply, the 2-man Soyuz 21 crew successfully docked and boarded 14 days after launch. Details of their 63-day flight and unexpected return can be found under Soyuz 21. Indications were that, like Salyut 3, this was mainly for military, rather than civil purposes. As misfortunes dogged Salyut 5's career, Soviet cosmonauts visiting America let it be known that it had only one docking port.

Following the failure of the Soyuz 23 docking attempt, Salyut 5's orbit steadily decayed, and re-entry was predicted for Mar 1977. However, trajectory corrections were made on Jan 14 and 18, raising the orbit to 275 × 256 km. By then, 3419 revolutions had been completed. Moscow announced that an Earthlike atmosphere was being maintained with an atmospheric pressure of 830–840 mm, and 21–23°C temperature. Then, on Feb 7, came the successful boarding by the Soyuz 24 crew. Controlled re-entry Aug 8 1977.

SOYUZ   Manned Spacecraft

History   Soyuz ('Union') is a long-term project aimed at the development of sustained flight, orbital manoeuvring, docking, scientific and technical investigations in Earth orbit, and ultimately at the assembling of manned orbiting stations. It was designed by Chief Design Engineer Sergei Korolev, and his assistant, Voskresensky, shortly before their deaths (*see* Voskhod spacecraft). The spherical and cylindrical shape of the Soyuz modules appear to be logical developments of their earlier vehicles. Soyuz is equipped for Earth-orbit missions up to altitudes of about 1300 km, and can be used either automatically or manually for both manoeuvring and docking. The system was successfully tested with an automatic docking of the unmanned Cosmos 186 and 188 satellites on Oct 30 1967, which was the first Soviet docking of any sort to be achieved. This was successfully repeated with Cosmos 212 and 213 in Apr 1968. The series has encountered many setbacks as well as successes; the first trial flight of Soyuz 1, in Apr 1967, ended in disaster; the Soyuz 10 crew, after achieving Russia's first manned docking, were unable to enter the Salyut space station; and the Soyuz 11 crew, after what was then a record 23 days in space, were killed by a catastrophic depressurization. As a result of this the crew was reduced from 3 to 2, so that spacesuits could be worn during lift-off and re-entry, and during all other critical manoeuvres. Until then, spacesuits were not considered necessary; comfortable leather jackets and helmets containing headsets were used. In any case, the re-entry module was not large enough for 3 men in suits. Various types of suit, for use during EVA and other experiments, had presumably been stowed in the orbital module and jettisoned with it until Soyuz 11. Despite the setbacks, there had been many achievements: the development of long-duration flight, the first space-welding experiments and the first transfer of crews from one vehicle to another, in a simulated orbital rescue—all well ahead of America. But there were more setbacks when Soyuz 15 went out of control during docking attempts with Salyut 3, and when in mid 1975, what should have been Soyuz 18 suffered man's first launch abort—happily with crew surviving an emergency return. As usual, Soviet scientists continued undaunted with development of Soyuz as what Gen Shatalov described as 'a universal spacecraft', for use both independently, and for carrying both freight and crews to space stations with multi-docking ports. Cosmos 772, a 3-day flight by an unmanned Soyuz type vehicle in Sep 1975, was thought at the time to be in preparation for a return to 3-cosmonaut crews.

Spacecraft Description

Overall Length:	10·36 m
Diameter:	2·29 to 2·97 m
Docking Probe:	2·74 m
Solar Wingspan:	10·06 m
Total Weight:	6000 kg
Total Volume:	9m³

There are 3 compartments: a Re-entry, or Landing Module, in the centre, flanked on one side by an Orbital Compartment, used as a workshop and for resting while in orbit; and on the other side by an Instruments Compartment. The Orbital and Instrument Compartments are jettisoned just before re-entry.

Re-entry Module   This is shaped like a car headlight, with

Interior view of control panel in Soyuz descent module simulator used for Apollo/Soyuz mission

an exterior coating as protection against re-entry heat, and interior insulation to provide both heat and sound protection. During re-entry interior temperatures do not exceed 25–30°C. The shape ensures aerodynamic lift, with deceleration normally producing no more than 3–4 G, compared with 8–10 G endured by cosmonauts during the ballistic re-entries of Vostok and Voskhod. The cosmonauts take their seats 2 hr before launch. The panel in front of the commander contains instruments and switches for the spacecraft systems, a TV screen, and optical orientation view-finder set up on a special porthole next to the panel. Orientation controls are on the commander's right, and manoeuvring controls on his left. There are portholes to starboard and port for visual and photographic observations. There are 4 TV cameras (2 mounted externally and 2 inside), which provide 625-line transmission, 25 pictures per second. The atmosphere regeneration system contains alkali metals which absorb carbon dioxide and simultaneously release oxygen, maintaining a 1 kg/cm² conventional nitrogen/oxygen atmosphere (compared with Apollo's all-oxygen atmosphere). Heat-exchange units condense excessive moisture and direct it to special moisture collectors. Water, as well as food, is carried in containers, since Soyuz does not use fuel-cells providing water as a by-product as in Apollo. Multi-channel telemetry systems store information in onboard memory units, and transmit it to Earth during regular radio sessions. The hatch, in the upper part, or nose, is used for entry before launch and for transfer after launch to the orbital compartment. The re-entry procedure is started by a retrograde firing lasting about 146 sec. The single main parachute, for which a back-up is available and which is preceded by a drogue, is deployed at 8000 m; solid-fuelled retro-rockets, fired at about 0·91 m above the ground, ensure that landing velocity does not exceed 3 m/sec. A direction-finder transmitter sends out one signal during descent, and another after landing, to aid search parties if necessary.

Orbital Module Mounted on the nose of the completed craft, with an estimated volume of 6·26 m^3, this provides sufficient room for the cosmonauts to stand up, and an area for work, rest and sleep. It has 4 portholes for observation and filming, controls and communications systems, a portable TV camera, photo and cine-cameras. A 'sideboard' contains food, scientific equipment, medicine kit and washstand. Communications and rendezvous radar antennas are mounted on the exterior. It is used as an airlock by closing and sealing the hatch communicating with the Re-entry Module; when depressurized, the external hatch can be opened for egress into space. For entry into a Salyut space station, a hatch in the nose is used. In the case of an 'active' spacecraft, a 2·74m docking probe is added to the nose; on a 'passive' craft, an adapter cone is fitted to receive the probe.

Service Module, Instrument Compartment Cylindrical, about 2·97m dia, this is at the rear, and like the Apollo Service Module, cannot be entered by the cosmonauts. It has a hermetically sealed instrument section housing the thermo-regulation system, electric supply system, orientation and movement control systems with a computer, and long-range radio communications and radio-telemetry. In the non-sealed section are 2 liquid-propelled rocket motors (the main and stand-by engines), each providing a thrust of 400 kg. These are used for orbital manoeuvres up to a height of 1300 km, and for braking purposes to start re-entry. A separate, low-thrust engine system provides attitude control. Mounted at the rear are 2 wing-like solar cell panels, each 3·66 m long; when deployed in orbit they provide a span of over 10·06 m, and 14 m^2 of solar cell area to gather power for the spacecraft system. A single 'whip' antenna extends forward from the leading edge, near the tip of each panel.

Escape Tower For lift-off and launch, an escape tower is mounted on top of the Orbital Compartment, capable of lifting the Orbital and re-entry Compartments clear of the rocket in the event of an emergency either on the launchpad or immediately after lift-off. It contains 3 separate tiers of rocket motors for boost, trajectory bending, and vernier control, to ensure that the spacecraft could be quickly removed from the trouble area.

Soyuz 1 Russia's first manned flight for just over 2 yr. Vladimir Komarov was launched on Apr 23 1967, into a 224km apogee, 201km perigee orbit, with 51·7° incl. Unofficial reports from Moscow had forecast that 2 spacecraft of a new type would be launched. The fact that for the first time a Soviet cosmonaut was being given a 2nd flight,

Mockup of 2 Soyuz spacecraft docked, as shown at 1973 Paris Air Show. Solar panels give this version longer flight capability

supported speculation that there might be attempts at docking and/or spacewalks. After the launch, the objects were said to be to check the systems and components of the new vehicle, to hold extended scientific and physical/technical experiments, and to continue medical and biological studies of the effects of spaceflight on the human organism. There appear to have been flight problems with Soyuz 1, however, and re-entry was ordered on the 18th orbit. At a ht of 6·5 km, the main parachute harness twisted, the spacecraft crashed to the ground, and Komarov was killed. He was the first man to have been killed during a mission in 6 yr of spaceflight. Earlier in the programme, centrifuge tests had shown that Komarov had developed an 'extra systole', or irregular heartbeat, and he had been suspended from the flight programme. Only after strenuous appeals had he been allowed to continue with command of Voskhod 1. There is no evidence, however, that he had any health problems during either of his flights. (Donald Slayton, one of the original 7 Mercury astronauts, developed a similar heart condition; as a result, he was the only Mercury man not allowed to fly.)

Soyuz 2 and 3   Soyuz 2, used as an unmanned target vehicle, was launched on Oct 25 1968, into a 224 × 185 km orbit with 51° incl; it was followed by Soyuz 3, manned by 47-yr-old Giorgi Beregovoi, on Oct 26 into a 225 × 205 km orbit. The Soyuz 2 launch was not announced until after Soyuz 3 was in orbit. Soyuz 2 was believed to be equipped with a docking collar, and Soyuz 3 made an automatic approach and rendezvous to within 180 m, but there was no docking. The spacecraft then separated to 565 km, and Beregovoi then carried out a 2nd, manually controlled approach; no distance was given. On Oct 28 Soyuz 2 made an automatic re-entry, thus successfully testing the parachute system. Beregovoi continued his flight, making regular TV reports, his work including photography of Earth's cloud and snow cover, and study of typhoons and cyclones. During orbit 36 an automatic manoeuvre changed the orbit to 244 × 199 km; then Beregovoi manually oriented the spacecraft, and switched on the automatic system for re-entry. The retro-rocket was fired for 145 sec. The descent went so well that the search party was able to see and photograph the final descent, at the end of the 64-orbit flight lasting 94 hr 51 min. With Soviet recovery ships stationed in the Indian Ocean, it was expected that Beregovoi would come down at sea, especially as Russia had made her first sea recovery of an unmanned spacecraft (Zond 5) in that area a month earlier. These preparations, however, were probably for use in emergency; after the successful landing of Soyuz 2 Beregovoi followed Russia's established landing procedures, and came down in a snowdrift in Kazakhstan. Asked afterwards whether his age had made it difficult for him to get accepted for this flight, Beregovoi said his height of 180 cm had proved more of a problem than his age.

Soyuz 4 and 5 Russia's first manned docking, followed by the spacewalk-transfer of 2 cosmonauts from one craft to another, was achieved on this mission. Soyuz 4, piloted by Vladimir Shatalov, was launched on Jan 14 1969 at an incl of 51°, and its initial orbit of 225 × 173 km was corrected on the 4th orbit to 237 × 207 km. This was in preparation for the launch of Soyuz 5 on Jan 15 into a 230 × 200km orbit; on board were 3 cosmonauts, Boris Volynov (Commander), Yevgeny Khrunov (Research Engineer), and Alexei Yeliseyev (Flight Engineer). On Jan 16, while Soyuz 4 was completing its 34th orbit and Soyuz 5 its 18th, the 2 vehicles were automatically brought within 100 m; Shatalov then took over manual control, and docked Soyuz 4 with 5. Outside TV cameras transmitted the docking process to Earth. The 2 craft were coupled mechanically, and electrical connections and telephone communications established. Volynov oriented the joined craft, which Russia claimed to be 'the world's first experimental space station', so that the solar arrays were exposed to the Sun; 4 compartments, it was pointed out, had become available to provide comfortable conditions for work and rest for the combined crew of 4. Immediately after the docking Khrunov and Yeliseyev donned a new type of spacesuit with self-supporting life systems, and egressed into space through the hatch of Soyuz 5's orbital module. Life-sustaining packs containing oxygen supplies and air-conditioning systems, were attached to

their legs so as not to get in the way as they used external handrails during their 37-min spacewalk into the Soyuz 4 orbital compartment.

The 2 craft were undocked after 4 hr, and the following day Soyuz 4, on its 48th orbit after 71 hr 14 min of flight, successfully re-entered and landed. Helicopters sighted the orange parachute as the spacecraft touched down on target in the Karaganda area, in a strong wind and temperature of −35°, and warm clothes were hurried to the site for the cosmonauts. Volynov, alone in Soyuz 5, successfully re-entered and landed on Jan 18, after 50 orbits and 72 hr 46 min. In effect, the mission had rehearsed the first emergency rescue in space. Leonov, the first spacewalker, said afterwards he had advised Khrunov and Yeliseyev, when spacewalking, 'to think 10 times before moving a finger, and 20 times before moving a hand', since abrupt and hurried movements built up heat in the spacesuit and could make a cosmonaut unfit for work. This mission, following shortly after America's first flight around the Moon with Apollo 8, saw a new style of Russian coverage. TV pictures, though still not released live, were shown only an hour after the launches, and were followed by TV tours around the spacecraft, conducted by Shatalov in Soyuz 4.

Soyuz 6, 7 and 8 These spacecraft were launched on successive days, and each remained in orbit for 5 days; the group flight, therefore, the first time 3 manned vehicles had been in orbit simultaneously, covered a total of 7 days. Soyuz 6, launched on Oct 11 1969 into an initial orbit of 185 × 222 km, carried Georgi Shonin (Commander) and Valeri Kubasov (Flight Engineer). The incl in the case of all 3 craft was 51·7°. Soyuz 7, launched on Oct 12, carried Anatoli Filipchenko (Commander), Viktor Gorbatko (Research Engineer) and Vladislav Volkov (Flight Engineer), into an initial orbit of 206 × 225 km. Soyuz 8, on Oct 13, carried Vladimir Shatalov (Overall Commander), and Alexei Yeliseyev (Flight Engineer), into an initial orbit of 204·5 × 222 km. Both crew members of Soyuz 8 were making their 2nd flight.

Early in the mission it was stated that Soyuz 6 carried extra scientific, instead of docking equipment; the other vehicles were believed to be equipped for docking, but none took place, despite much speculation during the flight that 7 and 8 would dock, and spacewalks would follow. Some observers thought there had been manual control problems with the improved autonomous navigation systems on Soyuz 7 and 8. At the subsequent Moscow news conference, however, Filipchenko said no spacesuits had been taken. Shatalov said docking would have been possible, but was not included in the tasks; he did, however, say there had been 'difficulties, as in every spaceflight'.

Objectives given included 'mutual manoeuvring' to test the complex system of controlling group flight by 3 spaceships. For the first time Soviet ground tracking stations were supplemented by 8 Academy of Science research ships on station in various parts of the world. A total of 31 orbital changes were made, mostly manually, by the crews, in a wide variety of rendezvous manoeuvres. On Oct 15 Soyuz 7 and 8 conducted a manual rendezvous to within 488 m of one another, observed by Soyuz 6 from a distance of several kilometres. This position was maintained for about 24 hr. Continuous radio and radar contact was maintained among the spaceships, and with tracking ships and ground stations, through Molniya satellites. Crew tasks carried out included scientific and photographic studies of near-Earth space, and Earth photography for geological, geographic and meteorological purposes. The activities which attracted most attention, however, were experiments in remote-control welding; for the first time in their space programme, Soviet scientists announced in advance that they were to take place. When assembling space stations, it was explained, cosmonauts would have to work outside the spacecraft for long periods. Welding processes were much preferable to joining parts with nuts and bolts, since the latter required revolving movements, which were difficult for weightless cosmonauts to achieve. On their 77th orbit, Shonin and Kubasov, in Soyuz 6, sealed themselves in their re-entry module, and depressurized the orbital workshop, creating the conditions known to be ideal for welding. They used a remote control panel to operate the 'Vulkan' experimental equipment in the workshop. Weighing about 50 kg, this consisted of a welding unit, turntable with specimens of welded metals, instrument unit with power pack, and safety shield to cover the welding unit. Kubasov tested 3 methods: compressed arc welding (low-temperature plasma); electron beam; and fusible (or consumable) electrode welding. The last 2, it was stated afterwards, were the most promising methods because solar energy could be used to heat the components. After repressurizing the orbital compartment, Kubasov transferred samples of the experiments to the re-entry module, and Soyuz 6 landed after completing 80 orbits and 118 hr 42 min. Soyuz 7 landed 1 day later, on Oct 17, after 80 orbits and 118 hr 41 min; and Soyuz 8 on Oct 18 after 80 orbits and 118 hr 41 min.

Soyuz 9 This established a new long-duration record of nearly 18 days' (424 hr) spaceflight, surpassing the previous record established 5 yr earlier by Gemini 7. It was the first manned launch made at night. The 2-man crew was Andrian Nikolayev, Commander, who had to wait 8 yr for his 2nd flight; and Vitali Sevastyanov, a civilian, as Flight Engineer.

Launch on Jun 1 1970, was into a 222 × 207·5km orbit, with 51·7° incl, increased by manual firings on the 5th and 17th orbits to 267 × 248 km. The primary aim was to study the physical effects of prolonged weightlessness; other work, however, included the study and photography of ocean behaviour, including coastal currents and surface temperatures, aimed at developing fish-searching techniques employed by Soviet fishing fleets. During the flight cabin pressure was maintained at an average of 1·12 kg/cm², or slightly higher than sea level, with 23·4% oxygen and the remainder nitrogen and other gases; average temperature was 22°C. Final descent was shown on Soviet TV—the first time the final stages of a Russian recovery had been seen in this way. Soviet doctors' concern about long-term effects appeared justified by the fact that both cosmonauts had difficulty in re-adjusting to Earth conditions. For 5 to 8 days they said they felt as if they were in a centrifuge, being subjected to 2 G (that is, feeling as if they were twice their actual weight); they found difficulty in walking, and sleep in bed was uncomfortable for 4 or 5 days. During the flight Nikolayev's daughter, Elena, was brought to Ground Control to talk to him to celebrate her 6th birthday; she was born within a year after Nikolayev married Valentina Tereshkova, the first woman in space, following their original flights.

Soyuz 10   L 3 days after Salyut 1 (*qv*). Russia's first experimental space station, Soyuz 10 had a 3-man crew: Vladimir Shatalov (Commander), and Alexei Yeliseyev (Flight Engineer), both making their 3rd flights; and Nikolai Rukavishnikov (Test Engineer). It was a pre-dawn launch on Apr 23 1971 into a 248 × 209km orbit, and incl of 51·6°. Mid-course corrections, transmitted out of range of Soviet territory by the research ship *Sergei Korolev*, started the

Soyuz 9 on its transporter just before pad erection at Tyuratam

automatic rendezvous procedure. Yeliseyev said the cosmonauts first saw Salyut on an 'optical instrument' at a range of 15 km. When the automatic procedures had brought Soyuz 10 to within 180 m of Salyut, the crew took over manual control, and Shatalov completed the final approach and docking during the 12th orbit of Soyuz 10 and the 86th orbit of Salyut. Cosmonaut Dr Boris Yegorov (who flew on Voskhod 1) observed later that the manoeuvre 'led to a considerable emotional load on the cosmonauts'.

The 2 vehicles remained docked for 5½ hr, and official statements said the principles of closing and berthing the craft, the operation of the new coupling units, and the complex radio engineering equipment were tested. The crew did not enter Salyut, and Western speculation was that either they had been unable to open the docking tunnel, or that the flight had been cut short because Rukavishnikov became ill. (Rukavishnikov had said during a communications session that the presence and advice of Shatalov and Yeliseyev 'helped him to get accustomed to weightlessness, and to overcome the unusual and rather unpleasant sensations arising as a result of the increased flow of blood to the head'.) Soviet reports, however, emphasized that Shatalov had carried out more than 10 complex manoeuvres during rendezvous, berthing, docking and undocking, and that docking had had to be done independently while the crew was 'out of radio contact with ground stations on Soviet territory'. (Ships were apparently not mentioned on this occasion.) It was also pointed out that docking a small spacecraft with a vehicle of much greater mass was more difficult than docking 2 Soyuz or Cosmos craft. Soyuz 10 remained in orbit for 16 hr after undocking; then, after 30 orbits, re-entered and landed in darkness (02.40 local time) 120 km NW of Karaganda on Apr 25. At a subsequent news conference Rukavishnikov, asked about Western reports of his health, said 'work in a weightless state was a joy'; after the 2nd day he had been too busy even to notice any unpleasant sensations. Reasons for the cosmonauts not entering Salyut, which continued its flight after Soyuz 10 re-entered, were apparently not discussed.

Soyuz 11 The 3-man crew made history by spending nearly 24 days in space—23 of them working in Salyut 1— but they died during re-entry owing to the failure of an air valve on the spacecraft. Soyuz 11, crewed by Georgi Dobrovolsky (Commander), Vladislav Volkov (Flight Engineer), making his 2nd flight, and Victor Patsayev (Test Engineer), was launched on Jun 6 1971, and following a 4th orbit correction, given an apogee of 217 km and perigee of 185 km, and $51 \cdot 6°$ incl. The purpose of the mission was 'to continue comprehensive scientific and technical studies in joint flight with the Salyut orbital scientific station'. On entering orbit, Soyuz 11 was 3000 km behind Salyut; the following day, on the 16th orbit, a 2nd manoeuvre initiated rendezvous, and automatic approach was carried out from 6 km to 100 m. The crew then took over, completing the docking manually, with the 2 craft in a 217×185 km orbit. The total weight of the 2 craft was over 25 tonnes. The vehicles were rigidly coupled, and electrical and hydraulic communications connected. The hatches of the airtight passage were opened and the cosmonauts entered the scientific station. It was announced that Soyuz spacecraft had been assigned the role of ferry-craft for orbital stations, with the task of carrying relief crews and supplies, and returning the results of experiments to Earth.

The preparation of Salyut as a manned orbital station took 2 days, during which the cosmonauts complained about having too much work. Soyuz 11 was then powered down for 23 days while work continued aboard Salyut (*qv*). On Jun 29 after, it was stated, completing their flight programme in full, the cosmonauts transferred their research materials and logs to Soyuz 11, and prepared to return to Earth. Undocking took place at 21.28, the crew reporting that separation was normal, with all systems functioning. At 01.35 on Jun 30 (a record 23 days 17 hr 40 min, and 380 orbits since launch), a normal firing of the braking engine, usually 146 sec, was carried out. From that moment, all communications with the crew ceased; but the automatic re-entry procedure of aerodynamic braking, parachute deployment, and soft-landing engines continued smoothly, with an on-target touchdown. A helicopter-recovery crew, landing simultaneously with the spacecraft, opened the hatch and found the crew lifeless in their seats. Although telemetry had shown the cosmonauts to be well after their long period in space, they had reported feeling fatigued on the day of their return; consequently many newspapers carried banner headlines speculating on the possibility that the debilitating effects of their long period of weightlessness had led to the cosmonauts being unable to withstand the sudden return to G-forces during re-entry. A Soviet Government Commission, however, announced 12 days later that a rapid depressurization in the re-entry vehicle, as a result of a loss of sealing, had led to the cosmonauts' deaths. They added that an inspection of the vehicle showed there were no failures in its design. Nearly 3 yr later, after NASA had insisted that they must have a full report on the accident before the joint ASTP flight could take place, it was learned that it was not a hatch seal (as had been assumed in the West) but a faulty valve, which caused the fatal decompression. Modifications, since successfully tested on Cosmos 496 and 573, and on Soyuz 12, included not only an improved valve system, but a return to the wearing of spacesuits by cosmonauts, with a direct oxygen supply. These changes took up so much room that it meant only 2, instead of 3 men could be carried in the existing version of Soyuz, with serious effects on the Salyut programme. On Jul 2 the cremated remains of Lt-Col Dobrovolsky, aged 43, of Victor Patsayev, who celebrated his 38th birthday aboard Salyut, and Vladislav Volkov, aged 36, were placed in the Kremlin Wall, following a Moscow funeral attended by Brezhnev, Kosygin and other Soviet leaders.

Soyuz 12 This was announced as a 2-day flight as soon as launch had taken place on Sep 27 1973, into a 194 × 249km orbit, and the usual 51·6° incl. Only 2 crew were carried, Lt-

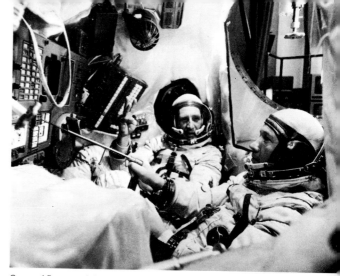

Soyuz 15 crew giving inflight demonstration of operating controls with extending arm when strapped in

Col Vasily Lazarev, who is a physician as well as test pilot (and was no doubt chosen for that reason, following the Soyuz 11 disaster), and Oleg Makarov, Flight Engineer, both making first flights. There is almost certainly no room for a 3rd crew member in the command module now that spacesuits must once again be carried; though in this case it seems they were worn only for re-entry. The objects were stated to be testing of structural modifications, improved flight systems and controls, and Earth spectrography to obtain data 'for the solution of economic problems'. Details of a press conference released shortly after launch showed that Shatalov had announced the flight 1 day beforehand, a few hours after the 2nd Skylab team had splashed down, offering Bean and his crew congratulations as he did so. 1st day manoeuvres changed the Soyuz 12 orbit to 326 × 345 km. After soft-landing 400 km SW of Karaganda, Soyuz 12 was said to have functioned faultlessly.

Soyuz 13   The immediate result of this launch on Dec 18 1973, was that for the first time Soviet cosmonauts and American astronauts were in space simultaneously. Major Pyotr Klimuk, Commander, and Valentin Lebedev, Flight Engineer, both making first flights, joined the 3rd crew of 3 Skylab astronauts in Earth orbit, then in the 32nd day of their 84-day mission. Differing frequencies did not permit direct exchanges between the 2 crews, nor, during the 8-day Soyuz mission, did either crew report sighting the other spacecraft. Soyuz 13's initial orbit was not given; but after 5th orbit manoeuvres was stated to be 272 × 225 km, incl 51·6°. Although Cosmos 613, launched on Nov 30, was believed to be a Salyut-type vehicle, the orbits were not compatible for docking, and there is no evidence that this was ever intended. On their 2nd day, the cosmonauts were said to be 'degreasing' an Orion-2 telescope system, first tested in Salyut 1, and carried outside Soyuz 13's command module, with instrument panel inside. This was used both for studying selected stars for space navigation purposes, and for surveying Earth's resources, perfecting long-term weather forecasting, and predicting natural calamities. The orbital compartment contained an Oasis-2 'greenhouse'—an experimental biological system to grow nutritive protein during prolonged spaceflights. This was a follow-up of a 6-month ground experiment during which 3 researchers lived

in a closed ecology. Oasis-2 consists of 2 connected cylinders, one of which breaks down waste products, while the other provides the ability to cultivate water-oxydizing bacteria grown on hydrogen. The first 'harvest' of protein was reported to have been gathered only 2 days after launch. Later Konstantin Feoktistov, former cosmonaut and now Doctor of Technological Sciences, disclosed that the Soyuz 4/5 docking in 1969 was the first step towards developing 'a universal spaceship', which Soyuz 13 was continuing to do. Re-entry and landing on Dec 26 was admitted by Shatalov to have been an anxious time because of a heavy snowstorm in the landing area, 200 km SW of Karaganda. But recovery followed touchdown within a few minutes, and both cosmonauts were said to be in good health.

Soyuz 14   A night launch, at 21.51 Moscow time, on Jul 3 1974, 8 days after Salyut 3 (*qv*). The Commander, Col Pavel Popovich, was making his 2nd flight after an interval of 12 yr; the Flight Engineer, Lt-Col Yuri Artyukhin, was making his first flight, despite having been a cosmonaut for 11 yr. The veteran crew was probably an indication of the importance attached to the mission, with Soviet scientists anxious to demonstrate to American astronauts, at that time training in Star City for the 1975 ASTP, that the Soyuz technical troubles had been overcome. The initial orbit, followed by a trajectory correction, placed Soyuz 14 (wt approx 6570 kg), in a 255 × 277km orbit, at 51·5° incl, and 3500 km behind Salyut 3. The Flight-Director, Alexei Yeliseyev, said the crew caught up with the minimum fuel expenditure for orbital corrections. At 1 km, the automatic approach unit started functioning, and at 100 m the crew took over manual control for docking, which took place about 26 hr after launch. Russia's first docking since the Soyuz 11 tragedy, it was only the 4th manned docking carried out by Russia. Major solar eruptions, which began unexpectedly the day after launch, resulted in some Soviet physicists suggesting that an emergency return should be made. But a careful watch on the radiation gauges on the cosmonauts' suits showed no significant increase. On a polar or lunar flight however, such intense solar activity might have made it necessary to abort. Popovich and Artyukhin quickly settled down to a routine which included much time spent on Earth observations—mapping the Soviet Union for mineral deposits, soil resources, glacier movements, etc. Code words used by the cosmonauts suggested that military reconnaissance was among their activities; special targets were reported by America (presumably having been seen by their 'spy' satellites) as having been laid out near the Tyuratam launch site for the cosmonauts to observe and photograph, in order to measure the reconnaissance potential of Salyut space stations. 4 daily exercise periods, totalling 2 hr, and involving the use of special suits and elastic belts harnessing them to a running track, helped to keep this crew fitter than any previous Soviet crew. During running and jumping exercises, and tests of man's load-carrying abilities during weightlessness, pendulum instruments measured the vibration effects on Salyut; high-precision instruments had given incorrect readings on previous flights as a result of such vibrations. Speculation that the flight would end with Russia's first manned 'splashdown' began during the mission when a pre-recorded interview with Popovich disclosed that they had rehearsed sea recoveries. However, 3 hr 18 min after undocking on Jul 19, the usual land descent was made on the 16th day, 140 km SE of Dzkezkazgan, Kazakhstan, only

View of Soviet Flight Control Centre near Moscow during Soyuz 16—released during planning for Apollo/Soyuz

2000 m from the target point. The crew egressed without waiting for the recovery team to arrive.

Soyuz 15   Another disappointing failure. Instead of making a 2nd visit to Salyut 3, and probably staying in it for about a month, Soyuz 15 was unable to achieve what should have been only Russia's 5th manned docking, and after about 36 revolutions had to make an emergency return to Earth at night. Launch was also at night (22.58 Moscow time), on Aug 26 1974. Official announcements did not specify the orbit. The Commander, Lt-Col Gennady Sarafanov, 32, and Flight Engineer, Col Lev Demin, 48, the first grandfather in space, were both making their first flights.

Initially placed in a 230 × 180km orbit, Soyuz 15 was expected to have rendezvoused with Salyut 3 in its 16th revolution in a 259 × 249km orbit. At that time, however, Soyuz 15 was observed to be 120 km ahead of and below Salyut 3. Official reports merely said that during their 2nd working day Soyuz 15 approached Salyut many times, and made observations and inspections as it did so. Over a year later, Gen Shatalov revealed that the failure was associated with a fully automatic docking system being tested for later use on Soyuz 'tanker' spacecraft to replenish space station consumables and thus give them a much longer orbital life. Twice the system went wrong, throwing Soyuz 15 out of control, when the distance between it and Salyut 3 was only 30–50 m. Following what must have been a major emergency, Sarafavov and Demin returned to Earth.

Soyuz 16 L Dec 2 1974, this 6-day flight was a near-perfect rehearsal for ASTP. Flown by the ASTP 2nd Prime Crew, with Anatoli Filipchenko as Commander and Nikolai Rukavishnikov as Flight Engineer, Moscow announced the orbit as 177 × 223 km, with 51·8° incl, after a course correction on orbit 5. NASA officials later said that their tracking data showed an incorrect initial orbit, and an unusual sequence to get out of that orbit. According to NASA, the initial orbit for an ASTP practice mission should have been about 230 × 187 km. Soyuz 16's initial orbit was in fact 254 × 169 km; the spacecraft then went to a lower orbit to correct, and used a double sequence to get back to the acceptable ASTP orbit. The incident was reassuring, however, to the extent that it demonstrated the Russians' ability to manoeuvre into a proper rendezvous orbit. NASA had been told 3 months before that 2 unmanned Cosmos flights had been flown as ASTP rehearsals earlier in the year, and that a manned Soyuz rehearsal would follow. No advance date was given,

Soyuz 16: helicopter views of recovery procedure. Note dust raised by spacecraft retrorockets. This was Soviet rehearsal for Apollo/Soyuz

however, and first information received by NASA was a telephone call from Moscow to Houston at 04.40 local time, 1 hr after the launch, and $\frac{1}{2}$ hr before the public announcement. Dr Glynn Lunney, NASA flight director for ASTP, and others were then summoned from bed to Mission Control. At the start of the flight, Dr Konstantin Bushuyev, Soviet technical director for ASTP, said the spacecraft was identical with the ASTP vehicle, and the mission was intended to test modifications—mainly the Soviet part of the docking mechanism, and spacecraft ability to change its atmosphere. Additional tasks included Earth observations and photography. Dr Bushuyev pointed out that America was using left-over spacecraft and rockets from their Apollo programme, with changes concentrated in the combined docking and transfer compartment, which did not need flight testing. On the other hand, Soyuz spacecraft and rockets were still in production, so that modernization was possible. Soyuz 16's improved life-support system, able to cater for 4 men at a time, contained 'additional air regeneration blocks'. Soon after launch, pressure was apparently reduced to 530 mm of mercury (10 psi) to make it more compatible with that of Apollo. According to Rukavishnikov, this would enable spacemen transferring from Apollo to Soyuz or vice-versa to spend only 1 hr, instead of 2 hr, of acclimatization time in the docking tunnel. Soyuz 16 also carried 'an imitation docking ring', apparently simulating part of the Apollo mechanism; this enabled Filipchenko and Rukavishnikov to carry out about 20 simulations of the retraction and other procedures involved in docking and undocking, before they finally jettisoned the docking ring. NASA technicians also believed that modifications to the Soyuz control system were checked: one involved a method of disconnecting the Soyuz control system, for safety purposes, while Apollo was carrying out the actual docking; another involved reversing the normal control system, giving Soyuz the ability to roll only to the right while under automatic control. Had this not been done, Soyuz would have been required to roll 300° instead of only 60° just before docking, to match an Apollo roll manoeuvre needed to align the US vehicle's high-gain antenna with ATS-6. Before re-entry preparations, the crew also succeeded in rehearsing the 'artificial solar eclipse' manoeuvres planned for ASTP just after undocking. On Dec 8, Soyuz 16's braking engine was fired for 166·5 sec. The parachutes opened at a height of 'less than 10 km'; a recovery aircraft spotted the spacecraft at 5 km. Touchdown was on ground frozen to a depth of 40 cm, 'which meant the landing would be soft', 30 km NE of Arkalyk in Kazakhstan. Only 5 min after landing, Dr Bushuyev phoned Dr Lunney in Houston to tell him that the flight was successfully concluded. The crew, who were recovered in good condition, had not made a single mistake. There had been slight US disappointment that Soyuz 16 had not carried the radio equipment to enable the crew to talk directly to Houston; in an oblique reference to this, Yeliseyev said that all that remained after this flight was to rehearse the Moscow-Houston communication line 'in the next few months'.

Soyuz 17 Another night launch, made at 00.43 Moscow time (21.43 GMT) on Jan 11 1975, stated afterwards to have been 'put into the calculated orbit of an artificial Earth satellite', this appears to have gone well from the start. It followed Salyut 4 by 16 days. After correction the orbit was 293×354 km, with $51·6°$ incl. Both crew members, Lt-Col Alexei Gubarev, Commander, and Georgi Grechko, Flight Engineer, were 43, and making first flights. The latter helped to design the original Luna soft-landers, and this flight was possibly in part a reward for that achievement. Docking with Salyut 4 was achieved apparently without difficulty, after about 30 hr. A detailed official description said that approach was made automatically up to 100 m; with Salyut in a circular 350 km orbit. The Soyuz onboard automatic search and approach systems picked up Salyut at a distance of about 4000 m. The propulsion system was switched on, and approach rate increased up to 12 m per sec. At 100 m, the cosmonauts asked permission to switch to manual control. The spacecraft went out of ground contact as the Soyuz docking latches were ready to grip the station's docking unit, and ground control had to wait for the next orbit to learn of its success. The crew, without spacesuits, opened the hatchway, floated through the transfer hatch, switched on Salyut's lights, energized the radio transmitters, and began inspecting their equipment. When they opened the Salyut hatch, they found a notice left by Soviet engineers where a normal home would have left the doormat: 'Wipe Your Feet'. By 13.00 Moscow time, on Jan 12, they had occupied their Salyut berths. It took Gubarev and Grechko rather longer than expected to adapt fully to weightlessness—from 5 to 8 days—probably because of their 'irresistible and avid urge for work'; during the first week they worked between 15 and 20 hr a day, and as a result became fatigued and had some difficulty in sleeping. By the 8th day, Grechko had recorded over 4·8 km of weightless movement around the 2 compartments. Like Skylab, Salyut was equipped with a teleprinter, used to send up instructions in connection with the stiff programme of experiments. It was also used to send up personal letters and messages; and once, instructions to enjoy some leisure time when the crew, instead of having 1 day off in 6, continued to work. Each night, 1 hr was allowed for leisure before bed, and there were 4 small meal breaks each day.

A major experiment not yet tackled by America's Skylab crew was recovery of evaporated water produced by the body—up to 1 litre per man per day could be recovered by Salyut 4's regeneration unit, it was claimed. The crew used this water for washing, and also tasted it, reporting that 'it was the same as running water'. Skylab, it was pointed out, had carried 3 tonnes of drinking water; the Salyut 4 experiment would make it possible to reduce stocks of water to a minimum in future flights. Another indication of Russia's steady progress towards long flights to the planets was the onboard 'cultivator', in which green peas were grown. On some days, over 6 hr were devoted to studying the Sun with the orbital solar telescope in Salyut's funnel-shaped housing (in which ground reconnaissance equipment is also located). The Sun's rays are reflected on to a flat mirror and then on to a spectograph, which is registered on a film camera inside the cabin; the rays then pass on to another 2 mirrors which the crew can observe. Other instruments enable them to record the details of what is being studied. 2 X-ray telescopes and an infrared telescope were also used to study the Crab Nebula and the planets Mars, Saturn and Jupiter. Looking inwards, Gubarev and Grechko measured the transparency of the Earth's upper atmosphere, in experiments to check the water vapour and ozone which protects man from the Sun's ultraviolet radiations. Several cameras were also used to study and photograph the surface; particular attention was paid to selected areas of Russian territory in southern Europe, and areas of Central Asia, Kazakhstan, and the Far East.

Before returning, the crew used a remote-control system to spray new reflecting layers on to the lenses of the solar telescope, to counteract the dust and condensed vapours which had gathered on them. This was caused, it was explained, by micrometeoroids which knock particles out of the space station skin, and create a fine dust around it. It seems possible that this pollution was added to by the fact that the Salyut 4 crew ejected their waste matter through an airlock—something avoided, at considerable inconvenience, in Skylab. The Russian view is that the waste packages will drift away, and ultimately be burnt during re-entry.

In Western countries, so little interest was taken in this flight that most people forgot the crew was in orbit until their return was reported—even though it beat Skylab 1 for duration, and became man's 3rd longest flight. This may have been partly because the crew themselves preferred it that way; on one occasion, when ground controllers turned on the TV, Grechko paused briefly to comment: 'Publicity isn't the best thing during our work'. But on the 19th day, a rest day, the crew did volunteer to do an additional TV report, during which they showed Salyut's 'Red Corner', with banners and souvenirs on display.

Conversation monitored at Kettering, in England, indicated that, 3 days before re-entry, the crew had already begun the transfer of scientific materials, cassettes with still and cine film, and log books, from Salyut 4 into Soyuz 17. No EVAs, or spacewalks, were required to recover film or other samples. Although the crew had kept fit during over 500 orbits by conscientious use of Salyut's ergometric bicycle and running track, they were advised to wear gravity suits for re-entry. With Salyut then in a 361×334km orbit, undocking was at 09.08 Moscow time on Feb 9. Landing took place 'in complex weather conditions', with wind velocity of 20 m per sec, cloud ht of 250 m, and visibility 500 m, 110 km NE of Tselinograd, Kazakhstan. Rapid detection and recovery of the cosmonauts was carried out by the search-and-rescue team. The mission had lasted 29 days 14 hr 40 min. Gubarev was found to have lost 2·7 kg and Grechko 4·5 kg. Both recovered this weight about a week after landing, but both were said to be still tiring easily more than a fortnight after the landing.

Soyuz 00 What was supposed to be Soyuz 18, sending a 2nd crew to Salyut 4 for a stay of up to 60 days, ended dramatically after about 9 min. According to a Soviet announcement, issued $1\frac{1}{2}$ days after the launch on Apr 5 1975, 'on the 3rd-stage stretch the parameters of the carrier rocket's movement deviated from the pre-set values, and an automatic device produced the command to discontinue the flight under the programme, and detach the spaceship for return to Earth'. It was the first time on either a Soviet or US manned flight that an abort became necessary between lift-off and orbital insertion. But the 2-man crew, Col Vasily Lazarev and Flight Engineer Oleg Makarov (who flew together on the 2-day Soyuz 12 mission) landed safely near the Siberian town of Gorno-Altaisk, in the foothills of the rugged Altai mountain range, N of a point where Mongolian, Soviet and Chinese frontiers meet. The cosmonauts were later said to be 'feeling well', but it seems certain they had a very uncomfortable time after landing. Subsequent US reports said that the upper stage of the SL-4 rocket ignited for only about 4 sec before cut-off occurred. During the emergency, the cosmonauts, who, unlike Apollo crews, have no control over the automatic abort sequences, 'indicated substantial concern' about whether the system was working properly; they also sought repeated reassurance that they were not going to land in China. For short periods during the descent, the spacecraft was pulling up to 14–15 G. The delay in announcing the launch and abort was probably because it took about a day to locate and rescue the crew. Western observers, after noting that Salyut 4, then in orbit, had been manoeuvred into a suitable docking orbit, had worked out the exact launch time; but because no orbit was achieved, failed to detect the launch. It would, however, have been fully monitored by America's military satellites and radars; it was probably this that prompted the official Soviet announcement about a mission that they would have preferred to forget. Later to allay American concern about ASTP, a scientific attaché at the US Embassy was told that the Soviet booster being used was an early version, 'less diligently checked' than the rocket to be used on the joint mission. The crew chosen for this flight was an indication of its importance: Col Lazarev is also a doctor; he and Makarov had flight-tested the improved spacecraft after the Soyuz 11 disaster. On this occasion he was no doubt intended to provide Russia with medical progress reports on the effects of long-term weightlessness.

Soyuz 18 This flight was also successful from the start; 63 days in space (61 of them in Salyut 4) was a record for Soviet spaceflight, and 2nd only to the 84 days of the final US Skylab. Shortly after the launch, at 14.58 GMT on May 24 1975, Moscow was able to announce that 'continued experiments with Salyut 4' were intended. Commander was Lt-Col Pyotr Klimuk, with Flight Engineer Vitali Sevastyanov, both making 2nd flights. According to the Novosti Press Agency, Sevastyanov became the first journalist in space; not only did his wife, Alevtina, work at Novosti, but he himself was a commentator of a popular science series, 'Man, Earth, Universe' on Soviet TV. Gen Shatalov, director of cosmonauts' training, aroused curiosity by announcing that the aims included the 'search for new possibilities for using both an individual spacecraft and a group of spacecraft to carry out research and applied scientific tasks in near-Earth orbit'. Soyuz 18's initial orbit was 247×193 km, with the usual $51 \cdot 6°$ incl. By orbit 19, when docking operations began, at least 3 course corrections had been made, 2 by firing the main engine. Docking was achieved on May 26, apparently without difficulty, with Salyut 4 then on its 2379th orbit, with 344km perigee and 356km apogee. As on previous successful dockings, the crew switched to manual control at a distance of 100 m; the manoeuvre was completed on Earth's dark side. Sevastyanov reported that 'not one gramme of fuel' was wasted in absorbing the shock. As they entered Salyut they found a notice 'Welcome to our communal house', left by the previous crew, Gubarev and Grechko, which had been intended for Lazarev and Makarov. Having checked the space station's life support systems and found them in order, the crew settled down to studies of the Sun, planets and stars in various bands of the electro-magnetic radiation spectrum; geological and morphological studies of the Earth, physical processes in the Earth's atmosphere, and biological research on themselves. To the usual collection of fruit flies, frogspawn and chlorella was added flour beetles, with a supply of dry yeast for food, to see whether, in weightlessness, they could still maintain their reputation for rapid development. On the 2nd day Klimuk and Sevastyanov loaded both cine and still cameras with film; the latter switched on dust filters to clean the atmosphere. About 2 tonnes of scientific equipment had to be checked and reactivated—a task needing 48 hr. 'A few problems' were reported, but nothing serious. One of the 6 gas analysers was malfunctioning; also 1 of 6 pumping condensers had to be replaced by a hand pump. By the 4th day both cosmonauts had adapted to weightlessness, and their health remained good throughout the flight; afterwards Klimuk said that the body 'remembers' weightlessness, and 'knows ways' to adapt itself. By the 5th working day they had settled down to the routine of taking one another's blood samples, and regular exercise periods etc. The running track introduced on Salyut 3, together with the special suits worn by the Soyuz 14 and 17 crews, to put loads on the muscles to simulate gravity, had been so successful that both the loads and the exercise periods were increased on this flight. The suits were only taken off at night. On Jun 1 and 2 the characteristics were studied of the constellations of Scorpio, Virgo and Cygnus with a mirror X-ray telescope. Studies of ripening crops etc, on Earth, provided a contrast with the earlier winter photographs taken by Gubarev and Grechko. Biological experiments included observations of the 2nd generation of drosophila (fruit flies)—descendants of those taken aboard by the first crew—and flour beetles taken by this crew, because this insect is a rapid developer. In the 'Oasis' system for plant growing, the first crew succeeded in getting only 4 out of 28 peas to sprout in 29 days. But although 13 of the 16 peas planted by the 2nd crew sprouted within 4 days, the young plants died 4 weeks later; Sevastyanov thought this was because they had been scorched by the lamps switched on for a live TV transmission from Salyut 4. By the 3rd week the outside world had almost forgotten that the Soyuz 18 crew was still in orbit. But Moscow announced details of their new 'Kaskad' (Cascade) navigation system, first tried out by the previous Salyut 4 crew. This had reduced fuel consumption 'by several times' by orientating the space station with great precision and accuracy for photography over the Soviet Union, Central Asia, the Kurile Isles, etc. A 10-day period in Jun, during which Salyut was passing over Russia in daylight, was devoted to Earth resources; during 60 orbits, 2000 photographs were obtained. Another series of experiments called 'Emission' studied the distribution of energy in the upper atmosphere and 'the mechanism of its intensive heating'. Russia's determined progress towards planetary spaceflight was emphasized by studies of the effect of the space environment on Salyut 4's hull. As Klimuk and Sevastyanov entered their 5th week in space, Moscow made only casual references to the fact that they had broken the 30-day record established earlier by Soyuz 15; by Jun 25 they had covered 21M km, and completed 500 orbits. Salyut 4 itself had completed 6 months in space, and the living conditions had worked so well that not once had the warning light come on. Salyut 4's energy resources were described as 'practically inexhaustible'. Sevastyanov celebrated his 40th birthday on Jul 8, their 45th day in space. A celebration dinner included spring onions which had been grown on board.

When Leonov and Kubasov were launched on Jul 15 into an orbit 125 km below that of Salyut 4, Klimuk and Sevastyanov had just completed their 52nd day in space. Both were in excellent health and spirits; their only problem seemed to be that Soyuz 18 was going through a period when it hardly entered the Earth's shadow at all; extra fans were needed inside to keep the temperature down. As the 2 Soviet crews passed the next day, Leonov congratulated the 'old timers' in space—the first of 2 such exchanges. As preparations began for their return, a Soviet Forestry Commission official said the Salyut 4 crew had opened up new possibilities for plotting forests, running them economically, distinguishing coniferous from deciduous trees, and detecting any that were diseased. Major scientific achievements claimed for the mission included 600 pictures of the Sun provided by the solar telescope; recordings of X-ray radiations from 10 sources in deep space; photography of 8·5M sq km of Soviet territory; and 2 scientific 'firsts': spectographic investigation of the polar lights and of noctilucent, or luminescent clouds. Data had also been obtained showing that the star Cygnus X-1 was 'a black hole'.

While Apollo was landing on Jul 24 at the end of the joint Apollo/Soyuz 19 flight, Klimuk and Sevastyanov began preparations for their own landing in Soyuz 18. The Salyut/Soyuz orbit was changed to 349 × 369 km, and about 50 kg of film tape and other research material packed into the descent module. They also prepared Salyut 4, which by then had been in orbit for over 7 months and completed 3352 revolutions, for continued automatic operation under ground control. About 24 hr before re-entry, a short test-burn was made on the main Soyuz propulsion system, to ensure it was operational after 2 months in space. (Soyuz capability is still a major limiting factor on long-duration flights. Before this mission an unmanned Soyuz, designated Cosmos 613, was flown for 60 days to ensure that it would remain operational for that length of time.) Undocking occurred 3 hr 22 min before landing, which was at 14.18 GMT on Jul 26, well within the target area, 56 km E of Arkalyk, in the same area as the Soyuz 19 landing 5 days earlier. For the 2nd time, live TV was shown of the descent and landing. Doctors had asked the cosmonauts to agree to being carried out of the spacecraft on stretchers after their long period of weightlessness; but both refused and insisted on walking out. It was 2 days, however, before Klimuk took a 10-min walk, and Sevastyanov called for a tennis racket. During the flight, Klimuk lost 3·8 kg in wt, and Sevastyanov 1·9 kg. The latter recovered his lost weight in his first day

Soyuz 19: rear view taken by Apollo crew during joint flight

back on Earth. Opening a space exhibition in Belgrade a few weeks later, Sevastyanov said that one result of their flight was that the whereabouts of even the smallest ore deposits under 22M sq km of Soviet territory (more than one-third of the country) was now known. They had brought back so much new data that no new crews would have to be sent into space for the next 6 months; an unexpected revelation that Russia, as well as America, was to have a period without manned flights.

Soyuz 19   This first joint Soviet/US flight, launched on Jul 15 1975, with Alexei Leonov and Valeri Kubasov as crew, is fully described in the International Section.

Soyuz 20   Launched unmanned on Nov 17 1975, in an apparent rehearsal of both a space station re-supply mission, and a 90-day manned flight, Soyuz 20 docked automatically with Salyut 4 on Nov 19 'with the aid of onboard radio devices and computer installations'. It was the first time an unmanned spacecraft had been given a Soyuz designation, and the first time Russia has achieved 3 dockings with a Salyut. The docked orbit was 367 × 343 km. Konstantin Bushuyev, Apollo/Soyuz technical director, in America at the time, disclosed that no transfer of propellant was involved, and a decision on whether to send a 3rd crew to Salyut 4 depended on the success of this mission. This mission picked up the 'tanker' experiments, aimed at prolonging space station life, started on Soyuz 15, when the automatic docking system failed. There was speculation too, that Soyuz 20 might be the test flight of a bigger spacecraft, able to carry the 3-man crew really needed for Salyut missions, and also tying in with the Saturn 5-type launcher, known to be nearing operational readiness. Soyuz 20, with its load of tortoises and other livestock, gladiola bulbs etc, returned to Earth on Feb 16 1976 after undocking and making 2·6 revolutions before re-entering.

Soyuz 21   With Col Boris Volynov as Commander (making his 2nd flight) and Lt-Col Vitaly Zholobov as Flight Engineer, Soyuz 21 was launched on Jul 6 1976, on what became Russia's 2nd longest flight, 14 days after Salyut 5 into an orbit of 253 × 193 km; Salyut 5's orbit at that time was 275 × 220 km. After 1 day in orbit, during which its orbit was corrected to 254 × 280 km, Soyuz 21 docked with Salyut 5; the operation took only 10 min, with minimum use of fuel and was probably Russia's smoothest docking so far. Gen Shatalov, probably giving warning that no dramatic space walks or spectaculars should be expected, said no 'basically new' engineering tasks had been set; the main trends in space research for the current 5-yr plan (presumably starting in Feb 1976) were 'tasks of an applied nature'. The combined weight of Salyut 5/Soyuz 21 was given as 25 tonnes, and the length as just over 26 m. By Jul 10, when Salyut 5 had completed 284 revolutions, the orbit had been raised to 269 × 281 km. The crew were studying an aquarium which included 'a very pregnant guppy'; the melting and hardening of molten metals in weightlessness with an ingot of bismuth, lead, tin and cadmium; and, by means of a hand spectograph rather like an amateur cine camera, were studying aerosol and industrial pollution of Earth's upper atmosphere. The instrument could be used for spectography of layers of atmosphere in the rising and setting Sun, and against various types of Earth's land or water surface. An 'active day of rest' on Jul 11 included exercises, medical checks and TV reports. Another experiment operated by Zholobov, demonstrated how liquid in a sphere, connected to another empty container above it, could be transferred from below to above, without the use of pumps or other equipment 'under the effects of surface tension'. Zholobov said the method could be used for spacecraft refuelling. During a 2nd rest day on Jul 17, details were given of Soyuz 5's 'sports ground' which differed from earlier ones. This time there was no bicycle ergometer; the cosmonaut strapped himself on to a moving pad, so that he had to move—either walking or running—at the speed of the pad. With body sensors transmitting their heart rates to Earth, it was said the cosmonauts 'enjoyed a physical workout for a couple of hours a day'. Weightlifting and exercises with dumb-bells were included (though presumably in space it would become 'weightpulling' once the weights had been given momentum). At this stage it was merely stated that the cosmonauts felt well. By Jul 30 the cosmonauts were reported to be taking Earth pictures of areas likely to contain mineral deposits, studying seismic activity, 'appreciating the danger of mud streams in mountains' and exploring areas on which hydro-engineering structures were to be built. They were said to 'feel well' several times, with no further details. By Aug 6, having completed a month in orbit, they had also carried out studies of the solar corona; by Aug 9 they had extracted from the crystallizer one of the crystals they had grown in space over a period of 26 days, placed it in a container for return to Earth, and started a 2nd experiment with a dye added, to study the diffusion of the mixture into the crystals grown. Discussing their medical condition at this point, there was a note of anxiety in a report by Dr Igor Pestov, who said the doctors were 'sometimes worried' because, due to non-stop work, the cosmonauts did not always manage to get proper rest; they had each lost about 1·5 kg in weight, and had higher blood pressure; there was nothing exceptional or dangerous about it, but steps were being taken to 'normalize' these reactions. During the 5th week, 2 days were devoted to medical experiments, studying the cosmonauts' hearts, body masses etc, following which they were said to be 'feeling fine'. First details were also given of an instrument aboard Salyut 5 designed to test the cosmonauts' reactions. It consisted of a small unit with 3 coloured lamps and 3 key switches, a master switch, starting button and control lamp, with a transfer switch, for work regimes. Cosmonauts had to answer lamp signals by pressing the appropriate keys; and their reactions were said to be 'somewhat slower' in space. The 6th week included exercises in manual control of the space station, and more medical checks. The Salyut systems were reported to be functioning normally, and the cosmonauts to be feeling well. Aug 14 was 'the most festive and joyous day of their flight', when schoolchildren from Vietnam, 'Congo', Cuba, Poland, Russia, the US, Czechoslovakia, Sri Lanka and Chile travelled from an international camp on the Crimean coast to the Yevpatoria Control Centre. They talked to the cosmonauts and were given 'exhaustive replies to all their questions'. By Aug 18, Salyut 5 had completed 900 orbits, of which 700 had been with the crew on board; by then they were studying the Sun with an infrared telescope, and working with enthusiasm. In an interview Sevastyanov said on Aug 18 he did not know whether Volynov and Zholobov would break the 63-day Soviet record set up by Klimuk and himself on Soyuz 18: 'the time of the flight is of secondary importance, though it is not to be ignored'. On Aug 24 Volynov and Zholobov were said 'to be starting their 7th week in "good health and spirits" and to be carrying out infrared studies of the Earth's land and ocean surfaces; then unexpectedly on the same day, only 10 hr ahead of landing, another report said the flight was nearing completion; the crew were putting Salyut 5 into automatic regime, and preparing Soyuz 21 for return with their film, samples and materials. Usually at least 2 days are devoted to preparations for return; this time touchdown was announced on Aug 25, 200 km SW of Kokchetav in Kazakhstan. A smooth landing was reported but the fact that it was made in darkness, at 21.33 Moscow time the previous night, seemed to support Western theories that there had been an element of emergency about the return. A report (which the author has been unable to identify) just before landing said the crew were suffering from 'sensory deprivation', and Izvestia had reported a week earlier that mission psychologists monitoring their health had asked for music to be played to them to counter the effects of their prolonged isolation. There was no official confirmation that the decision to return had been sudden; but US reports several weeks later stated that an emergency return became necessary because of an acrid odour flowing from the Salyut

5 environmental control system. Zholobov and Volynov endured it for some time but were unable to locate the cause before it became unbearable. Thus it became the 6th time in 9 manned missions that it was not possible to complete the flight plan. The cosmonauts were said to be 'walking a little shakily' on landing, and suffering from the usual post-flight effects, but they quickly recovered. Salyut 5, it seems, was basically a military mission similar to that of Salyut 3; the orbit regularly took it over 'Operation Sevier' ('North') in Siberia, where massive military and air and sea manoeuvres were being carried out. The crew would thus be able to assess the reconnaissance potential of a manned spacecraft in monitoring and participating in such operations. About 770 of their 789 revolutions were spent in Salyut 5.

Soyuz 22   Launched on Sep 15 1976, and at first expected to be a 2nd crew for Salyut 5, Soyuz 22 was in an inclination (64–75°) not used for a Soviet manned flight since Vostok days, ruling out any such docking. The Commander, Col Bykovsky, aged 42, was making his 2nd flight 13 yr after his first; Flight Engineer Vladimir Aksenov, aged 41, more fortunate, was getting his first trip only 3 yr after becoming a cosmonaut. Orbital parameters, noticeably omitted from the official announcements, were initially 200×281 km, circularized on revolution 16 to 257×251 km. The docking mechanism had apparently been replaced with a multiple-spectral camera developed and built by the GDR company Karl Zeiss Jena—'the first occasion on which foreign instruments had been installed in a Soviet manned spaceship'. (ASTP was obviously excepted.) During their 8-day flight, individual sections of Soviet Union and GDR territory were photographed—part of a co-operative study of geological and geographical features in the interests of the 2 national economies. The cosmonauts were reported as 'enjoying a

Soyuz 24 almost ready for lift-off. Note huge concrete basin in which launchpad stands, and patches of snow

large measure of independence in controlling the craft and making various investigations'. This, coupled with the unusual orbit, taking it over the main areas of activity in a massive NATO land, sea and air exercise stretching from Norway to Turkey, was one of the many factors which suggested that this was a 'first' which Russia was unlikely to claim: the first manned 'spycraft'. Other factors were the seniority of Bykovsky, said to have 'graduated' at the Zhukovsky Military Air Engineering Academy in the 13-yr interval since his first flight; and the unexpected choice of Aksenov (also with a specialized air force background, with no rank given) rather than any of the other cosmonauts with greater seniority waiting for first or 2nd flights. Their experience and classified knowledge might well enable them to select and point their new cameras at NATO activities of reconnaissance interest much more quickly than ground control could have achieved. During the flight it was revealed that it was the start of increased manned spaceflight co-operation between USSR and Soviet bloc countries. Selections of non-Soviet cosmonauts would begin in 1977, with flights for them in the 1978–83 period. All missions would be commanded by a Soviet cosmonaut, with a foreign country providing the flight engineer. The landing on Sep 23 (10.42 MT) was 150 km NW of Tselinograd, Kazakhstan. Both cosmonauts were in good health; they had taken 2400 detailed photographs, in 6 different bands of the spectrum, of Soviet and GDR territory.

Soyuz 23   L Oct 14 1976, with the announced aim of continuing 'the scientific-technical research and experiments in conjunction with orbiting scientific station Salyut 5', this was obviously intended to lead to a 60–90 day stay. Soviet scientists were presumably satisfied that the fault in Salyut 5's air-conditioning system had either been corrected automatically, or could be corrected by the new crew. But the flight proved to be the 7th Soviet failure in 11 attempts to complete space station missions, and the 2nd successive Soyuz failure. The Commander, Lt-Col Vyacheslav Zudov and the Flight Engineer, Lt-Col Valeri Rozhdestvensky, were both making first flights after a 10-yr training; the latter, a former diving unit commander, somewhat prophetically said in the pre-launch Press conference: 'I think my pre-cosmic profession may yet come in handy'; it did, since the emergency return, 48 hr 06 min after lift-off, ended in the first Soviet splashdown. After being placed in a 275 × 243km orbit, with 51·6° incl, Soyuz 23 began automatic rendezvous with Salyut 5 during its 17th revolution,

25 hr after launch. The space station's orbit was 253 × 268 km. Soviet sources said later that Soyuz's rendezvous approach electronics malfunctioned; the spacecraft never got to the point, about 100 m from Salyut, at which the crew would have been able to take over manual control for final docking. Since this version of Soyuz is flown without solar panels, to make it lighter and more manoeuvrable, and the internal battery power limits flight independent of Salyut to $2\frac{1}{2}$ days, what should have been Russia's 9th manned docking had to be cancelled, and preparations begun immediately for return to Earth. Retrorockets were fired 44 min before landing, and Gen Shatalov warned the crew to stay in their seats after touchdown, because of wind conditions. He congratulated them on working calmly and confidently and said their reports were very clear and good. Possibly because high winds dragged the main parachute in the final 7 km of descent, the spacecraft came down in darkness and a snowstorm in the 32km wide Lake Tengiz, 195 km SW of Tselinograd, thus fulfilling Rozhestvensky's prediction. Aircraft located Soyuz 23, and helicopters dropped frogmen with flotation devices. The official announcement said 'evacuation was under difficult conditions at night in heavy snowfalls'. After recovery, announced 12 hr later, the cosmonauts were flown to Arkalyk, 140 km W of the lake, and reported to be in good condition.

Soyuz 24   L Feb 7 1977 'to continue the experiments started by Soyuz 21 with Salyut 5', which had then been in orbit 8 months, it was clearly too late to break America's long-duration record. The Commander, Victor Gorbatko, on his 2nd flight and Flight-Engineer Lt-Col Yuri Glazkov his first, had been the stand-by crew for Soyuz 23. It had taken 4 months to launch their back-up mission. From a corrected orbit of 281 × 218 km, they successfully docked with Salyut

Soyuz 24 crew, Victor Gorbatko and Yuri Glazkov, in training simulator

5 the next day, and took a sleep period before boarding. Gen Shatalov had warned at the start that the flight would be 'routine'; in fact it covered the now familiar routine of exploration of Earth's atmosphere and surface, crystal growing and technological experiments and biological research—involving plant growing, and cultivation of fish and other livestock. On Feb 23 the crew began preparations for re-entry, set up Salyut 5 for continued operation, and landed 37 km NE of Arkalyk on Feb 25.

SPUTNIK   World's First Satellites

History   When Sputnik 1, the world's first artificial satellite, soared into orbit on Oct 4 1957 it acted as the starter's pistol in the Soviet-American race to put men on the Moon. In America its launching brought bitter disappointment; earlier the same year a US Army plan, supported by Dr Wernher von Braun, who wanted to use his Redstone rocket to put up a satellite in Sep, was turned down. Had it been successful the US would have been 1 month ahead of Russia with man's entry into space. As it was, Russia was to put up Sputnik 2 before America joined the contest with Explorer 1 on Jan 31 1958.

The whole series of 10 appears to have been devoted to developing the ability to place men in Earth orbit, using at least 6 dogs for the purpose. All were launched from Tyuratam over a period of $3\frac{1}{2}$ yr. Cosmos 1 was also Sputnik 11, and for a time the numbers of the 2 series overlapped. Sputniks 19, 20 and 21 were believed to be early attempts at launching Venus probes. Sputniks 22 and 24 were attempts to launch towards Mars; and Sputnik 25 (Jan 4 1963) was a Luna failure. From that point, Russia herself allotted Cosmos numbers to any launch failures occurring in non-Cosmos programmes.

Launchers   The standard Vostok vehicle ('A'), based on the original Soviet ICBM, providing lift-off thrust of 509,840 kg was used for Sputniks 1–3; for most of the subsequent launches an upper stage, providing 90,260 kg was added to the Vostok vehicle.

Sputnik 1   L Oct 4 1957. Wt 83·6 kg. Orbit 947 × 228 km. Incl 65°. The world's first artificial satellite. It consisted of a

Man's first satellite, Sputnik 1

polished metal sphere, dia 0·58 m, with 4 whip-type aerials from 1·5 m, to 2·9 m long. Instrumentation included radio telemetry and devices for measuring the density and temperature of the atmosphere and concentrations of electrons in the ionosphere. It transmitted 21 days; dec. after 1400 orbits in 96 days.

Sputnik 2   L Nov 3 1957. Wt 508 kg. Orbit 1671 × 225 km. Incl 65°. The world's 2nd satellite; it contained a spherical pressurized container to carry the dog Laika, the first living creature in space, to obtain data on weightlessness effects on living organisms. Transmissions lasted for 7 days, when Laika presumably died painlessly when her oxygen supply ran out. Spacecraft re-entered and dec. after 2370 orbits in 103 days.

Sputnik 3   L May 15 1958. Wt 1327 kg. Orbit 1880 × 217 km. Incl 65°. Cone-shaped, 1·73 m dia at base, and 3·57 m long, it carried instruments to study Earth's upper atmosphere, solar radiation, etc; dec. after 690 days.

Sputnik 4   L May 15 1960. Wt 4540 kg. Orbit 368 × 312 km. Incl 65°. First Sputnik for 2 yr, with Lunas 1–3 intervening. This was believed to be a test flight for a manned Vostok; recovery failed when the cabin went instead into a higher orbit and finally re-entered over 5 yr later on Oct 15 1965.

Sputnik 5   L Aug 19 1960. Wt 4600 kg. Orbit 305 × 339 km. Incl 80°. 2nd Vostok trial. 2 dogs, Belka and Strelka, were ejected and recovered by parachute after 18 orbits.

Sputnik 6   L Dec 1 1960. Wt 4563 kg. Orbit 186 × 265 km. Incl 65°. 3rd Vostok trial; recovery failed, and canine passenger was killed 1 day later.

Sputnik 2 as launched in Vostok's nose cone. Note spring for ejection

Sputnik 7   L Feb 4 1961. Wt 6482 kg. Orbit 327 × 223 km. Incl 65°. Apparently a test flight in preparation for Sputnik 8, the first attempt at a Venus flyby. Dec. Feb 25 1961.

Sputnik 8   L Feb 12 1961. Wt 6474 kg. Orbit 318 × 198 km. Incl 65°. Launched Venus 1 Russia's first Venus probe, from Earth parking orbit. Sputnik 8 dec. Feb 25 1961.

Sputnik 9   L Mar 9 1961. Wt 4700 kg. Orbit 183 × 250 km. Incl 65°. 4th Vostok trial; dog Chernushka successfully recovered on same day.

Sputnik 10   L Mar 25 1961. Wt 4695 kg. Orbit 178 × 246 km. Incl 65°. 5th Vostok trial; dog Zvezdochka successfully recovered after 1 orbit. Yuri Gagarin, Russia's first man in space, went into orbit 18 days later.

STATSIONAR   Communications System

History   Russia's stationary-satellite system, on present plans, is likely to duplicate and rival the 91-nation Intelsat system by the end of the decade. Starting 12 yr later than America with her first geostationary satellite, Russia had registered the Statsionar project with the International Telecommunication Union on Feb 3 1969, and further details on Dec 1 1970. But it was over 3 yr later, on Mar 26 1974, before Cosmos 637, her first stationary experiment, went into orbit, followed by Molniya 1S on Jul 29 1974. Technical difficulties it seems, led to Statsionar being 5 yr late. There was considerable concern, just before Statsionar 1 was launched on Dec 22 1975 when the Soviet Union notified (but did not consult) the International Frequency Registration Board (part of the ITU) that it intended to launch a global satellite communication system, consisting of 7 Statsionars positioned over the Indian, Pacific and Atlantic Oceans. The problem was that they would operate in the 4-gc and 6-gc frequency bands used by Intelsat and were likely to create frequency interference problems not only with Intelsat, but with Franco-German Symphonie over the Atlantic, and 2 planned Indonesian spacecraft for domestic communications. Examples are that Statsionar 5 is planned for 58°E, only 2° from the operational Intelsat 4; and that the new 4a, with more sensitive antennas, is likely to pick up more interference. It may be the Soviet attitude was influenced by the fact that in 1964 Soviet officials had suggested a joint venture, but Comsat and European countries then forming Intelsat, were not interested. Later the Soviet view was that Intelsat was US-dominated.

Statsionars 1 & 2 L Dec 22 1975 & Sep 11 1976 by D1e from Tyuratam, are 3-axis and spin-stabilized, derived from Molniya. They were placed in 35,800 and 35,000km synchronous orbits over the Indian Ocean at 35°E and 85°E. With no wt given, and nicknamed Raduga (Rainbow), they relay radio and TV programmes to most of Russia and Soviet-bloc countries. Statsionar operates with cheap Earth terminals, equipped with 10m dia antennas, compared with Intelsats' antennas of 27–30 m dia, but the latter can carry 2-way telephone traffic.

Statsionar-T L Oct 16 1976 by D1e from Tyuratam. Also sited over the Indian Ocean, at 99°E for direct TV broadcasting to community aerials in Russia, particularly in high northern latitudes. Also called Ekran (Screen).

Statsionars 4–10 With launches spread between 1978–80, will form the global system as follows: Statsionar 4 over the S Atlantic at 14°W, covering the eastern half of US and Canada, Africa, S America and USSR. Statsionar 5 over the Indian Ocean at 58°E, covering most of the USSR, Europe, the Middle East and Africa. Statsionar 6 over the Indian Ocean at 85°E covering one-third of the world, from the Polar regions through Central Africa and all Australia. Statsionar 7 over the Pacific at 140°E to cover eastern USSR and down to the S Polar region. Statsionar 8 over the Atlantic at 25°W covering eastern USSR to mid Canada and US. Statsionar 9 over Indian Ocean at 45°E covering most of USSR, Europe, Africa and Middle East. Statsionar 10 over the Pacific at 170°W giving transpacific coverage.

VENUS Planetary Explorers

History After more than 14 yr of persistent work, and many disappointments in the early years, Soviet scientists have achieved remarkable successes in penetrating the mysteries of Venus. Their efforts were rewarded in Oct 1975, when Venus 9 & 10 sent back man's first pictures from the surface of another planet. Venus 3 was man's first spacecraft to reach a planet (Mar 1 1966); but it plunged into the hostile atmosphere without returning any planetary data. Russia was able to claim the first transmissions from a planet's surface following the successful descent of Venus 7 on Dec 15 1970; while there may be some doubt whether this reached the true surface, or fell on mountain peaks, with the result that it tumbled, and abruptly ceased transmissions after 23 min, there can be no doubt at all that Venus 8

Venus 4 Descent Capsule, showing parachute attachments

transmitted from the surface for about 1 hr on Jul 22 1972, before its resistance to the temperatures of up to 500°C and atmospheric pressures up to 100 times those on Earth, was finally overcome.

The painstaking persistence of Soviet scientists in using the knowledge gained on each mission to improve both the techniques and spacecraft on each succeeding flight can be followed by a study of the missions detailed below. The information has gradually been gleaned as facts about past missions have been released during the project's progress.

With its clouds reaching an altitude of up to 60 km and surrounding the planet with what has been described as 'a heavy oily smog', Venus presents much greater technical problems than Mars for exploring spacecraft. This may be one explanation for the extent of Soviet interest in Venus, which sometimes seems disproportionate to the likely rewards. By the time it was announced that Venus 8 was on

course in Mar 1972, at least 16 launching attempts had been made. Venus designations are given to spacecraft only when they are safely on course; failures are merely given the next available Cosmos number. Evidence of these failures is to be found in the tables of space launches compiled by Western observers. Russia prefers to operate her planetary probes in pairs, so that transmissions can be compared. This was achieved with Venus 2 and 3, and with Venus 5 and 6; but what should have been Venus 8, accompanying Venus 7 in Aug 1970, failed to achieve escape velocity, and was written off as Cosmos 359. Similarly, what was clearly intended to be Venus 9, accompanying Venus 8 in Mar 1972, and launched 4 days later, had to be written off as Cosmos 482 when the escape stage fired only partially, and left it stranded in an elliptical 205 × 9800km orbit with a 6-yr life. Venus 9 was intended to land on the planet's dark side, to provide comparative readings with those sent by Venus 8 from the sunlit side.

However, as Soviet scientists developed the use of structural strengthening and ablative materials, and found ways of adjusting the descent rates to achieve greater penetration before their spherical descent craft were destroyed, confidence grew. Following the successful Venus 8 launch, advance information was given that a descent was intended—almost certainly the first time in Soviet space history that flight plan details were published in advance. The information gained on that flight was used, in the 3-yr gap that followed, for a complete redesign of Venusian spacecraft. That, coupled with a near-doubling of their size and wt, brought the remarkable successes, provided by Venus 9 & 10. From Oct 1975, man could work from facts, not speculation, in his studies of Venus, its origin and development.

Spacecraft Description Venus 1–8 consisted of a cylindrical main section, containing mid-course propulsion engine, telemetry, attitude control, guidance, sensory and power systems with solar panels. The spherical descent module was about one-third the usual payload of around 1180 kg. Total wt did not rise much as the series progressed; but increasing knowledge made it possible to use less of the wt on protection from pressure and temperature, and more on scientific content. Centre of gravity was offset to assist self-orienting during descent, with separate hermetically sealed compartments for the instrumentation and dual-parachute descent system. Launch into Earth parking orbits was from Tyuratam by 43m high Voskhod rockets providing lift-off thrust of 509,840 kg. A 2-engine 2nd stage adds 140,000 kg thrust, and there is a 3rd 'escape' stage. In the case of Venus 8 it was announced that this burned for 243 sec and accelerated the spacecraft to 11·5 km per sec (41,433 kph), 'somewhat greater than second cosmic velocity'.

Experience with Venus 4, 5 & 6 showed that the design of the main spacecraft was satisfactory; but the landing vehicle was redesigned to withstand external pressures of up to 180 atmospheres and 530°C, increasing its wt over Venus 5 & 6 by about 100 kg. The parachute canopy was made of heat-resistant cloth able to withstand temperatures up to 530°C. Published wt figures vary slightly; the total given for Venus 7 was once again 1180 kg.

Western suggestions that Venus 7 ended its life on mountain peaks were not fully accepted by Soviet scientists, who considered then that all mountains would have been eroded into sand and dust by the hot atmosphere. But a light wind, they conceded, could have raised the dust into the atmosphere, and caused echoes in radio frequencies and discrepancies in radio-altimeters. The performance of Venus 7, and the knowledge it added to that gained on earlier missions, made another major redesign possible for the Venus 8 lander.

Since Venus 4–7 had all studied the planet's night side it was decided that Venus 8 should aim for the sunlit side, to check the amount of light reaching the surface through the cloud layers; this, it was hoped, would indicate how far the solar rays penetrated the thick atmosphere, and perhaps explain how the heating of the atmosphere to such high temperatures took place.

A major advance on Venus 8 was a dual antenna, which enabled it to transmit 60-min data when it succeeded in reaching the sunlit side on Jul 22 1972. Additions like this were possible because Venus 7 experience showed that weight-consuming insulation and protection against the elements could be reduced. A refrigeration unit was added to cool the interior below freezing during the early part of the descent; the dual-parachute system, first tried on Venus 7, aided the effectiveness of this by enabling Venus 8 to drop rapidly through the upper atmosphere and then slow down as it approached the surface (*see* page 323). At touchdown a separate antenna, tripod-mounted with large pads and able to withstand heavy winds, was thrown off, coming to rest a few metres away. This was to ensure transmission even if the main craft rolled or fell so that its primary antenna was not pointed to Earth. In the event, it was possible to use both antennas; the first 13 min of data came from the primary antenna, and included temperature readings, atmospheric pressure and light levels; 20 min of data concerning the nature of the hard surface then came from the secondary antenna, with the final 30 min again from the main antenna.

Venus 9 & 10, described as '2nd generation' Venusian explorers, were completely re-engineered. At the time of writing, no precise weights and details have been given. But Soviet engineers began by redistributing weights between the orbiter and lander. It was also decided that all transmissions back to Earth—on this occasion 80M km—should be passed via the orbiter, thus lightening and simplifying the lander's transmitter. As before, the lander started as a sphere. The need to protect it against over 2000°C and pressure equivalent of 300 tonnes on its front part during entry into the Venusian atmosphere, was provided by 2 jettisonable semi-spheres. An astonishingly elaborate procedure—which nevertheless worked perfectly—was devised to ensure that the craft touched down safely, and yet passed through the cloud layers quickly so that its brief surface life was not diminished. Details, given under Venus 9, apply equally to Venus 10. To help it survive, the interior was chilled beforehand to minus 10°C. The sequence of retro-rockets, parachutes, parachute-jettisoning, followed by aerodynamic braking by means of a circular metal disc, and in the final seconds, retro-rockets and a metallic ring to absorb the impact, reduced the entry speed from 38,520 kph to 24–27 kph.

Venus 1 L Feb 12 1961 by Sputnik 8 from Tyuratam, wt 643·5 kg. Radio contact lost at 7·56M km, but spacecraft passed 99,800 km from planet. This was the first Soviet planetary flight.

Venus 2 L Nov 12 1965 by A2e from Tyuratam, wt 963 kg. Passed 23,950 km from Venus on Feb 27 1966, but failed to return data.

Venus 3 L Nov 16 1965 by A2e from Tyuratam, wt 960 kg. After 105 days' flight impacted on Venus on Mar 1 1966, man's first spacecraft to reach a planet. It failed, however, to return any planetary data.

Venus 4 with solar panels and high-gain antenna folded. Spherical descent capsule is at bottom

Venus 4   L Jun 12 1967 by A2e from Tyuratam, wt 1106 kg of which 383 kg was the entry vehicle. After 128·4 days' flight, presumed to have impacted on Oct 18 1967, 1 day before US Mariner 5 flyby. Descent capsule transmitted data during 94-min parachute descent. Highest temperature sent back was 540°F; also sent measurements of pressure, density and chemical composition of atmosphere before ceasing transmissions, possibly as a result of striking 1 of 3 mountain ranges recently detected by radar bounces. This flight enabled later spacecraft to be redesigned so that they could penetrate more deeply and survive longer.

Venus 5   L Jan 5 1969 by A2e from Tyuratam, wt 1130 kg, of which 405 kg was the entry vehicle. After 131 days' flight, the descent capsule separated at a distance of 37,002 km from Venus. Initial entry velocity of 11·2 kps, was reduced by atmospheric braking to 756 kph before deployment of main parachute one-third the size of that on Venus 4. The capsule entered the Venusian atmosphere on the planet's dark side on Mar 16 1969, and transmitted data for 53 min, travelling 36 km into the atmosphere before being crushed.

Venus 6   L Jan 10 1969 by A2e from Tyuratam, wt 1130 kg. After 127 days' flight, the descent capsule separated at a distance of 25,000 km, and 24 hr after Venus 5, entered the Venusian atmosphere on the planet's dark side, and transmitted data for 51 min, descending 37·8 km into the atmosphere before being crushed.

Joint Results   Data from both spacecraft revealed that the concentration of nitrogen with inert gases in the atmosphere of Venus is from 2–5%; oxygen does not exceed 0·4%; carbon dioxide represents 93–97%. Water vapour content was very low. During descent, temperature readings ranged from 77°F–608°F. Pressure readings ranged from 0·5 kg sq cm to 28 kg sq cm. Extrapolation of data from both spacecraft suggested surface temperatures on Venus might range from 400–530°F, while the pressure would be from 62–144 kg sq cm. Both spacecraft were believed to have been crushed when the ambient pressure exceeded 27 atmospheres (28 kg sq cm); and the temperatures exceeded 600°F.

Venus 7   L Aug 17 1970 by A2e from Tyuratam, wt 1180 kg. After a 120-day flight, entry into the planet's atmosphere began on Dec 15. Distance from Earth at that time was 60·6M km. After separation, the descent craft's speed was reduced aerodynamically from about 41,400 kph, to 724 kph. It was subjected to 350G, and temperature differences between the shock wave and vehicle reached 11,000°C. The parachute was opened 60 km above the surface, when external pressure was about 0·7 atmospheres. The parachute canopy, of cloth designed to withstand temperatures up to 530°C, was bound by a Kapron cord, so that a fast descent was made through the upper layers; the cord was made to fray and tear apart in the lower layers, so that the parachute canopy would then open fully, slowing the descent to allow fuller study of the lower layers. During this period only the gradually increasing temperatures were transmitted to Earth; the signals took 3 min 22 sec to reach Earth. According to a Tass report, changes in descent velocity signals showed that the probe, which carried pennants with a picture of Lenin and the Soviet Union's coat of arms, had landed. Moscow radio said that at this point the radio signals dropped to 100th of descent strength, probably because the antenna's axis had deviated from the direction of Earth after landing. Signals continued for 23 min after landing; temperatures were 475°C plus or minus 20°C; pressures were 90 atmospheres plus or minus 15.

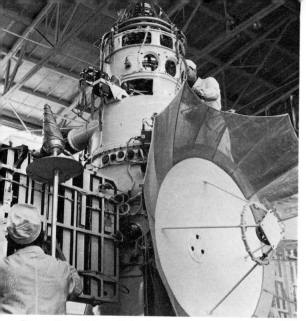

Preparation of Venus 8. Its domed capsule (not visible) reached the Venusian surface and transmitted for 60 min

During the flight of Venus 7, its instruments sent back readings of the 'powerful chromospheric flare' which began on Dec 10 1970; it was possible to compare Lunokhod 1 readings received simultaneously from the lunar surface, and from satellites and ground observatories. When Venus 7 landed Soviet scientists were able to claim that for the first time they were receiving information simultaneously from 2 celestial bodies.

Venus 8   L Mar 27 1972 from Tyuratam, wt 1180 kg. After 117 days' flight, and having travelled 300M km, Venus 8 reached the planet on Jul 22, when the latter was 107·8M km from Earth. 86 communication sessions had taken place during the flight to control the course and check the on-board systems. It was manoeuvred to a touchdown in the narrow, crescent-shaped sunlit portion of the planet visible from Earth; the landing site, about 2896 km from the site of Venus 7, was selected to minimize the Earth-distance. Earth was relatively low on the Venusian horizon, so that Soviet scientists could obtain readings from the sunlit portion; it was early morning at the landing site, with the Sun equally low on the local horizon. As the spacecraft entered the upper atmosphere the descent module separated, while the service module went on to burn up in the atmosphere. The entry speed of about 41,696 km, was reduced by aerodynamic braking, and the parachute deployed at 900 kph. During descent, a refrigeration system, involving a compressor and heat exchange unit, was switched on to offset the 500°F temperature found on earlier missions. Transmissions of temperatures, pressures, light levels and descent rates were interrupted for 6 min during descent for calibration of on-board instruments; touchdown was at 09.29 GMT. Soviet scientists promised that details would be issued when information obtained had been processed, and said so much heat was stored in the dense and heavy atmosphere that night and day temperatures levelled out, even though each lasted about 2 terrestrial months. Surface light levels were considerably lower than Earth's. Surface temperatures probably varied only 1°F; they were such that tin and lead

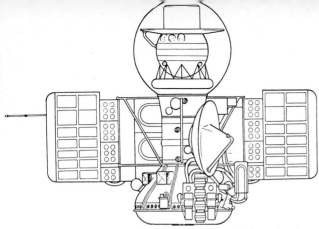

Drawing by D. R. Woods of Venus 9 and 10 shows lander at top inside protective sphere. With this 2nd generation, the main spacecraft went into Venusian orbit after ejecting the lander

would melt, and iodine, mercury, bromine and sulphur would evaporate.

Venus 9 L Jun 8 1975 by Proton from Tyuratam, wt ? 4000 kg. The first spacecraft to send to Earth a picture from the surface of another planet. Its launch and course were compared by Soviet experts to firing a rifle at a flying coin at a distance of about 1 km. Having covered over 300M km in 136 days it also provided man's first Venusian orbiter on Oct 22, with a perigee of about 1500 km and a 2-day revolution period. During its flight, 90 communications sessions took place, though only 2 trajectory corrections were necessary. The last of these took place 7 days before the landing.

Separation of the lander and orbiter took place 2 days beforehand. Since all the lander's signals were to be transmitted to Earth via the orbiter, the orbit had to be perfectly phased so that as the landing occurred, the vehicles were approaching one another from opposite directions. At 09.58 GMT on Oct 22, the lander entered the Venusian atmosphere at 38,558 kph. Aerodynamic braking reduced this to 899 kph; then the elaborate new parachute system, using a total of 6 chutes was automatically switched on by an overload detector. First the pilot parachute pulled out the deflection parachute, which pulled aside the upper part of the heat protective sphere. That in turn brought out the braking chute, which cut the speed until it was possible for the 3 main chutes to be deployed. At 50 km the main chutes were jettisoned in their turn, with the final descent and landing once more being dependent on the aerodynamic landing shield. This elaborate system ensured a relatively fast descent through the cloud layers, in which it was found earlier landers had tended to 'hover', with the result that they had been overwhelmed by the intense heat. The Venus 9 & 10 landers were apparently identical spheres, made of strong heat-resistant material, with thick insulation added outside. First signals were received 5 min after the Venus 9 lander entered the atmosphere and continued for nearly 2 hr 53 min from the surface itself. The descent itself took 1 hr 15 min. The single but remarkable panoramic picture arrived 15 min after touchdown; it made nonsense of centuries of theorizing about what lay below Venus's dense cloud cover. It was expected that diffused, dim light, would result in a poor picture. Instead, the scanning photo-tele-photometers sent back a picture of startling sharpness revealing that, instead of a sandy desert resulting from wind and heat erosion, Venus 9 had landed on a steep slope, possibly of a

Historic first view of Venusian surface, sent by Venus 9 shows rocky surface. Vertical stripes indicate transmission sections

volcano, 2500 m above Venusian mean level, among a scattering of large, perhaps recently erupted rocks—possibly on top of one of them. In addition to the surface close-up picture, a 2nd panoramic view showed a horizon estimated at 200–300 m and angular rocks of up to 1 m; one, nearby the lander, was about 30–40 cm across.

Venus 10   L Jun 14 1975 by Proton from Tyuratam, wt ? 4000 kg. Identical with Venus 9, this spacecraft operated equally well. Course corrections were made on Jun 21 and Oct 18. The lander was separated from the orbiter on Oct 23; the descent procedure first used on Venus 9 was again successful, and this time took 75 min. During it, measurements of the Venusian atmosphere and details of the physical and chemical contents of the cloud-layers, were transmitted to the orbiter 1500 km above, and from that, 80M km to Earth. This landing on Oct 25 was 2200 km from the first, much lower, and showed a surface temperature of 465°C, atmospheric pressure 92 times that of Earth, and a wind velocity of 3·5 m per sec. While the Venus 9 picture had showed a scattering of large rocks on what Boris Nepoklonov, Russia's chief Venusian topographer, described as 'a typically young mountainscape', the single Venus 10 area showed 'a landscape typical of old mountain formations'. The rocks were not sharp, but resembled huge pancakes with sections of cooled lava or debris of weathered rock in between. Neither lander, when it hit the surface at a speed of 7–8 m per sec, raised any dust, either of the lunar or Martian type; a Soviet planetologist said that while Venus 9 had landed on a high plateau, Venus 10 had landed in a stony desert. Venus should therefore be classed as a young, still evolving planet. In the month following the landings, the Venus 9 & 10 orbiters made 15 and 13 revolutions around the planet, and Venus moved a further 33,500,000 km from Earth, making the total distance 119M km. During that period 40 radio communication sessions took place with Venus 9 and 35 with Venus 10. Soviet scientists began to build up a large-scale composite picture of the planet's cloud blanket. Credit was given to the French-made ultraviolet spectrometer for measurements of the upper atmosphere, with its ratios of hydrogen and deuterium.

VOSKHOD   Manned Spacecraft

History   Controversy as to the origin and purpose of these 2 flights has never been completely resolved. One view is that the first resulted from pressure brought by the then Soviet Prime Minister, Mr Krushchev, on the Chief Designer, Sergei Korolev, to perform a 3-man flight before America flew a 2-man Gemini spacecraft. This would certainly explain why, although Voskhod is officially described as 'different from Vostok in both structure and equipment', no pictures have ever been released; it would also line up with suggestions that it was so difficult to cram 3 cosmonauts into a stripped-down Vostok that it was necessary to fly them without spacesuits. However, these arguments cannot detract from the value and courage of Leonov's first spacewalk during Voskhod 2. The international sensation it caused—coupled no doubt with Leonov's demonstration of its relative safety and feasibility—led to a NASA meeting at Houston 11 days later (Mar 29 1965). There it was decided that on Gemini 4, the following Jun, preparations should be made for Edward White to carry out a similar spacewalk, although up to that stage the plan had merely been for the spacecraft hatch to be opened so that White could stand up, with head and shoulders protruding into space, without actually leaving the vehicle. The official Soviet description of the objects of the Voskhod missions were 'to test the new multi-seater spaceship, to investigate the work capacity of a group of spacemen specialized in different spheres of science and engineering, to make physical and technical experiments, and to perform an extensive medico-biological investigation programme.'

The Voskhod flights proved to be the last made under the leadership of Sergei Korolev, Russia's Chief Design Engineer for rockets and spacecraft throughout the development of orbital flight. He died on Jan 15 1966, 10 months after Voskhod 2, aged 60; only after his death was his identity publicly revealed. Korolev's health, like that of his deputy, Voskresenky, who died aged 52 after the Voskhod 1 mission, had been undermined by 6 years of imprisonment under the Stalin regime. Posthumous recognition came with the public burial of his ashes in the Kremlin wall. Cosmonauts present included Komarov, whose ashes joined those of Korolev less than a year later as a result of the Soyuz 1 accident.

Spacecraft Description

Weight *Voskhod 1*:	5320 kg
Voskhod 2:	5682 kg
Size:	Not given

No exterior pictures of Voskhod ('Sunrise') have ever been issued; all the evidence, including the weights, suggests that they are basically Vostok spacecraft modified for multi-man flights. The main change is the addition of a soft-landing system, providing retro-rockets to assist the parachutes just before touchdown. This makes it possible for the cosmonauts to remain in the spacecraft for the landing, instead of ejecting and making their final descent on individual parachutes.

Voskhod 2, although it carried only 2 men instead of 3, was 560 kg heavier than Voskhod 1; this would no doubt be due to the fact that this time spacesuits were essential, and in addition, a telescopic inflatable airlock had been fitted to the outside. This was tube-shaped about 1·83 m long and 0·91 m in dia, to form a chamber which the cosmonaut,

Leonov makes man's first spacewalk—filmed by Belyayev

having crawled inside, could depressurize before opening the external hatch. Voskhod's interior was said to have 2 comfortable armchairs, upholstered in white, with 2 instrument panels overhead; one of them was for the airlock chamber.

The spacecraft's control panel includes 'a long handle to operate the manual orientation system'. A red metal hood covers a small black button marked 'Descent TDU'—the Russian initials for the retro-engines. To the left of the cabin, an instrument board includes a revolving globe, which continually indicates the spacecraft's exact location. The lenses of the cine and TV cameras are trained down from above.

Voskhod 1   The first 3-man flight, and the first flight on which a medical man had flown. Vladimir Komarov (pilot), Konstantin Feoktistov (scientist), and Boris Yegorov (physician), were launched on Oct 12 1964, and completed 16 orbits in 24 hr 17 min. The apogee was 409 km, perigee 178 km, and the inclination 65°. The higher orbit was possible because a reserve retro-rocket had been added; Vostok orbits were planned so that atmospheric resistance would ensure re-entry within 10 days in the event of retro-rocket failure. For the first time, the cosmonauts had no spacesuits; they wore 'light grey sports suits and special white space helmets'. The cosmonauts emphasized that they were much more comfortable; but the absence of spacesuits was almost certainly due to the need to save weight and space when fitting 3 men into the spacecraft. Dr Yegorov's seat was apparently set above and in front of the other 2. Despite the cramped conditions, however, Dr Yegorov, using the detachable biosensors, was able to study the effects of flight on his companions, while Feoktistov carried out some experiments on the behaviour of liquids in weightlessness, in addition to geophysical and astronomical observations. The usual telephone conversation between the spacecraft and the Soviet Prime Minister was notable because it proved to be Mr Krushchev's last public statement; while talking to them he observed that Mikoyan was 'pulling the receiver out of my hand'. Mr Krushchev was displaced the day after the landing, and the usual 'Hero's Welcome' in Red Square was delayed so that his successors, Mr Brezhnev and Mr Kosygin, could attend.

Voskhod 2   Pavel Belyayev and Alexei Leonov were

327

launched on Mar 18 1965 into the highest apogee yet obtained in manned flight; apogee was 495 km, perigee 172 km; incl the usual 65°. The flight lasted 26 hr 2 min, and 17 orbits. Leonov later described how, over the USSR on the 2nd orbit, he donned his spacesuit, with the help of Belyayev, entered the airlock and inflated his suit. The pressure he gave later as 0·4 atmospheres, or about 0·42 kg/cm². When he emerged he pushed himself away from the spacecraft, and he could clearly distinguish the Black Sea with its very black water and the Caucasian coastline. He began rotating on his 4·88m tether 10 times a minute, but did not lose orientation, and was able to maintain communications both with Earth and spacecraft. His pulse rate was 150–160 when he left the spacecraft, and peaked to 168. When he pulled too vigorously on the tether he had to put out his hands to avoid collision with the spacecraft. After 10 min, during which his manoeuvres were watched by TV, Leonov was instructed to return to the airlock; unofficial reports say he then ran into difficulties for the first time. His spacesuit had 'ballooned', an effect predicted by British aerospace scientists, with the result that it took 8 min of struggling before he was able to force his way in. Later in the flight the cosmonauts reported sighting an unidentified Earth satellite.

A major crisis occurred on the 16th orbit, when the automatic re-entry system failed. For the first time it became necessary for a Soviet spacecraft to make a manual re-entry; an extra orbit was flown, while Belyayev made preparations to fire the retro-rocket himself; when he did so, he was at first uncertain about the spacecraft's attitude, but it proved to be correct. The landing was made in deep snow in a forest near Perm, about 2000 km N of the planned area; the cosmonauts had to wait 2½ hr for the first helicopter, and

Inside Voskhod 2's cramped quarters. Note telescopic airlock for Leonov's spacewalk in centre

then had to stay overnight at Perm before being flown back to base. Leonov suffered no ill effects from his spacewalk; and so far as is known, neither did Belyayev, but he died as a result of internal trouble, on Jan 11 1970, aged 37.

VOSTOK   Manned Spacecraft

History   Manned spaceflight began with Yuri Gagarin's 1-orbit flight in Vostok 1 on Apr 12 1961. The name Vostok ('East') applies both to the spacecraft and the launch vehicle. Although the spacecraft were designed for automatic operation throughout the series, the cosmonauts' flight programme involved numerous astronomical and geophysical studies, in addition to the development of manned spaceflight techniques. Observations of constellations, photographs of the Sun and the disc of the Earth both at daybreak and sunset, were included. On Vostoks 3–6 seeds of higher plants, bacteria and human cancer cells were carried, for later study of the effects of spaceflight.

Soviet preparations for the first manned flight really started with Sputnik 2 in Nov 1957, when the dog Laika was placed in orbit and his behaviour monitored for 7 days. Recovery was not attempted and he died in space; Sputnik 4, launched on May 15 1960, saw the launch of the first, unmanned Vostok prototype; the recovery attempt failed, and the re-entry section remained in space until its orbit decayed and it was burnt-up on re-entering 5 yr later. Sputnik 5, the 2nd Vostok trial in Aug 1960, was successful, and 2 dogs, Belka and Strelka, were ejected and recovered in a parachute-borne container after 18 orbits; the 3rd Vostok trial (Sputnik 6), in Dec 1960, was successfully orbited, but again the recovery system failed. The trials were repeated with 2 more Vostoks (Sputniks 9 and 10) in Mar 1961, and this time dog passengers were successfully recovered on both occasions. 18 days after the 2nd of these dogs, Zvezdochka, had flown, Russia's first man went into space.

Yuri Gagarin inside Vostok 1, looking tense before the first spaceflight

Spacecraft Description

Orbit Payload (Spacecraft & Last Stage:	6170 kg
Spacecraft Weight:	4730 kg
Re-Entry vehicle:	2400 kg
Spacecraft Diameter:	2·3 m
Length (Spacecraft & Instrument Cylinder):	7·35 m

Vostok consists of a relatively simple re-entry sphere, heat-shielded all round, attached to a cylindrical instrumentation section and re-entry rocket engine. Designed for operations up to 10 days, it has manual as well as automatic control, though basically intended to be controlled from the ground. The retro-rocket engine is used to reduce the orbital speed and start the vehicle on a descent trajectory. Environmental control maintains an oxygen-nitrogen atmosphere at sea-level pressure.

Soviet concern about the possible effects of weightlessness from the start of spaceflight was demonstrated by the fact that in Vostok 3, cosmonaut reactions were monitored by 2 television cameras, one transmitting a full-face view,

Vostok 1, in which Gagarin made the first flight, was 7 m long, including the final rocket stage; re-entry sphere only 2·3 m dia

and the other a face view in profile. Also starting with Vostok 3, electro-encephalogram, electro-oculogram and galvanic skin reactions were monitored throughout the flights.

Vostok 1   Gagarin was launched from Tyuratam (though official records still give its starting point as Baikonur), 370 km to the SE, at 09.07 Moscow time on Apr 12 1961, and landed 1 hr 48 min later at the village of Smelovka, Saratskaya. His orbit had a perigee of 181 km, and apogee of 327 km. Gagarin appears to have remained inside the spherical capsule for the landing, although the 5 subsequent Vostok cosmonauts all used ejection seats at 7000 m, presumably because of the danger of the spacecraft bump-ing too heavily when it hit the ground. It was 4 yr before pictures of Vostok were released, and the secrecy surround-ing the early Soviet flights, never fully relaxed, makes it difficult even today to establish with certainty precise details of the flight.

The handsome Gagarin, with an engaging personality, ready smile and quick sense of humour, had clearly been chosen for the first-class public relations job he carried out following the flight. His first Western visit was to England, where he lunched with the Queen at a time when East-West relations were particularly strained, and went on to make appearances in many other countries. He was only 34 when he was killed, with another pilot, in what seems to have been a routine training flight in a jet aircraft.

Vostok 2   The first 1-day flight. Herman Titov, launched on Aug 6 1961, completed 17 orbits in a flight lasting 25 hr 18 min. Apogee was 244 km, perigee 183 km, and incl 65°; the orbit and inclination were similar for all the Vostok flights. A major factor in the Soviet decision to go straight from 1 orbit to 17 was the problem of a suitable landing site on Soviet territory if fewer orbits were done. Spacecraft improvements were said to include a more advanced air-conditioning system. Titov, who was 26, emerged with credit from man's first full day in space, and from the subsequent news conferences and hero's parade in Red Square. Only later was it disclosed that he had suffered serious disorientation during the flight, and had inner ear trouble for some time afterwards.

Vostoks 3 and 4   The first 'group', or double flight, followed 1 yr later. Andrian Nikolayev, launched in Vostok 3 on Aug 11 1962, completed 64 orbits in 94 hr 27 min; for the last 3 days he was accompanied in orbit by Pavel

Popovich, launched on Aug 12 in Vostok 4, who completed 48 orbits in 70 hr 29 min. Popovich, although he became the 4th Soviet cosmonaut to fly, had in fact been the first to be appointed. At one time the 2 spacecraft were within 5 km of one another. This has been attributed to the reliability of the Vostok launcher rather than to the sophistication of the spacecraft. Though at the time Western speculation was that the flight indicated that Russia was already moving towards rendezvous and docking techniques, there is no evidence that the Vostok spacecraft had any independent manoeuvring capability. The flight was notable for the first TV transmissions from space, and the first demonstrations of weightlessness seen by the public. Popovich was later said to have suffered some disorientation. The 2 cosmonauts, having used their ejection seats following simultaneous re-entries, finally landed only 193 km apart within 6 min.

Vostoks 5 and 6   The 2nd group flight. Valeri Bykovsky, launched on Jun 14 1963 in Vostok 5, established a space record of nearly 5 days (119 hr 6 min) and 81 orbits which stood for more than 2 yr until broken by Gemini 5. World attention, however, was focused on Vostok 6; launched 2 days later, this carried the world's first spacewoman, Valentina Tereshkova, aged 26, who completed 48 orbits in 70 hr 50 min. According to some reports, Tereshkova was substituted at the last moment for a much more highly trained woman pilot who became indisposed. It is known that she suffered some disorientation and space sickness, but regular TV pictures relayed to Earth, and exchanges in which she took part with Bykovsky and the ground did not fully accord with one report (by Vladimir Leonid) that 'when she landed in the Southern Urals she was in the most pitiful condition'. Tereshkova had limited space experience,

and said herself soon after the flight that at the time of Titov's flight she was 'not even dreaming of becoming a cosmonaut'; she had, however, made 126 parachute jumps before Vostok 6.

Five months after her flight she married Nikolayev in Moscow, with the Soviet Prime Minister, Mr Kruschev, leading the wedding festivities. The healthy daughter born to the 2 cosmonauts within a year was an added bonus for Soviet scientists, still much concerned with possible radiation damage as a result of spaceflight. Apart from the considerable prestige associated with the first spaceflight achieved by a woman, this mission appears to have differed little technically from that of Vostok 4 and 5; however, the spacecraft were reported to have improved control systems, and further experience was gained in orbital rendezvous, and the effects of extended periods of weightlessness.

Biosatellite Flights   Vostok spacecraft were used for 2 biosatellite missions in 1973 & 74 (Cosmos 605 & 690) to test the long-term effect of weightlessness on various mammals, insects and bacteria, etc. Russia invited America to participate in at least one of a further planned series. (*See* International Section—Biosatellite.)

ZOND   Lunar and Planetary

History   When the series began in 1964, the spacecraft were described as 'automatic interplanetary stations and deep spaceflight technology development tests'. For Western observers, this series has been the most puzzling of all Soviet programmes. President Keldysh, of the Soviet Academy of Sciences, confirmed in Jan 1969, following Russia's first manned docking with Soyuz 4 and 5, that Zond spacecraft were 'adapted for manned flight'. However, he added a warning that such flights should not be expected 'in the next 2 or 3 weeks'. At that time, it seems, Russia was still undecided upon the division of space exploration between manned vehicles and automatic apparatus. But Zond 5 was the world's first spacecraft to make a circumlunar flight and return safely to Earth in Sep 1968; 2 months later, the feat was repeated by Zond 6. Tortoises, insects and seeds carried on board, showed no ill effects. Spurred on by the fact that the Russians were apparently making final circumlunar tests with a manned spacecraft, America took a chance and sent Apollo 8 round the Moon at Christmas 1968; though they did not land, the first manned flight *around* the Moon was an achievement ranking 2nd only to the first landing. Perhaps one day we shall learn why the Russians, so conscious of the prestige value of space 'firsts', failed to achieve this one when they clearly had the capability. Probably their rockets did not have sufficient lifting-power for a fully manned Zond. One theory is that they could have sent a single cosmonaut, but felt it was too risky to sent one man alone on the first circumlunar flight; if he became ill, or technical problems led to survival depending on manual control of the spacecraft, the workload might well have proved too much for one man.

Zond 8's activities suggest that this spacecraft may well be undergoing tests for the first manned flyby of Mars, perhaps towards the end of this decade.

Cosmos 21 on Nov 11 1963; Cosmos 27 on Mar 27 1964; and Cosmos 379, 382, 398 and 434 in 1970 and 1971, were probably unsuccessful Zond missions.

Zond 1 L Apr 2 1964 by A2e from Tyuratam. Wt 950 kg.

Zond 3 sent back 25 photos of the Moon's farside

Believed to be a Venus probe, but communications were lost. It finally passed Venus at a distance of 99,779 km, and went into solar orbit, after course corrections at 563,270 km, and 14M km from Earth.

Zond 2   L Nov 30 1964, by A2e from Tyuratam. Wt 950 kg. Passed less than 1609 km of Mars, but failed to return any data, and went into solar orbit.

Zond 3   L Jul 18 1965 by A2e from Tyuratam. Wt 950 kg. Took 25 photographs of lunar farside at distances of 9219–11,568 km, and transmitted them to Earth 9 days later from distance of 2,200,000 km. Transmission was repeated from 31·5M km, presumably in a test of photographic systems for Venus 2 & 3 later that year.

Zond 4   L Mar 2 1968 by D1e (Proton) from Tyuratam. Wt 2500 kg. Flight tested new systems 'in distant regions of circumterrestrial space'.

Zond 5   L Sep 15 1968, by D1e (Proton) from Tyuratam. Wt 2500 kg. First spacecraft to circumnavigate the Moon and return to Earth; fired out of Earth-parking orbit before completion of first orbit into a free-return translunar trajectory. A Soyuz-like vehicle, it consisted of a heat-shielded re-entry module, with camera, scientific package, radio and telemetry, re-entry control and parachute systems; an instrument compartment with both major course correction and low-thrust vernier engines, and solar-powered batteries. Tortoises, insects, plants and seeds were carried. After passing around the Moon with the closest point 1950 km, it took Earth photos at a distance of 90,000 km (no Moon photos were mentioned). During the ballistic trajectory return to Earth, a tape-recorded Russian voice was heard calling out simulated instrument readings.

Zond 5 being winched aboard Soviet ship after Russia's first sea recovery. Expectations that Zond would be used to send men to the Moon were never realized

333

Parachutes were deployed at 7 km; splashdown, at the end of a 7-day flight, was in the Indian Ocean, Russia's first sea recovery. The capsule was taken to Bombay, and flown back from there in a Soviet aircraft.

Zond 6 L Nov 10 1968, by D1e (Proton) from Tyuratam. Wt 2720 kg. 2nd circumlunar flight, described by Tass as intended 'to perfect the automatic functioning of a manned spaceship that will be sent to the Moon'. As it passed around the Moon at a minimum distance of 2420 km, the farside was filmed. Re-entry was made by a skip-glide technique; the spacecraft's aerodynamic lift was used to bounce it off the atmosphere, at a shallow angle, thus reducing speed and gravitational forces. Recovery, after another 7-day flight, was on land inside the Soviet Union.

Zond 7 L Aug 7 1969 by D1e (Proton) from Tyuratam. Wt 2720 kg. 3rd circumlunar flight. Farside was photographed from distance of 2000 km; colour pictures of both Earth and Moon were brought back, following re-entry, again by skip-glide technique, with recovery on Feb 14.

Zond 8 L Oct 20 1970 by D1e (Proton) from Tyuratam. Wt ?4000 kg. 4th circumlunar flight, with more colour pictures of Earth and Moon. The spacecraft itself was photographed by an optical telescope at distances up to 277,000 km from Earth; the telescope was pointed with the aid of an onboard laser. The main variation in this flight was re-entry from the Northern Hemisphere, as opposed to the normal Southern Hemisphere approach for touchdown on Soviet territory. The result was Russia's 2nd sea-recovery; splashdown was on Oct 27 in the Indian Ocean. Photographs taken by Zonds 3, 6, 7 & 8 were used to compile a 3-volume atlas of the Moon's farside.

'Earthset' as seen on Aug 11 1969 by Zond 7

LAUNCH VEHICLES

History In 20 yr of spaceflight the Soviet Union has issued few details about her launch vehicles. Detailed information, still far from complete, has been painstakingly assembled by Western observers over a period of many years. Their task has at least been simplified by the fact that there have been few completely new Soviet launchers, especially in recent years; developments have consisted of additions and improvements to the original military missiles. It is interesting to compare this process with the much faster development of rocketry techniques which resulted from the much more varied launchers produced by the competing aerospace companies in the United States. Throughout the *Observer's Spaceflight Directory*, I have followed the identification system originated by Dr Charles Sheldon, of the US Library of Congress, and combined information provided by him with explanations given by Dr J. A. Pilkington, former Director of the Scarborough Planetarium.

By 1969 6 basic Soviet launch vehicles had been identified, classified as A, B, C, D, F and G. X was used for some launches which were difficult to classify. Added upper stages have been designated 1, 2, 3 etc; where there is doubt whether the stage used is 1 or 2, it is marked 1/2. The escape rocket, often a 4th stage, is labelled 'e'; manoeuvrable stages thus become 'm'; and a re-entry rocket 'r'. G was for 'Webb's Giant', named after the former NASA Administrator, who said in 1967 that Russia was developing a huge new launcher, bigger than Saturn 5. Reports that 3 test-firings, beginning in 1969, had ended in disastrous explosions, are still not fully confirmed. But if the Russians are to send men to the Moon, and assemble large space stations in orbit, they will need much greater lifting power than they possess now; the chances are, that by the time this book is published, Webb's Giant, with 10 yr of development behind it, will have been successfully fired.

Current Soviet rockets go back to the early 1950s, and her first ICBM, the SS-6 (code-named Sapwood by NATO), which used the RD 107 rocket engine, burning kerosene and liquid oxygen, with 4 combustion chambers and exit nozzles, plus 2 steering rockets. Dr Sheldon's Table of Soviet Launch Vehicles (page 336) shows how its initial ability to place Sputniks 1–3 in Earth orbit has been improved with the addition of more powerful stages, so that it now launches Soyuz spacecraft. The smallest Russian launcher, used for small Cosmos and Intercosmos satellites, is derived from the

Soyuz Launch vehicle and Soyuz spacecraft being installed at Tyuratam

SS-4 missile (NATO code-name Sandal). The SS-5 missile (code-named Skean), has been developed for military missions. The SS-9 missile (code-name Scarp) has been developed to launch inspection-and-destroy satellites, as well as the Fractional Orbit Bombardment (FOBS) nuclear bombs,

TABLE OF USSR LAUNCH VEHICLES

Designa-tion	Derivation	Orbital Stage(s)			Maximum Payload Wt to Low Orbit	Typical Payloads
		Dia	Lgth	Empty Wt		
A	SS-6 (Sapwood)	2·95	27·5	4000	2000	Sputnik 1–3
Am	SS-6 + manoeuvrable stage	2·0	5·0	700	2300	Polet (unitary payload and stage)
A1	SS-6 + Luna stage	2·6	3·8	1440	4750	Luna 1–3, Vostok, Cosmos suborbital
A1m	SS-6 + Luna stage	2·6	3·8	1440		Cosmos 102, 125 (unitary payload and stage)
	+manoeuvrable stage	2·0	5·0	700	4050	
A2	SS-6 + Venus stage	2·6	7·5	2500	7000	Cosmos, Voskhod, Soyuz
A2e	SS-6 + Venus stage	2·6	7·5	2500		Molniya, Luna 4–14, Venus, Prognoz
	+ probe rocket	2·0	2·0	440	7500	
A2m	SS-6 + Venus stage	2·6	7·5	2500		Cosmos 379, 398, 434
	+manoeuvrable stage	2·0	5·0	700	7500	
B1	SS-4 (Sandal) + Cosmos stage	1·65	8·0	1500	450	Cosmos, Intercosmos
C1	SS-5 (Skean) + restart stage	2·0	6·0	2000	1000	Cosmos, Aureole, Intercosmos
D1	Original	4·0	12·0	4000	20,000	Proton, Salyut, Cosmos
D1e	Original	4·0	12·0	4000	22,000	Luna 15–, Mars 2–, Molniya 1S
	+ probe rocket	3·9	4·0	1900	22,000	
D1m	Original	4·0	12·0	4000		Cosmos 382
	+manoeuvrable stage	3·9	7·5	2500	22,000	
F1m	SS-9 (Scarp) + stage	2·5	6·0	1500		Cosmos (ocean surveil., inspector)
	+manoeuvrable stage	2·0	5·0	700	4550	
F1r	SS-9 + stage	2·5	6·0	1500		Cosmos (fractional orbit bombardment)
	+ retro stage	2·0	8·0	700	4550	
G1	Original	?	?	?	135,000?	Future orbital stations
G1e	Original	?	?	?		
	+ probe rocket	?	?	?	140,000?	Future manned lunar flights

(*data in metres and kilogrammes*)

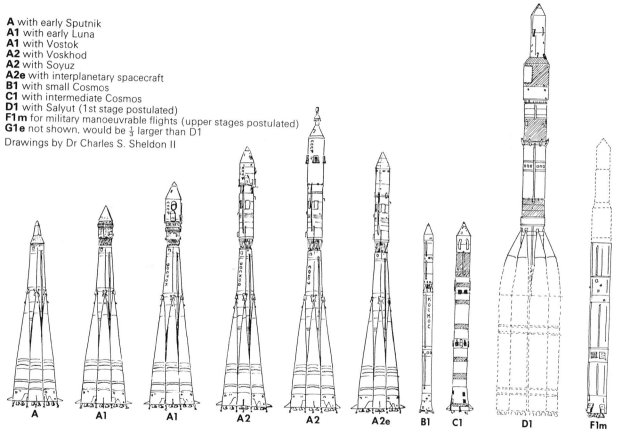

A with early Sputnik
A1 with early Luna
A1 with Vostok
A2 with Voskhod
A2 with Soyuz
A2e with interplanetary spacecraft
B1 with small Cosmos
C1 with intermediate Cosmos
D1 with Salyut (1st stage postulated)
F1m for military manoeuvrable flights (upper stages postulated)
G1e not shown, would be ⅓ larger than D1

Drawings by Dr Charles S. Sheldon II

A   A1   A1   A2   A2   A2e   B1   C1   D1   F1m

337

which go into orbit but re-enter to strike their target before completing one revolution. The only launcher, it seems, with no missile background, is the Proton, or D-type, introduced in 1965 for heavy Earth-orbit and planetary missions. Some further details, for use in conjunction with Dr Sheldon's table, are to be found below.

A1 Vostok core + Luna stage. This early vehicle consists of the original Soviet (SS-6) ICBM, plus 4 tapered, strap-on boosters to make the first Vostok launcher, with a total lift-off thrust of 509,830 kg. The 'cluster design' ('A') is regarded as 2 stages; the 3rd, Luna stage ('1') provides a single engine giving 90,260 kg thrust. It was so named because it first appeared with the Luna 1 launch, and subsequently in the Vostok, Elektron and Meteor projects. Payload capacity of this combination is 4720 kg in low Earth orbit, with launchpads at Tyuratam and Plesetsk.

A2 Vostok core + Venus stage. This standard vehicle consists of the Vostok core ('A') described above, plus a 2-engine 3rd stage, ('2') giving 140,160 kg thrust. It was so named because it first appeared in Feb 1961, to launch Venus 1, and was later used in the Voskhod and Soyuz projects. Payload capacity is 7480 kg in low Earth orbit, with launchpads at Tyuratam and Plesetsk. For the Molniya, planetary, and later Luna programmes, an additional 'escape' stage was added (thus 'A2e').

B1 Sandal (SS-4) + Cosmos stage. This small, 2-stage launcher, first used for Cosmos 1 in Mar 1962, has as a 1st stage a modified Sandal IRBM (B), with thrust approx 74 tonnes. The single engine upper stage (1) has a thrust of approx 11 tonnes. The complete assembly is called the Cosmos launcher, first used for Cosmos 1 in Mar 1962, has as a 1st stage a

Explorers, weighing between 129–453 kg. Launchpads are at Kapustin Yar and Plesetsk.

C1 Skean (SS-5) + Restart stage. This intermediate launcher is believed to use as 1st stage a modified MRBM (C) of unknown thrust, with an unidentified upper stage (1), dubbed a restart stage because it is used to inject satellites into medium-height circular orbits. This requires 1st- and 2nd-stage firing into elliptical orbit, coast to apogee, and then 2nd-stage re-ignition to circularize the orbit. C1 first appeared in Aug 1964 for Russia's first triple-launching (Cosmos 38–40); subsequent use has included Cosmos navigational satellites. Payload capacity up to 1451 kg in low orbit, with launchpads at Tyuratam and Plesetsk.

D1 Proton core + Proton stage. This heavy launcher consists of a multi-engined 1st stage (D) with 1,496,880 kg thrust, which can be used with or without an upper stage (1) and an escape (e) stage. The final stage was first used for the big Proton satellite launchings beginning in Jul 1965, and later in the Zond programme. D1e was used for the Venus 9/10 lander-orbiters in 1975. Payload capacity up to 22,680 kg for low Earth orbit, 4536 kg to the Moon, or 3175 kg to the planets. Launchpads only at Tyuratam.

F1r Scarp (SS-9) + FOBS stage. This military vehicle consists of either the 3-stage Scrag ICBM, or more probably the newer 2-stage Scarp ICBM (F1), with a re-entry FOBS (r). Both Scarp and Scrag are capable of orbital delivery of nuclear bombs to any point on Earth. The final stage, dubbed FOBS stage (for 'Fractional Orbital Bombardment System'), because it has only been used in military FOBS tests, first appeared in Sep 1966 in the unannounced Cosmos U1 launch. Launchpads only at Tyuratam.

MANNED LAUNCHERS

SOYUZ

Lift-Off Thrust:	509,840 kg
Upper Stage:	140,160 kg
Height:	42·65 m

General Description   Soyuz figures are estimates. This launcher appears to be a development of the Vostok and Voskhod rockets, consisting mainly of the insertion of 11·8 m of additional upper staging, with strengthening of the inter-stage truss to safeguard against bending. An emergency escape tower is mounted above the spacecraft, with 3 separate tiers of rocket motors; in the event of a launchpad abort, these systems boost the spacecraft clear of the launchpad, bend the trajectory and control the recovery. The Soyuz spacecraft usually weigh just over 6000 kg.

Perhaps the most remarkable thing about Soviet manned spaceflight is the small amount of rocket development which appears to have taken place between the first Vostok mission and the Soyuz series, when compared with the 166,460 kg thrust of the Atlas rocket used for the first US orbital flight, and the 3,400,000 kg thrust of the Saturn 5 used for Apollo.

VOSKHOD

Estimated { Total Thrust: 650,000 kg
              Height: 42·65 m

General Description   A developed version of the Vostok launcher. The first stage is probably almost identical with Vostok's although there is some evidence that the propellant tanks have been enlarged.

The upper stage has been lengthened to about 6 m compared with 1·98 m on Vostok. The Russians say that it contains 7 engines instead of 6. The 2nd stage, therefore, is believed to have a twin-chamber rocket engine, developing about 140,000 kg thrust.

Since ejection seats were omitted from the Voskhod spacecraft, an emergency escape tower was added to the rocket for the launch of Voskhods 1 and 2.

VOSTOK

Lift-Off Thrust:	509,840 kg
Upper Stage Thrust:	90,265 kg
Total Height:	38 m
Upper Stage and Fairing:	10 m
Strap-on Boosters:	19 m
Overall Diameter:	10·30 m
Diameter of Boosters:	3 m

General Description   All figures are necessarily estimates. The Vostok rocket, not shown publicly until 6 yr after it had been used to place Yuri Gagarin in orbit, is in a cluster, or pyramid design, and consists of 6 units. The central unit with 4 primary nozzles and 4 verniers (small swivelling nozzles), is supported by 4 clip-on boosters, each with 4 primary and 2 vernier nozzles. Thus, no fewer than 32 rocket chambers must be fired simultaneously for lift-off. The upper stage (referred to as the 3rd stage by Soviet scientists) has 1 primary and 4 vernier, or steering nozzles. The central unit is a slim cylinder which flares out into a larger diameter at the top, thus enabling the clip-on units to fit snugly against it; they are fastened by 2 belts of explosive

Vostok rocket at the Paris Air Show

bolts at the top and bottom, and are jettisoned when burnt out. The central unit continues firing at full thrust (102,000 kg), and, when through the densest layers of atmosphere, the protective nose cone is jettisoned. The upper stage ignites at burn-out of the central unit, which then separates. This cuts out when the required orbital velocity is achieved, and separates from the spacecraft. All the primary nozzles are fixed, control during launch being exercised by the vernier nozzles, supplemented by 4 small aerodynamic control surfaces (officially described as 'rudders'), at the base of each clip-on booster. Each rocket unit is stated to use an individual engine firing LOX (liquid oxygen) and hydrocarbons. Vostok can place a payload of about 4715 kg in Earth orbit, not counting the upper stage, the burn-out wt of which is about 1450 kg. Vostok is assembled in a horizontal position attached to a 'strongback', then taken by rail to the Tyuratam launchpad, where it is raised to a vertical position over a flame detector pit. It is steadied by supporting arms which swing up from the pad and remain there until lift-off, when they swing back again.

The Vostok rocket was designed by the late Sergei Korolev.

THE SOLAR SYSTEM

Escape Velocities The speed a visiting spacecraft would have to achieve to leave a planet to return to Earth or visit another—are as follows:

	mph	kph
MERCURY	9272·5	14,919·4
VENUS	22,908·5	36,859·7
EARTH	25,022·0	40,260·3
MOON	5318·0	8556·6
MARS	11,386·0	18,320·0
JUPITER	134,314·5	216,112·0

PLANETARY PRIORITIES

The exciting pattern of future exploration of the Solar System and beyond begins to emerge quite clearly from the recent successes of Russia and America in unravelling the secrets of the inner planets. As recorded in the *Observer's Space Directory*, the exploration of the Outer Planets is already well underway. By the end of the century, man should be reaching out to Pluto, 9th and furthest of all the planets. And long before then, the Space Telescope will be studying the planetary systems of other suns, where it seems inevitable that sooner or later a planet will be found with life forms at least as advanced as our own. So far Russia has shown no enthusiasm for NASA suggestions that the 2 countries might embark on joint planetary missions. NASA's 'Planetary Priorities' are as follows:

Jupiter The 2 Mariner spacecraft (recently renamed Voyager) launched in Aug and Sep 1977, should swing by Jupiter in Mar and Apr 1979, and provide close-up views of the inner planets as well as Jupiter itself. In Dec 1981 a Jupiter Orbiter Probe (JOP) will be launched on a 1000-day journey; it will drop a probe into the atmosphere, and repeatedly encounter the moons Ganymede and Callisto in its Jovian orbit.

Saturn The 2 Voyager spacecraft should pass Saturn at 209,000 km in Nov 1980, passing within 6430 km of Titan, and providing close-ups of 5 inner satellites. The rings are said to be 'potential treasure houses of information'.

Uranus/Neptune If the first Voyager spacecraft is successful at Jupiter, the 2nd will use Jovian gravity to swing towards Uranus, and possibly Neptune.

Venus 2 Pioneer spacecraft will be launched in mid 1978. One will drop 4 probes and then enter the atmosphere itself. The 2nd spacecraft will orbit within 150 km of the surface.

Mars A roving vehicle, able to operate for 2 yr and travel up to 500 km is being designed for launch in 1984, possibly accompanied by a Sample Return Vehicle.

Moon A close look at the lunar poles, with a spacecraft in polar orbit, is considered long overdue. Apollo astronauts had been hoping to do this as their final mission.

ASTEROIDS

The Asteroid Belt was discovered at the end of the 18th century by astronomers who thought there might be an undiscovered planet between Mars and Jupiter, since they were separated by such a disproportionate distance. Instead they found thousands of tiny planets and fragments, estimated to contain enough material to make a planet about one-thousandth the size of Earth. Astronomers have so far identified and calculated orbits for 1831 asteroids. They have been given both numbers and names. There may be a total of 50,000, ranging from the largest, Ceres, which has a diameter of 770 km, to bodies only 2 km in dia; there are hundreds of thousands of asteroid fragments below that size. The belt of debris forms a doughnut-shaped region. In its centre the asteroids orbit the Sun at about 61,200 kph The orbit is elliptical, ranging from 300–545M km from the Sun, and about 245M km wide. A few asteroids stray beyond the Belt; Hermes can come within 354,000 km of Earth, or closer than the Moon. Icarus comes within 14M km of the Sun. Many meteorites which penetrate Earth's atmosphere and reach the surface are believed to be asteroidal material. Most are stony, but some are iron; some contain large amounts of carbon and inorganic chemicals; and occasionally diamonds.

Scientists believe the asteroids either condensed individually from the primordial gas cloud which formed the Sun and planets; or (and they think this is less likely) they are the debris from the break-up of a very small planet. Pioneers 10 and 11, on their way to Jupiter and Saturn, passed safely through the Asteroid Belt in 1973 and 74. Fears that they were likely to be destroyed by collisions were disproved. Pioneer 10 took a few glancing blows from tiny fragments, but was undamaged. The debris is very widely dispersed, and Soviet scientists have calculated that a collision with a large body is likely only once in 600 yr. In Nov 1975, the Russians identified an asteroid with a 10km dia and named it after the late Sergei Korolev.

With NASA now seriously making long-term plans for the colonizing of space, much greater attention is being paid to the asteroids. They are a ready-made source of material for space colonies; it might be possible for men to select say, an asteroid of pure nickel, attach some low-powered ion engines to it, and leave it to make its way, over a period of about 5 yr, into a suitable near-Earth orbit. There it could be mined for its materials and also provide a convenient base while that is being done.

EARTH

Third planet from the Sun, at an average distance from it of 149·6M km, Earth has an equatorial dia of 12,723 km; the polar dia is 42 km less. It is not therefore a true sphere, but is flattened at the poles with an equatorial circumference of 40,054 km. Earth is believed to be about 4·5B yr old, with man himself having evolved about 1M yr ago. Its interior is least known; probably consisting of a solid inner core with a 1206km radius, with a temperature of about 10,000°F, surrounded by a liquid outer core of nickel-iron. Of the surface, 70·8% is covered with water to an average depth of 3810 m. Average land elevation is 820 m. The average annual temperature on the surface is 60°F. The atmosphere consists of 78% nitrogen and 20·9% oxygen (the balance being made up of minute quantities of various other gases) for 100 km from the surface, and above that there are successive layers of atomic oxygen, helium and hydrogen. Planet Jupiter probably has the same type of reducing atmosphere that Earth did when life originated here.

Apollo 17 crew took this Earth view, showing S Pole icecap. It shows most of the African coastline, Arabian peninsula and Asian mainland

Travelling at 30 km per sec, Earth orbits the Sun in 365 days 5 hr 48 min 45·51 sec; it completes 1 rotation on its axis in 23 hr 56 min 4·09 sec.

Earth's Moon With an elliptical orbit 356,334 × 406,610 km, the Moon takes 27 days 7 hr 43 min to revolve around Earth. Because its rotation about its axis is equal to its period of revolution, the Moon always shows the same face to Earth, although 59% of the lunar surface is visible at different times. With a dia of 3475 km, the Moon's mass is 1/81 of Earth's. Man's conquest of the Moon may be followed by a study of America's Lunar Orbiter, Ranger and Surveyor projects, followed by the manned Apollo programme; in the case of Russia, much has been learned from the Luna and Lunokhod projects.

JUPITER

By far the largest of the 9 planets, Jupiter contains three-quarters of the material in the solar system. Its nearest approach to Earth is 627M km. There are at least 13 moons—the 13th having been discovered in Sep 1975. Accurate measurements and details of its atmosphere were finally provided by Pioneers 10 and 11 (qv), which flew past the planet in Dec 1973 and Dec 1974. Circling the Sun at a mean distance of 778M km (compared with Earth's 149·6M km) and taking nearly 12 yr to complete 1 orbit, Jupiter has an equatorial dia of 142,071 km and a polar dia of 133,508 km. The equatorial 'bulge' is caused by the speed at which Jupiter spins—35,400 km at the surface, completing a revolution or 'day' in 9 hr 55 min, making it the fastest planet in the solar system.

The Great Red Spot, in the southern hemisphere, which has been visible through telescopes for several centuries, was photographed by Pioneer 10, and studied in detail by

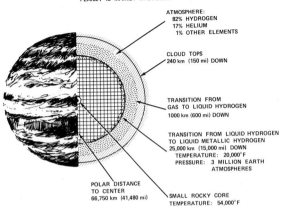

INTERIOR OF LIQUID JUPITER
PLANET IS MAINLY HYDROGEN

ATMOSPHERE:
82% HYDROGEN
17% HELIUM
1% OTHER ELEMENTS

CLOUD TOPS
240 km (150 mi) DOWN

TRANSITION FROM
GAS TO LIQUID HYDROGEN
1000 km (600 mi) DOWN

TRANSITION FROM LIQUID HYDROGEN
TO LIQUID METALLIC HYDROGEN
25,000 km (15,000 mi) DOWN
TEMPERATURE: 20,000°F
PRESSURE: 3 MILLION EARTH
ATMOSPHERES

POLAR DISTANCE
TO CENTER
66,750 km (41,480 mi)

SMALL ROCKY CORE
TEMPERATURE: 54,000°F

Pioneer 11. It is now thought to be the vortex of an intense hurricane, which has been in progress for hundreds of years. A mass of whirling clouds, 2000 km wide, and towering about 8 km above the surrounding cloud deck, it could easily swallow 3 Earths. The Pioneer pictures also revealed that there were other, smaller red spots in the Jovian atmosphere. The atmosphere circulates much less rapidly at the poles than at the equator. As well as spectacular thunderstorms, dwarfing anything known on Earth, there is 'blue sky' visible at the poles.

Analysis of the seething, coloured cloud bands surrounding the planet suggests that the grey-white zones are cloud ridges of rising atmosphere—crystals of frozen ammonia—about 19 km above the darker, orange-brown belts, which are troughs of descending atmosphere, 19 km deep. This is thought to consist of icy particles of ammonium hydrosulphide. As sunlight reaches this material, it turns yellow, then orange, then brown, providing a partial explanation of much of the colour we see on Jupiter. Below that, a 3rd cloud-layer consists of liquid-water droplets, suspended in hydrogen-helium. It is in this area, in the relatively stagnant polar regions, with pressures and temperatures within the limits that terrestrial organisms can tolerate, that some form of life may one day be found, since it contains hydrogen, methane and ammonia—the ingredients in which life on Earth probably originated thousands of millions of years ago. Further down, however, about 200 km below the uppermost cloud layers, temperatures reach 800°F and 100 times Earth's atmosphere pressure at sea level. Beneath this, hydrogen is probably squeezed into a dense, hot fluid under the increasing pressure. About 2900 km down, pressure has reached 100,000 atmospheres and temperatures 12,000°F higher than the Sun's surface. At 24,000 km the liquid hydrogen becomes metallic, a form not known on Earth. Jupiter may have a small rocky core with temperatures rising to 54,000°F, and the pressure to 40M atmospheres.

Unlike any other planet, Jupiter radiates 2·7 times the energy it receives from the Sun, thus behaving more like a star than a planet. it is probably 'left-over heat', from gravitational contraction after the planets condensed some 4·6B yr ago.

Pioneer 11 also showed that Jupiter's magnetic field (Earth and Jupiter are the only planets known to have a substantial magnetic field) is much more complex than expected. Such fields are believed to be produced by motions of the liquid material in the interior, similar to the mechanisms of electric dynamos. Several 'generators' may be contained inside Jupiter, creating a wobbling, tilting magnetic field which sometimes stretches 14·5M km into space, and at other times shrinks by three-fourths or more. Inside this pulsating field are belts of intense radiation; electrons trapped there are 10,000 times more intense than those in the Van Allen radiation belts around Earth.

Jupiter's Moons   There are at least 13. Working inwards towards the planet, and with the year of their discovery in brackets, they are as follows:

The 4 small, outermost moons, with diameters ranging from 24–129 km, are 'maverick' bodies orbiting in a retrograde direction at distances ranging from 24–11M km. They are Sinope (1914), Pasiphae (1908), Carme (1938) and Ananke (1951). Next come 3 inner moons, Lysithea (1938), Elara (1905) and Himalia (1904), at distances of 11·6–11·4M km from the Jovian surface. Most recently discovered is Leda (1975) at 11·1M km. The 4 large moons discovered by Galileo are of most interest. First comes Callisto (1610) over 4800 km dia and as big as the planet Mercury. It has an atmosphere, and its brightness varies. The only one outside the intense radiation areas, it is a possible site for man's first landing in the Jovian system, 1,800,000 km from Jupiter; Pioneer 11 photographed a well-defined south polar ice cap. Next comes Ganymede (1610) about 5280 km in dia, 1½ times the size of Earth's Moon, the largest moon in the solar system, about 1M km from Jupiter. It has a rocky surface, probably with a top layer of ice rubble, which causes its brightness. Then comes Europa (1610) equal in size to Earth's Moon, again believed to have water ice on its surface, orbiting about 600,000 km out. Most exciting of all is orange-coloured Io (1610) the most reflective object in the solar system. It probably has an atmosphere of methane or molecular nitrogen that freezes and snows down to the surface when it is in Jupiter's shadow; when it emerges, glistening into the Sun, it melts into gas again, and the brightness fades in about 15 min. Io is also known to affect low-frequency radio waves from Jupiter. Amalthea (1892) is the closest moon; the 5th to be discovered, it orbits 180,160 km from the surface and has about 160 km dia.

MARS

The Red Planet, named by the Romans after their god of war because of its blood-red hue even to the unaided eye, is much further from Earth than Venus, and is also further from the Sun. On average, Mars is 228M km from the Sun, travelling round it at 77,000 kph. In its larger orbit, Mars moves more slowly than Earth, so that a Martian Year is 687 days 23 hr. With a dia of 6786 km, Mars is about twice the size of the Moon and half the size of Earth. Maximum distance between the 2 planets, when they are on opposite sides of the Sun, is 398,887,000 km; but every 2nd year, the 2 planets come within about 55,784,000 km of each other, providing launch 'windows' approximately every 25 months. According to the initial velocity given to the spacecraft, Mars can be reached in periods varying from 259 to 105 days. Faster journeys mean higher fuel consumption by the launch rocket in order to achieve the greater velocity; and this of course can only be achieved by reducing the weight, or payload of the spacecraft.

Mars from 560,000 km as seen by Viking 1. *Top centre* Tharsis Mts, 3 volcanoes 20 km high; *top left*, Nix Olympica, Mars' largest volcano; *extreme right* Argyre impact basin

Viking 1 picture of Arsia Mons, one of Tharsis volcanoes, standing 19 km above surrounding terrain. Central caldera is 120 km across; top and south, vast amounts of lava have flooded out

Thanks to the remarkable series of unmanned spacecraft launched by Russia (in their 'Mars' series) and by America (in their 'Mariner' and 'Viking' series), in the last 20 yr man has been able to build up a detailed globe of the Martian surface. The gathering of that information is recounted in detail under the relevant spacecraft projects. But it was Galileo's invention of the telescope in 1608 which enabled man to begin his study of the surface; and Christian Huygens in 1659 who first sketched the dark region, Syrtis Major ('giant quicksands'), and from that established that Mars rotated on a north-south axis like Earth, with a day 30 min longer than ours. The dust storms—one of which in 1971 was to ruin Russia's Mars 2 and 3 missions—were identified in the 1700s. The discovery that Mars had seasons and polar ice caps, just like Earth, inevitably led to the belief that Mars was inhabited—strengthened by the observations of Lowell and Schiaparelli between 1877 and 1920. Their discovery of channels, or 'caneli' (mistranslated into 'canals') which Lowell thought were the work of 'intelligent creatures', led to speculation that there was life on Mars. The Mariner flybys, which disclosed details of the cratered surface with its gigantic volcanoes, established that there were no man-made channels or canals and established that at most, only the most primitive form of life was likely to exist.

However it is now known that Mars is geologically active, different from both Earth and Moon; details of the landscape are described under Mariner 9, as well as of the 2 cratered, potato-like moons, Phobos (dia 19 km) and Deimos (dia 10 km) which orbit Mars at distances of 6100 km and 19,900 km. The planet's magnetic field, polar ice caps (now known to contain much water), thin atmosphere (on the surface, no thicker than Earth's at 30,000 m), and dried up river beds, are also discussed in more detail under the entries which record their discovery. When the first Apollo men reached the Moon, there was optimistic talk of a 12-man expedition to Mars, departing on Nov 12 1981, arriving on Aug 9 1982, and arriving back on Earth on Aug 19 1983. With the cutback soon afterwards in America's space programme, that project was abandoned indefinitely. But man's growing shortage of energy is raising hopes that such an expedition will be mounted before the end of the 20th century.

MERCURY

Little was known about Mercury, the nearest planet to the Sun, until many of its secrets, and more than 10,000 pictures were provided by America's Mariner 10 in 11 remarkable days as the spacecraft flew past within 431 km at the end of May 1975. The smallest of the planets, with a 4876 km dia, it had always been difficult for astronomers to study because of its proximity to the Sun. It was not until 1965 that radar measurements by Cornell University established that Mercury's eccentric orbit took it, at its closest, to within 46,259,000 km of the Sun; it then retreats to

70,393,000 km. The long-held view that Mercury always keeps the same face to the Sun proved to be wrong; in fact it rotates on its axis in 58·6 days; the combination of that and its solar orbit makes a Mercury day equal to 180 days on Earth. It is a heavy planet for its size, like the Moon outside, but similar to Earth inside, with a density of 5·5 times that of water, the same as Earth's. Unexpectedly, it was found to have a magnetic field, though 100 times smaller than Earth's.

The Mariner 10 pictures revealed a heavily cratered, dusty surface like the Moon's; its instruments showed a large, heavy core of iron like Earth's. Like both Mars and the Moon, Mercury is 'two-faced'; the spacecraft showed that the 'in-coming' side (furthest from the Sun at that time) was heavily cratered, like the lunar highlands. Then, as it passed to the farside (nearest the Sun at that time), a less heavily cratered area was revealed, with extensive volcanic 'flooding' similar to the so-called lunar seas. Giant basins, the result of terrible meteorite impacts when the planet was young, include at least 18 which are more than 193 km in dia. Most prominent of these, given the name Caloris, is 1280 km across; the shock wave appears to have passed right through the planet to shatter the surface opposite, creating an area named 'Weird Terrain'. The basins show no sign of erosion from water or wind, leading US geologists to conclude that Mercury has had no appreciable atmosphere for at least 4000M yr.

The condition of Mercury, coupled with recent knowledge gained of the Moon and Mars, also suggested that all the planets from Mars inwards suffered heavy bombardment at the same time. On Earth, similar cratering has been largely erased by erosion, volcanic action, and movement of the crust. Other interesting features identified on the Mercurian surface include towering, curving cliffs, often 3·2

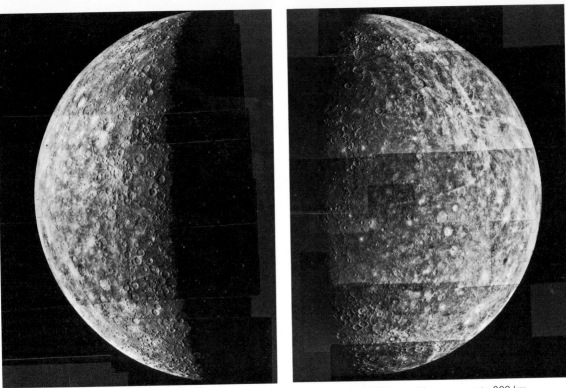

Mariner 10 mosaics of Mercury. *Left* shows $\frac{2}{3}$ of southern hemisphere, from 200,000 km, with craters up to 200 km dia; *right*, from 210,000 km, shows N Pole at top; Caloris Basin at top, with crater rays, is hottest place on Mercury

km high, running for hundreds of kilometres and cutting across crater walls and floors.

Mercury has 2 'hot poles', resulting from its elliptical orbit around the Sun, combined with its rotation. To a man on the surface, the Sun would appear to stop at Mercury's perihelion, or the point where it passes nearest to the Sun; the Sun then goes back more than 1°, and then resumes its forward course. This erratic action takes about one Earth week, and during it the surface directly beneath the Sun gets baked longer. One of these hot poles is near Caloris; on them, the temperature is believed to range from 800°F to −300°F—an incredible 1100°F variation, greater than any other planet.

NEPTUNE

The last but one of the outer planets, and at 4497 km, 8th in distance from the Sun. Finally identified as a planet in 1846 as a result of deductions from irregularities in the orbit of Uranus, it is 51,000 km in dia, but has a density only one quarter that of Earth's, and probably consists largely of hydrogen gas. It has 2 moons. NASA hopes to 'bounce' a Voyager (see Mariner) spacecraft to Neptune via the gravitational fields of Jupiter and Uranus; launched in 1977, it should arrive about 11 years later in 1989.

Neptune's Moons Nereid, discovered in 1949, has an elliptical orbit of 9·6M km × 1,287,400 km, and a dia of approx 320 km. Triton, discovered in 1846, orbiting at 355,000 km has a 4020 km dia.

PLUTO

The last and outermost of the 9 planets, Pluto is 5950M km from the Sun with approx 5900 km dia. It was discovered in 1930 as a result of a systematic search based on the sort of predictions that led to the discovery of Uranus. Pluto appears to be more like the inner planets, such as Earth and Mars, than its huge gaseous companions among the outer planets. Its density, probably much greater than Earth's, may be because it is so far from the Sun that its once-extensive atmosphere has condensed, and lies frozen on the surface, making it as smooth as a mirror, and reflecting the feeble sunlight. It seems unlikely that Pluto's secrets will be revealed by a spacecraft flyby until the 1990s. It has no known moons.

In Jul 1975, the Soviet Institute of Theoretical Astronomy in Leningrad announced that it was likely that there were 2 more planets beyond Pluto—'transplutonian planets'. They are invisible from Earth, but the first is calculated to be about the same size as Earth in volume and mass, 54 times further from the Sun. Evidence for their existence is provided by computer calculations of Comet 1862–3, showing perturbations of its orbit. This comet will be seen again in 1992, when it will be possible to test these theories.

SATURN

The most spectacular of the 9 planets, Saturn is 6th in distance from the Sun—1425M km compared with Earth's 150M km. With a dia of 120,000 km, it is second in size only

Route of Voyager (formerly Mariner) spacecraft from Earth to Saturn using Jovian gravity for 'slingshot' effect. If all goes well, Voyager 2 may be sent on to Uranus (to arrive 1986) and even Neptune (1989)—not shown here

The Moons
Working inwards towards the planet, with the year of their discovery in brackets, followed by their mean distance from the planet, they are approx as follows:

Phoebe	(1898)	12,950,000 km, with dia of 1600 km
Iapetus	(1671)	3,560,000 km, with dia of 480 km
Hyperion	(1848)	1,483,000 km, with dia of 480 km
Titan	(1655)	1,222,000 km, with dia of 5712 km is bigger than Mercury and the largest satellite in the solar system, and known to possess an atmosphere
Rhea	(1672)	527,000 km, with dia of 1850 km 2nd largest of Saturn's satellites
Dione	(1684)	378,000 km, with dia of 1500 km
Tethys	(1684)	295,000 km, with dia of 1200 km
Enceladus	(1789)	238,000 km, with dia of 740 km
Mimas	(1789)	186,000 km, with dia of 595 km
Janus	(1966)	159,000 km, with dia of 350 km

The 5 inner satellites, with dia ranging upwards from 350 km, appear to be covered with ice. The arrival of Pioneer 11 in 1979, which it is hoped will send back man's first close-up pictures of Saturn and its rings, is awaited with impatience in the astronomical world.

to Jupiter, and it has 10 moons. Like Jupiter, Saturn is banded, but the darker belts are fewer and wider, and present less contrast with the brighter zones. The atmosphere is known to contain methane and ammonia. Despite its size, Saturn's density is only just over one tenth that of Earth, and the equatorial 'bulge' is even more marked than Jupiter's. The rotation period at the equator is about 10 hr 15 min.

The Rings There are 2 bright rings separated by a dark line, and an inner dusky one, which is transparent enough to see the body of the main planet through it. Inclined at 27° to the planet's orbit, they are believed to consist of a vast swarm of orbiting ice particles.

SUN

The Sun has been likened to a very slow-burning hydrogen bomb. On average it is 149·6M km from Earth, and scientists call this distance an Astronomical Unit (AU) and use it for measuring the solar system. Because of Earth's slightly elliptical orbit, the Sun's actual distance varies from 147M km to 151M km. A gaseous body, twisting on its axis so that the equatorial regions rotate faster than the polar caps, its dia is 1,390,176 km compared with Earth's 12,723 km. Since the Sun's mass is 330,000 times that of Earth, a man on its surface (if that were possible) would weigh 2 tonnes. Daylight takes 8 min to reach Earth from the Sun's surface; every second, 4M tonnes of solar hydrogen transforms itself into radiant energy which eventually floods into space. The Sun has probably been consuming itself at this rate for 5000M yr; current thinking is that ultimately the thermonuclear explosions will move outwards as they consume unused hydrogen and the Sun will become a monstrous ball of red-hot gas, big enough to engulf and destroy its 4 nearest planets—Mercury, Venus, Earth and Mars; but this will probably not occur for another 5000M yr. It will then cool and shrink to a 'white dwarf' no bigger than Earth but weighing several tonnes per cubic cm. The main body of the Sun is surrounded by a thin shell, or layer of gas, about 322 km thick, called the Photosphere; most of our light comes from this. Outside this is a layer of flamelike outbursts of gas called the Chromosphere; outside that again is an almost endless outer atmosphere called the Corona. Men have been observing the Sun for more than 4000 yr, and as a result we know that its eruptions reach a peak every 11 yr, slowly quietening down and then increasing again between the peaks.

Solar outbursts, or eruptions (also called 'solar flares') send out streams of invisible radiation and clouds of solar gas. On Earth man is protected from the worst effects because our atmosphere acts as a shield; but men in spacecraft would have to be brought home during a big solar storm because the invisible particles could penetrate their craft and damage, and even destroy human cells. The biggest eruption ever recorded was on Nov 12 1960, when American astronomers detected a huge explosion on the face of the Sun. 6 hr later a gigantic cloud of hydrogen gas, 16M km across, travelling at 6400 km per sec, collided with Earth. Its effect on Earth's atmosphere was to start worldwide electrical storms, and black-out all long-distance radio communications for hours, interfering with the navigation systems of ships and planes.

In only 3 days, the Sun delivers to us as much heat and light as would be produced by burning Earth's entire oil, coal and wood reserves; so, now that those reserves are running short, an increasing number of spacecraft, both manned and unmanned, are being used to develop ways of collecting that energy in outer space and beaming it down to Earth. Major advances in studying the way the Sun works were made by astronauts aboard America's Skylab space station in 1973/4, and by Soviet cosmonauts in Salyut 4 in 1975.

URANUS

Third largest planet after Jupiter and Saturn, and at 2860M km from the Sun, the 3rd most distant. Pale green, with occasional white spots, when seen through a telescope, it is believed to be a huge ball of poisonous gases, spinning rapidly on its side. Spectroscopy suggests it has a large amount of methane, as compared with Jupiter and Saturn, which are primarily hydrogen and helium. It may be covered with ammonia clouds. Uranus has a dia of about

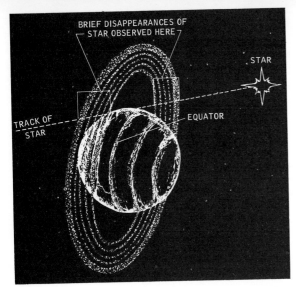

NASA drawing of Uranus's 5 rings, discovered by Cornell University on Mar 10 1977. Their existence was established by observing the light being blocked out 5 times on each side as the planet passed a selected star

In the image: BRIEF DISAPPEARANCES OF STAR OBSERVED HERE, STAR, TRACK OF STAR, EQUATOR

48,000 km and its rapid spin, at an angle of 90° to the ecliptic plane, gives it a day of only 10·8 hr. The unusual tilt also means that its poles have long periods either in the Sun or total darkness. It takes Uranus 85 yr to make 1 revolution round the Sun. It has 5 known moons.

Uranus's Moons Working inwards towards the planet, with the year of the discovery in brackets, followed by their mean distance from the planet, they are approx as follows:

Oberon   (1787) 585,600 km with dia of 1450 km
Titania   (1787) 438,000 km with dia of 1600 km
Umbriel   (1851) 266,900 km with dia of 644 km
Ariel   (1851) 191,600 km with dia of 960 km
Miranda   (1948) 130,000 km with dia of 240 km

Uranus's Rings 5 small rings were discovered in 1977, by an airborne Cornell University research team, to be encircling the planet about 18,000 km from the surface. About 7000 km wide, and much smaller than those around Saturn, the rings probably consist of some type of ice or frozen water, made up of fragments not much more than 1 km in diameter.

VENUS

With its mysterious thick clouds, racing around the equator at more than 321 kph. Venus is still the most puzzling planet in the solar system, despite recent remarkable discoveries by US and Soviet scientists. It has been a source of fascination to astronomers since Galileo first studied it through his newly invented telescope in 1610. Similar in size and mass to Earth, it has very different atmospheric and surface conditions. Of the 9 planets, it is 2nd nearest to the Sun, 108M km compared with Earth's 150M km. Dia is 12,400 km, compared with Earth's 12,900 km. Venus has no detectable magnetic field, and like Mercury and Pluto, no known satellites. It is often so bright in a clear night sky that when appearing just above the horizon, it has frequently started 'flying saucer' scares in England. This is the effect of the sunlight reflecting from the heavy cloud layers, found by Russia's Venus 8, 9 and 10 to be 30–40 km thick, with their base lying 30–35 km above the surface. It was Mariner 10's ultraviolet pictures which, in 1974, revealed that the cloud layers move around at different speeds. Nearest the surface the movement is insignificant, increasing at the top to 60 times that of the planet's own very slow rotation around its axis, and being fastest of all around the equator. Still unexplained is the mystery, discovered by American radar measurements in 1962, of why it is that, whereas Earth spins on its axis once a day, Venus turns lazily in the opposite direction once every 243 Earth days. At the same time it revolves around the Sun once every 225 Earth days—a combination which gives Venus a solar day (from one sunrise to the next) of 117 Earth days. The Sun rises in the west and sets in the east; and when Venus is closest, only the dark side faces us.

Following the 1975 landings, Soviet scientists concluded that the upper cloud zones, visible from Earth, probably consist of concentrated sulphuric acid, possibly mixed with hydrochloric and hydrofluoric acids. Because of high temperatures, causing various elements to vapourize, the lower cloud layers are probably equally inhospitable; bromide and iodine vapour may be present there. For a period it was thought that the thick clouds would result in a 'gloomy greenhouse' effect on the surface, like the dullest winter day on Earth. For this reason, Russia equipped her Venus 9 and 10 spacecraft with searchlights to illuminate their landing areas for the benefit of their TV cameras. In fact, the surface proved to be 'as bright on Venus as in Moscow on a cloudy June day'. Details could be clearly seen at a distance of 50–100 m. Neither Venus 9 nor Venus 10, when they landed 2200 km apart in Oct 1975, raised a dust cloud. In the hour before they were overwhelmed by the heat, the first provided a picture of a young landscape with large, sharp stones; the 2nd a picture of 'blighted scree', with smoother and more weathered rocks. Soviet geologists concluded that Venus was a young, still-living planet, and that it was in an early, cool-down phase of evolution rather than in a final stage of suffocation in a thickening atmospheric greenhouse. The big surprise at this stage was the lack of sand and dust, since the surface was expected to consist of eroded desert. At last an accurate surface measurement was obtained— $465°C$ by Venus 10, with a pressure of 92 Earth atmospheres (1352 psi) and wind velocity of $3·5$ m. With such a dense atmosphere, even slight winds create enormous forces; it was estimated that the wind at the Venus 10 landing site would have the effect, on Earth, of a sea wave pounding the coast in an 80 kph wind. After the landings, the orbiter sections of Venus 9 and 10 were used to build up a large-scale, composite picture of the planet's cloud blanket from an altitude of 1500 km. Launch opportunities for spacecraft occur approx every 19 months.

WORLD SPACE CENTRES

Successful Launch Total by Site 1957–1976

Place	Earth Orbit	Lunar or Escape	Total
Plesetsk, Russia	494		494
Tyuratam, Kazakhstan	370	49	419
Vandenberg AFB, California	395		395
Cape Canaveral, Florida	207	53	260
Kapustin Yar, Russia	64		64
Wallops Island, Virginia	16		16
Indian Ocean Platform, Kenya	8		8
Shuang Cheng-tzu, China	7		7
Uchinoura, Japan	6		6
Kourou, Guiana	6		6
Hammaguir, Algeria	4		4
Tanegashima, Japan	2		2
Woomera, Australia	2		2
Total	1581	102	1683

Australia

Woomera   About 430 km N of Adelaide, and running eastwards for 2000 km across the desert, the Woomera missile and rocket range was started jointly by Britain and Australia in 1946, and cost them well over £200M. Australian hopes that what was originally a test range for ballistic missiles and sounding rockets, and for shooting down pilotless planes, would have a future as a launch centre for satellites, were not fulfilled. During the period when Britain's Blue Streak was being developed as the main stage of ELDO's Europa rocket, a thriving township of 4500, with over 500 houses supplied with power and water across 160 km of desert, sprang up; 35,000 young trees were planted. But Woomera was not suitable for equatorial launches, and the French preferred to develop their own launch centre at Kourou. After successive cuts in its contributions to the range, Britain announced there would be no more work for Woomera after 1976; NASA also decided to close down its Carnarvon tracking station, which had played a major part in manned and unmanned spaceflights. A small test satellite, launched into polar orbit by an American Redstone in 1967, and Britain's one Black Arrow in 1971, are likely to be the only 2 satellites ever orbited from Woomera.

China

Shuang-Cheng-Tzu   About 1600 km W of Peking, and near Lop Nor, the nuclear test base. Pictures taken by Landsat-2 show this as a small facility, with little expansion in progress, even though China's first 5 satellites were all launched from here between Apr 1970 and Dec 1975. Orbital missions are launched SE, and military IRBMs are fired into the Lop Nor impact site 800 km W.

France

Guiana

France began her national space programme with launches from Hammaguir, in the Sahara Desert, but in 1964 decided to create a civil launch centre at Kourou, in French Guiana. Though hot and wet, its position 2°N of the Equator makes it ideal for launching synchronous satellites, and to the N and E rockets can travel 3000 km without passing over land. The first firing from Guiana was the first Veronique sounding rocket on Apr 9 1968, with the first satellite being launched in Mar 1970. With the decision in 1966 to transfer ELDO (European Launcher Development Organization) activities from Woomera in Australia to the more favourable Guiana site, it was developed into a space centre comparable with America's Cape Canaveral, and in some respects more modern. The installations now extend over a 30km strip of waterfront. They include a satellite control station (interferometer and telemetry) to the N; 3 launch pads originally for sounding rockets, Diamant and Europa II (the latter now being converted for Ariane); the radar and telemetry stations on a hill 20 km away (Montagne des Peres); and a tracking radar on Ile Royale, off Kourou (one of the 3 islands forming the notorious, but now disused penal colony, known as Devil's Isle). Between Nov 1965 and Sep 1975, 12 Diamants were fired, and 8 satellites placed in orbit. The final collapse of ELDO in 1973 left the Guiana Centre with only 1–2 Diamant B launches per yr, and the French having to bear the whole of the $5·7M per yr upkeep costs, instead of being able to rely on a 40% contribution from ELDO. The situation was saved by the formation of ESA, although it meant a gap of 4 yr between the last Diamant launch and the first Ariane test-launch in 1979. It also meant cutting the 500 personnel by about one-third.

Hammaguir

A military base at Colomb Bechar in Algeria's Sahara Desert used as a French rocket testing and development site from the mid 1950s until she was required to evacuate the base in 1967 following Algerian independence. It had an IRBM test corridor stretching 3000 km SE towards Fort Lamy. About 300 rockets were fired in upper atmosphere and missile tests, some of them detonating high explosive in the upper atmosphere. France developed the liquid propellant boosters, Emeraude, Saphir and Diamant at Hammaguir, and 4 small test satellites were orbited between 1965–67.

Italy

San Marco

Owned and operated by the Italian Government, this consists of 2 offshore platforms in Formosa Bay 4·8 km off the coast of Kenya, 2·9°S of the Equator. It became operational in 1966 and by mid 1976 8 satellites had been launched, 4 Italian, 1 British and 3 American. The launch platform has 20 steel legs sunk into the sandy seabed, linked by 23 cables to the Santa Rita control platform, 920 km away. This is a modified oil rig. Explorer 42, the Small Astronomy Satellite, launched by an Italian crew on Dec 12 1970, was the first US satellite to be launched by a foreign country. An example of the benefits of this equatorial site is that it was possible to launch the 195 kg Explorer 43 with a Scout rocket instead of the much larger booster needed to achieve the same orbit from Cape Canaveral. In addition to the launch ramp for Scouts, there is a ramp for US Nike-Apache and Nike-Tomahawk sounding rockets.

Sunrise on launch day for Britain's Ariel 5 from Italy's San Marco equatorial range off Kenya coast

Japan

Kagoshima, Uchinoura Japan's first 6 satellites were launched from these levelled hilltops facing the Pacific Ocean, on the southern tip of Kyushu Island. The site was completed in 1964 when the first Lambda-3 rocket was fired.

It was 6 yr later before the first successful orbital launch. The launchpads are 220 m above sea level, but local fishermen complained about the noise and hazards associated with the launches over their fishing grounds. The Government decided that improved and safe launch facilities should be provided at Tanegashima for the bigger 'N' rockets to launch applications satellites, although Kagoshima remained available to Tokyo University for launching scientific satellites.

Tanegashima The first phase of the Tanegashima Space

Japan's N Launchpad at the Tanegashima Space Centre, carved out of mountainous terrain

Centre was completed in 1974 in time for the launch of Japan's 7th satellite, ETS-1, by the new 'N' rocket. It is a 58km long island, about 100 km S of Kagoshima. Appropriately Tanegashima became the first Japanese word for 'firearm' because it was there that Fernao Mendes Pinto landed in 1542 carrying a musket. There are 2 launchpads, one for firing small rockets, the other for satellite launches; by mid 1977 3 had taken place.

US

Johnson Space Centre, Houston Control of all manned spaceflights passes from Launch Control Centre (LCC) at Cape Kennedy to Johnson Space Centre (formerly Manned Spacecraft Centre) 32 km SE of Houston, 10 sec after lift-off. Until 1961 it was undeveloped cattle country; in 3 yr it became the world's most sophisticated scientific centre. Unlike an airport, it was built as a whole, with no continuing process of additions and alterations. Its focal point, Mission Control Centre, has been the control point for all NASA's manned flights since Gemini 4, in Jun 1965. Flight controllers at 171 consoles, can check, at a touch of a button, the progress of the mission against the flight plan; 133 TV cameras and 557 receivers are used.

JSC is also responsible for design, development and testing of manned spacecraft and their systems, as well as the selection and training of astronauts, who live nearby in the Clear Lake and Key Largo areas.

Kennedy Space Centre, Cape Canaveral Cape Canaveral, a strip of sandy jungle on the Florida coastline, was

Cape Canaveral: Crawler transporter moves Saturn 5/Apollo from Vehicle Assembly Building towards the launchpad

originally set aside as a development area for ICBMs (Intercontinental Ballistic Missiles) in 1947. Since then it has been developed into the John F Kennedy Space Centre, and all US manned spaceflights have so far been launched from it. Originally known as Cape Canaveral, it was re-named Cape Kennedy after the late President's assassination; after several years of local protests and presentations to Congress, the Cape's original name was restored, but the Centre itself was named after the President. Cape Canaveral is also known as Eastern Test Range. About 40 Launch Complexes have been built, mostly by the US Air Force for their own military missile development and for unmanned satellite launchings which they carry out for NASA. The Merritt Island 'Moonport', Launch Complex 39, however, was developed and is operated exclusively by NASA for the Apollo Moonlandings, the subsequent Skylab missions, and the forthcoming Space Shuttle launches.

Bus tours are operated by NASA for visitors at the rate of about $2.50 for adults and half price for those under 18, and on Sundays visitors are allowed to drive their own cars to the launch sites. The centre attracts about 3M visitors per yr.

Launch Complexes involved in manned spaceflight are as follows:

5–6 Site of first 2 US suborbital manned flights, Mercury-Atlas 3 and 4. Complex now a museum piece.

14 Used for Mercury-Atlas 6–9, the four orbital flights, starting with John Glenn in Feb 1962. Blown up in 1976 when corrosion made it dangerous.

19 All 10 Gemini-Titan flights launched from here, between 1965–66. Site now inactive.

34 Site of first Saturn 1 launch. Altogether 7 Saturn 1 and 1B launches took place here, including Apollo 7, in Oct 1968, the project's first manned flight. The scene, too, of the Apollo 1 disaster; Grissom, White and Chaffee died when fire broke out in their spacecraft during what should have been a final rehearsal for the Apollo 7 launch. Site now inactive.

37 Site of 8 Saturn 1 and 1B launches, but no manned flights. Inactive.

39 Site of all manned Apollo flights except the first (Apollo 7), of Skylab launches and ASTP. Its construction reversed the fixed launch concept of earlier missions, with assembly, checkout and launch all being carried out on the pad. With the mobile concept employed on Complex 39, assembly is carried out in the huge Vehicle Assembly Building (VAB), which is 160 m tall, and so large that 4 buildings the size of the United Nations in New York would fit inside. The Launch Control Centre (LCC) nearby has 4 firing rooms, each with 470 sets of control and monitoring equipment, plus conference and display rooms etc. During launch 62 TV cameras at the pad supply 100 monitor screens. The Mobile Launcher 136 m high and 4808 tonnes in weight, serves as an assembly platform within the VAB and as a launch platform and umbilical tower at the launch site 5·6 km away. The Mobile Launcher is moved along a special roadway by the Crawler-Transporter, a double-tracked vehicle the size of a football field, 40 m long, and 35 m wide; the assembled Apollo-Saturn 5 vehicle which it carries stands 36 storeys high; the total load weighs 7711 tonnes.

Access to both the Apollo spacecraft and Saturn vehicle at the pad is provided by the Mobile Service Structure (MSS). 125 m high, this has 2 lifts and provides 5 platforms—the top 3 for spacecraft access, and the lower 2 for Saturn 5. About 11 hr before launch, the MSS is moved back 2134 m from Pad A. Pads A and B are almost identical, although Pad B was used only once, for Apollo 10, up to the end of the Apollo Project. They are 2657 m apart and roughly octagonal in shape. Below them are flame trenches 13 m deep and 137 m long. For emergency, a 61 m escape tube leads from the bottom of the lifts to a blast-resistant room 12 m below the pad; a cab on a slidewire also provides quick escape from the 98m level on the MSS to a revetment 762 m away.

Following the conclusion of the Skylab flights and ASTP, major modifications were carried out on Pad 39 and in the

VAB to adapt them for Shuttle assembly and launch. A 4500m runway was also laid down among Merritt Islands orange and grapefruit groves.

Vandenberg Air Force Base, California About 160 km N of Los Angeles, and by the end of 1975 had been the site of 382 launches, 2nd only to Russia's military launch site, Plesetsk. Since it is probably America's most important base for operational Minuteman ICBMs, sunk in deep underground 'silos' for protection, it is inevitably the most secret of America's 3 launch centres. The base has been created along 40 km of coastal desert and scrub, rather similar to the Cape Canaveral area. From here 'spy' satellites are regularly launched, frequently being recovered with the packages of reconnaissance film a few days later; others, like the sophisticated 'Big Bird', remain in orbit for 2–3 months at a time, passing over Russia and China every 90 min, transmitting TV pictures and close-up photographs of every military development as they pass over the American mainland. From here too ICBMs are test-fired under operational conditions over the Western Test Range, extending 8000 km across the Pacific to the tip of S America. About 10,000 Service personnel and their families live in the isolated area behind the base. From its southern tip, Point Arguello, spacecraft are launched into polar orbits, for both NASA and the USAF; polar launches are not permitted from Cape Canaveral, since this would involve firing them over the US mainland. Vandenberg will soon also become the 2nd US site for manned launches; the 20 Shuttle flights per yr expected to be flown for military purposes will be from Vandenberg; its runway has also been lengthened so that Shuttle landings can take place there as well. There are likely to be many occasions when Shuttle flights will begin at Vandenberg and

Rare view of Western Test Range at Vandenberg, with simultaneous test firings of 2 Minuteman missiles

end at the Cape, or vice versa. Vandenberg is also known as Western Test Range.

Wallops Flight Center Off Virginia's eastern shore, Wallops Island is America's 3rd launching site after Cape Canaveral and Vandenberg. Since 1945 more than 8000 rocket launchings have been made from the 2670-hectare facility, including the orbiting of 16 scientific satellites. On one occasion in 1974, 54 sounding rockets were launched in just over 24 hr (though by no means all were successful) in an elaborate upper atmosphere experiment. 6 launch areas and 24 launchers were used. Wallops employs over 400 personnel, and also carries out advanced aeronautical research.

White Sands Test site in New Mexico for sounding rockets and short-range vertical firings.

USSR

Baikonur (Tyuratam) The Soviet equivalent of Cape Canaveral, this is the cosmodrome from which all Soviet manned launches are made, as well as the Luna, Mars and Venus unmanned spacecraft. Its actual site was identified by a Japanese astronomer in 1957 as being near Tyuratam in Kazakhstan, E of the Aral Sea, and about 370 km SW of Baikonur. For at least 17 yr after that, however, the Russians continued to give its latitude and longitude as that of the town of Baikonur, though during the ASTP mission it was finally admitted that 'the highway passes near the Tyuratam railway station'. The site, nowadays much busier than Cape Canaveral, continues to be known as 'Tyuratam' in the West, and most of its details continue to be shrouded in mystery. America's 'Big Bird' spy satellites have no doubt provided much information for the military men, and photographs from Landsat, the Earth resources satellite, have given the public some idea of the vast sprawl of launchpads. (Some details of these can be found under 'ASTP—Preparations', thanks to Tom Stafford's refusal to go through with the joint flight unless the US crew went

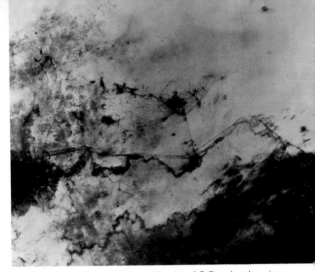

Landsat 1 view from 905 km altitude, of S Russia, showing Tyuratam (Baikonur) launch area, indicates how spy satellites can monitor events there

there.) Details issued by the Novosti Press Agency at that time said that the nearby town of Leninsk, which has grown to 50,000, is situated 2100 km SE of Moscow, a 3-hr jet flight. Founded 20 yr ago, 'it cannot boast a mild climate', being hot in summer and with violent snowstorms and 40°C frosts in winter. The whole city of Leninsk 'is entwined with a dense network of irrigation pipelines'. The cosmonauts have comfortable accommodation in cottages, and for pre-launch checks and activities there is a hotel complex with swimming pool and sports centre. The launchpad used for Sputnik 1 and Yuri Gagarin's historic flight is about 32 km from Leninsk; 2 small houses used by Gagarin and by Academician Korolev, 'who preferred to stay there during his frequent visits to Baikonur rather than at the hotel', are now memorial houses. Novosti added: 'The launching table for the Soyuz spacecraft is smaller and simpler than the American installations on Cape Canaveral. If one looks at the pad from a helicopter, it resembles a huge torch, with straight access routes being its handle and the roundish launching table its bowl.' Soviet journalists who visited its 'underground kingdom', found themselves in a 'several-score metre deep concrete gorge, under the nozzles of 20 rocket engines. Next to the launching pad deep underground is the command bunker whose periscopes make it look like a submarine. This is where all the key commands come from'. Though no Soviet reference to them has ever been made, it is known that 2 massive launch complexes have been built for the TT-5 (SL-14) 'super booster', which is bigger than Saturn 5, The pad had to be rebuilt when a booster exploded during a fuelling test in 1969, and 2 launch failures followed in 1971 and 1972. In mid 1976, testing was again underway in a 122m static rig, but there had still been no launch. By end 1975, Baikonur launches totalled 337.

Kapustin Yar   This was Russia's first rocket development centre, preceding the much more important centres at Tyuratam and Plesetsk. Thanks to a Landsat 2 picture obtained at the end of 1975, we now know that its facilities extend over a 96 × 72km area NE of the River Volga and 965 km SE of Moscow. It now mostly launches vertical probes as does White Sands, and small orbital payloads as does the Wallops Center. During its early years, Kapustin Yar was

used for testing captured V2 rockets, and for sounding rocket experiments which carried dogs and other research animals to heights of 500 km. Between 1962–70 orbital launches totalled 56. From 1971–5 only 7 launches were made, as work was switched to Plesetsk. Recent activities there have involved tests of ABM's (anti-ballistic missiles) which are fired through the atmosphere so that they fall back towards Russia's Sary Shagan missile base 2000 km to the E, near the Chinese border.

Plesetsk   Russia's most important military launch site, and the equivalent of America's Vandenberg, is S of Archangel. More payloads—over 450 by mid 1976—have been launched from here than from any other world space centre. Its existence, still not publicly admitted by the Russians, was first identified by Geoffrey Perry of Kettering Grammar School in England in 1966 when he and his science students noted that Cosmos 112 could not have been launched from either Tyuratam or Kapustin Yar. Landsat satellite pictures of the Plesetsk area, taken in 1973, showed it to be about 100 km long, with at least 4 launch complexes, and heavily defended by surface-to-air missiles. The scale of its activities (analysed by *Aviation Week*) can be judged by the fact that in 1973—probably a typical year—it supported 61 orbital launches compared with 23 from Tyuratam; 48 were military launches, with 27 involving recoverable photo reconnaissance craft, plus one other military-related satellite. Other Plesetsk launches, with Tyuratam's in brackets, were Scientific 2 (1); Weather 3 (0); Communications 7 (2); Biological 1 (1). In addition Russia launched a total of 128 test ICBMs from Plesetsk and Tyuratam in 1973.

SPACEMEN

US ASTRONAUTS

ASTRONAUTS' COMPARATIVE FLIGHT TIME TO ASTP

Astronaut	Flights	Hr., Min.	Mission
Carr	1	2017:16	SL-4
Gibson	1	2017:16	SL-4
Pogue	1	2017:16	SL-4
*Bean (4th)	2	1671:45	AS-12, SL-3
Garriott	1	1427:09	SL-3
Lousma	1	1427:09	SL-3
*Conrad (3rd)	4	1179:38	GS-5, GT-11, AS-12, SL-12
Lovell	4	715:06	GT-7, GT-12, AS-8, AS-13
Kerwin	1	672:50	SL-2
Weitz	1	672:50	SL-2
*Cernan (11th)	3	566:16	GT-9, AS-10, AS-17
*Scott (7th)	3	546:54	GT-8, AS-9, AS-15
*Young (9th)	4	533:34	GT-3, GT-10, AS-10, AS-16
Stafford	4	507:43	GT-6, GT-9, AS-10, AS(ASTP)
Borman	2	477:36	GT-7, AS-8
McDivitt	2	338:57	GT-4, AS-9
Gordon	2	315:53	GT-11, AS-12
Evans	1	301:52	AS-17
*Schmitt (12th)	1	301:52	AS-17
Schirra	3	295:13	MA-8, GT-6, AS-7
*Irwin (8th)	1	295:12	AS-15
Worden	1	295:12	AS-15
*Aldrin (2nd)	2	289:54	GT-12, AS-11
Collins	2	266:06	GT-10, AS-11
Mattingly	1	265:51	AS-16
*Duke (10th)	1	265:51	AS-16
Eisele	1	260:09	AS-7
Cunningham	1	260:09	AS-7
Schweickart	1	241:01	AS-9
Cooper	2	225:15	MA-9, GT-5
Brand	1	217:28	AS (ASTP)
Slayton	1	217:28	AS (ASTP)
*Shepard (5th)	2	216:17	MR-3, AS-14
*Mitchell (6th)	1	216:02	AS-14
Roosa	1	216:02	AS-14
*Armstrong (1st)	2	206:00	GT-8, AS-11
Anders	1	147:01	AS-8
Haise	1	142:55	AS-13
Swigert	1	142:55	AS-13
White	1	97:56	GT-4
Grissom	2	5:09	MR-4, GT-3
Carpenter	1	4:56	MA-7
Glenn	1	4:55	MA-6
Total	71	22503:49	31 Manned Missions

UNITED STATES FLIGHT SUMMARY

	Flights
4 men (4 flights)	16
3 men (3 flights)	9
10 men (2 flights)	20
26 men (1 flight)	26
Total 43 men	71

Source: NASA—Office of Manned Space Flight and Library of Congress
NASA—Executive Management Services Division

*Walked on the lunar surface (in order of walk)

John H. Glenn, now a Senator, at the time he became first US astronaut in orbit

Pete Conrad takes a shower after strenuous EVAs to repair Skylab launch damage. Picture on page 222 shows position of shower

By July 1977, only 27 of the 73 pilots and scientists selected in 7 groups since Apr 1959 were still available as flight crewmen; 24 remained on the active list, with 3 on other assignments but remaining available. Of those, 22 were pilot-astronauts and 9 were scientist-astronauts. While decisions have been made that all Shuttle test flights will be conducted by the current team of astronauts, NASA (at the time of writing) is already selecting an 8th group. Since Shuttle flying will be much less physically exacting, the tests and qualifications will be much less stringent for the new group. Over 1000 applicants sought 30 vacancies—15 for pilot astronauts, and 15 for a new category known as 'mission specialists'. Their job will be orbital management of the Shuttle's payload, including operation of the remote-control manipulating arm for deploying and retrieving spacecraft; if EVA becomes necessary, this too will be the mission specialist's job. Women and Blacks are expected to

be included in the new group. Later a 3rd category of 'payload specialists'—scientists, doctors etc, involved as experimenters—will be added to Shuttle crews.

Of the 46 original astronauts no longer on the active list, 8 are dead; 3 were lost in an Apollo fire during a ground rehearsal, but not one in space. By the end of ASTP, 43 astronauts had flown. Brief details of all 73 appear below, indicating whether they are active or inactive as astronauts.

Aldrin, Edwin E. (b. Jan 20 1930): Col USAF. Pilot-astronaut, selected Oct 1963. 2nd man to step on Moon, as pilot of Apollo 11's 'Eagle' on Jul 21 1969. Previously flew in Gemini 12 and spent $5\frac{1}{2}$ hr on EVA outside spacecraft. Left NASA 1971; ret. USAF 1972. Pres Research & Eng Consultants Inc Los Angeles. M; 2 sons, 1 dtr. INACTIVE.

Allen, Joseph P. (b. Jun 27 1937): Scientist-astronaut, selected Aug 1967; nuclear physicist. NASA

Virgil Grissom   James Lovell   Alan Shepard

A/Administrator for Legislative Affairs, 1975. Available for Shuttle. M; 1 son. ACTIVE.

Anders, William A. (b. Oct 17 1933). Lt-Col USAF (ret). Pilot-astronaut, selected Oct 1963. LMP on Apollo 8, 1st manned flight around Moon. Res NASA, Jul 1971, and joined Research Pilots' School, Edwards' AF Base. Atomic Energy Comm., 1973. Chmn Nuclear Regulatory Comm., 1975. US Ambassdr to Norway 1976. M; 4 sons, 1 dtr. INACTIVE.

Armstrong, Neil A. (b. Aug 5 1930): Civilian test pilot. Pilot-astronaut selected Sep 1962. 1st man to step on Moon, Jul 21 1969 as Cdr, Apollo 11. As Cdr, Gem. 8, on Mar 16 1966, performed 1st space-docking with Agena target. Prof Engineering, U of Cincinnati, Ohio, Oct 1971. M; 2 sons. INACTIVE.

Bassett, Charles A. (b. Dec 30 1931); Maj USAF; pilot-astronaut, selected Oct 1963. Was training as Gemini 9 pilot when he and Elliott See were killed in T-38 crash at St Louis in Feb 1966.

Bean, Alan L. (b. Mar 15 1932); Capt USN; pilot-astronaut,

selected Oct 1963. LMP on Apollo 12, Nov 1969, was 4th man on Moon. Cdr Skylab 3 mission, Jul 1973. M; 1 son, 1 dtr. ACTIVE.

Bobko, Karol J. (b. Dec 23 1937); Lt-Col USAF; pilot-astronaut, transferred to NASA when AF Manned Orbiting Laboratory programme was cancelled in Aug 1969. ASTP Support Crew. M; 1 son, 1 dtr. ACTIVE.

Borman, Frank (b. Mar 14 1928); Col USAF (Ret); pilot-astronaut, selected Sep 1962. Cdr Apollo 8, on first circum-lunar flight, Dec 1968. Cdr Gemini 7, in first RV with Gemini 6, on Dec 4 1965. Sen. Vice-Pres Eastern Airlines, Jul 1 1970. Pres Eastern Airlines 1975. M; 2 sons. INACTIVE.

Brand, Vance D. (b. May 9 1931). Civilian test pilot; pilot-astronaut, selected Apr 1966. Support crew member on Apollos 8 & 13. CMP ASTP Jul 1975. M; 2 sons, 2 dtrs. ACTIVE.

Bull, John S. (b. Sep 25 1934). Lt-Cdr USN (Ret). Pilot-astronaut, selected Apr 1966; withdrew, following illness, 1968. INACTIVE.

Carpenter, M. Scott (b. May 1 1925). Cdr USN (Ret).

Mercury astronaut, selected Apr 1959; made 2nd orbital flight in Mercury 7, May 24 1962. Res NASA 1967. Pres Sea Sciences Corp., Los Angeles. Later in wasp-breeding business. M; 4 children. INACTIVE.

Carr, Gerald P. (b. Aug 22 1932). Col USMC (Ret); pilot-astronaut, selected Apr 1966. Support crew member, Apollo 8 & 12. As Cdr, Skylab 4, Nov 1973, is joint holder of record for longest spaceflight of 84 days. Now on Shuttle development. M; 2 sons, 4 dtrs. INACTIVE.

Cernan, Eugene A. (b. Mar 14 1934) Capt USN (Ret); pilot-astronaut, selected Oct 1963. Cdr Apollo 17, and 11th man on Moon, Dec 1972. As pilot on Gemini 9, did 2-hr spacewalk, Jun 3 1966; as LMP on Apollo 10, descended to within 14·5 km of lunar surface in final landing rehearsal on May 18 1969. Res Jul 1976; Vice-Pres Coral Petroleum Co Houston. M; 1 dtr. INACTIVE.

Chaffee, Roger B. (b. Feb 15 1935). Lt-Cdr USN; pilot-astronaut, selected Oct 1963. Selected for Apollo 7, but died in launchpad fire, Jan 27 1967.

Chapman, Philip K. (b. Mar 5 1935). Civilian; scientist-astronaut, selected Aug 1967. Australian. Resigned Jul 1972, to become research scientist at MIT, Cambridge, Mass. M; 2 children. INACTIVE.

Collins, Michael (b. Oct 31 1930). Maj-Gen USAF Res.; pilot-astronaut, selected Oct 1963. As CMP of Apollo 11, in Jul 1969, remained in lunar orbit during 1st Moonlanding. Pilot on Gemini 10 on Jul 18 1966. Dir Air & Space Mus, Smithsonian Inst, Washington, Feb 1971. M; 1 son, 2 dtrs. INACTIVE.

Conrad, Charles (b. Jun 2 1930). Capt USN; test-pilot

Neil Armstrong, Michael Collins and Edwin (Buzz) Aldrin

astronaut, selected Sep 1962. As Cdr Apollo 12, 3rd man to land on Moon. Pilot of Gemini 5 on 8-day flight in Aug 1965; Cdr, Gemini 11, on Sep 12 1966. Head of Skylab Project, and Cdr, Skylab 2, May 1973. 3rd man to make 4 flights. Vice-Pres American TV & Comm. Corp, Denver, 1974. Vice-Pres McDonnell Douglas Corp 1975. M; 4 sons. INACTIVE.

Cooper, L. Gordon (b. Mar 6 1927) Col USAF (Ret); Mercury astronaut, selected Apr 1959. Made Mercury 9 flight, last of series on May 15 1963. Cdr Gemini 5, Aug 1965. Back-up Cdr Apollo 10. Ret Jul 1970; Vice-Pres Research, WED Enterprises Glendale, Calif. M; 2 dtrs. INACTIVE.

Crippen Robert L. (b. Sep 11 1937) Cdr USN; pilot-astronaut, transferred to NASA when A.F. Manned Orbiting Laboratory programme was cancelled in Aug 1969. ASTP Support. M; 3 dtrs. ACTIVE.

Cunningham, Walter (b. Mar 16 1932). Civilian pilot-astronaut, selected Oct 1963. LMP on Apollo 7 in Oct 1968. Worked on Skylab. Res Aug 1971. Vice-Pres of Operations, Century Development Co. then Pres Hydrotech Intl, Houston. M; 1 son, 1 dtr. INACTIVE.

Duke, Charles M. (b. Oct 3 1935): Col USAF, pilot-astronaut, selected Apr 1966. LMP Apollo 16; 10th man on Moon, Apr 1972. Res Jan 1 1976 to enter business in San Antonio, Texas. M; 2 sons. INACTIVE.

Eisele, Donn F. (b. Jun 23 1930): Col USAF (Ret); pilot-astronaut, selected Oct 1963. CMP on Apollo 7, Oct 1968; back-up CMP Apollo 10. Resigned NASA & USAF, Jul 1972; joined Peace Corps. M; 2 sons, 2 dtrs. INACTIVE.

England, Anthony W. (b. May 15 1942): Civilian, scientist-astronaut, selected 1967. Res Aug 1972 to take appointment with US Geological Survey, Denver. M; 1 dtr. INACTIVE.

Engle, Joe H. (b. Aug 26 1932): Col USAF, pilot-astronaut, selected Apr 1966. Support crew for Apollo 10; back-up LMP Apollo 14. 2nd Cdr Shuttle ALT tests 1977. M; 1 son, 1 dtr. ACTIVE.

Evans, Ronald, E. (b. Nov 10 1933): Capt USN, pilot-astronaut, selected Apr 1966. Support crew Apollo 7 and 11; back-up CMP, Apollo 14. CMP, Apollo 17, Dec 1972. Back-up CMP for ASTP, Jul 1975. M; 1 son, 1 dtr. INACTIVE.

Freeman, Theodore C. (b. Feb 18 1930): Pilot-astronaut, selected Oct 1963. Killed when his jet trainer collided with snow goose Oct 1964.

Fullerton, Charles G. (b. Oct 11 1936): Lt-Col USAF, pilot-astronaut, transferred to NASA when AF MOL programme cancelled in Aug 1969. Support crew, Apollo 14 & 17. 1st pilot Shuttle ALT tests 1977. M; 1 son, 1 dtr. ACTIVE.

Garriott, Owen K. (b. Nov 22 1930): Civilian, Scientist-astronaut selected Jun 1965. Science-pilot, Skylab 3, Jul 1973. Dir Science & Applications JSC. M; 3 sons, 1 dtr. ACTIVE.

Gibson, Edward G. (b. Nov 8 1936): Civilian, Scientist-astronaut, selected Jun 1965. Support crew for Apollo 12; as science-pilot, Skylab 4 Nov 1973, is joint holder of record for longest spaceflight of 84 days. Res Dec 74; now with ERNO, Bremen W Germany working on Spacelab. M; 1 son, 1 dtr. INACTIVE.

Givens, Edward G. (b. Jan 5 1930): Maj USAF. Died in car accident, Jun 1967.

Glenn, John H. (b. Jul 18 1921): Col USMC (Ret). Mercury astronaut, selected Apr 1959. First US man in orbit, on Mercury 6, on Feb 20 1962. Res NASA 1964. Senator (Democrat) for Ohio 1974. M; 1 son, 1 dtr. INACTIVE.

Gordon, Richard F. (b. Oct 5 1929): Capt USN (Ret); pilot-astronaut, selected Oct 1963. As CMP on Apollo 12, remained in lunar orbit during 2nd Moonlanding, Nov 1969. Pilot on Gemini 11, Sep 12 1966. Ret Jan 1972. Exec Vice-Pres, New Orleans Saints, professional football team. M; 4 sons, 2 dtrs. INACTIVE.

Graveline, Duane. Doctor; scientist-astronaut, selected Jun

Donald Slayton, Thomas Stafford and Vance Brand

1965; res for personal reasons 1965. Now with Dept of Health, State of Vermont. INACTIVE.

Grissom, Virgil I. (b. Apr 3 1926): Lt-Col USAF; Mercury astronaut, selected Apr 1959; flew 2nd sub-orbital Mercury mission, Jul 21 1961. Cdr Gemini 3, first US 2-man craft, Mar 23 1965. Died during Apollo launchpad rehearsal fire, Jan 27 1967.

Haise, Fred W. (b. Nov 14 1933): Civilian; pilot-astronaut, selected Apr 1966. LMP on Apollo 13, Apr 1970. Back-up Cdr Apollo 16, Apr 1972. Cdr Shuttle ALT tests 1977. M; 3 sons, 1 dtr. ACTIVE.

Hartsfield, Henry W. (b. Nov 21 1933): Col USAF; pilot-astronaut, transferred to NASA when AF MOL programme cancelled, Aug 1969. Working on Shuttle. M; 2 dtrs. ACTIVE.

Henize, Karl G. (b. Oct 17 1926): Astronomer; scientist-astronaut, selected Aug 1967. Support crew, Apollo 15. M; 1 son, 2 dtrs. ACTIVE.

Holmquest, Donald (b. Apr 7 1939): Doctor; scientist-astronaut, selected Aug 1967. On leave from 1971 as A/Prof at Baylor Coll of Medicine, Houston. Then Res NASA Assoc Dean of Medicine, Texas U. M. INACTIVE.

Irwin, James B. (b. Mar 17 1930): Col USAF; pilot-astronaut, selected Apr 1966. As LMP on Apollo 15, Jul 1971, was 8th man on Moon. Ret. USAF & NASA, Aug 1972. Now engaged in religious work in Colorado Springs. M; 1 son, 3 dtrs. INACTIVE.

Kerwin, Joseph P. (b. Feb 19 1932): Capt USN; scientist-astronaut, selected Jun 1965. Science-pilot for Skylab 2 May 1973. M; 3 dtrs. ACTIVE.

Lenoir, William B. (b. Mar 14 1939): Civilian; scientist-astronaut, selected Aug 1967. Back-up science-pilot for Skylabs 3 & 4 Jul and Nov 1973. M; 1 son, 1 dtr. ACTIVE.

Lind, Don L. (b. May 18 1930): Cdr USN (Res); pilot-astronaut, selected Apr 1966. Back-up pilot, Skylab 3 & 4 Jul and Nov 1973. Working on Shuttle. M; 2 sons, 3 dtrs. ACTIVE.

Llewellyn, John A. (b. Apr 22 1933): Civilian; scientist-astronaut, selected Aug 1967. From Cardiff, Wales, only British-born astronaut selected so far. Resigned for personal reasons, Aug 1968; Chemistry Prof, U of Florida, Tallahassee. M; 1 son, 2 dtrs. INACTIVE.

Lousma, Jack R. (b. Feb 29 1936): Lt-Col USMC; pilot-astronaut, selected Apr 1966. Support crew member, Apollo 9 & 13. Pilot, Skylab 3, Jul 30 1973. Back-up DMP, ASTP, Jul 1975. M; 2 sons, 1 dtr. ACTIVE.

Lovell, James A. (b. Mar 25 1928): Capt USN; pilot-astronaut, selected Sep 1962. 1st man to fly 4 missions, with total of 715 hr 5 min in space. Cdr Apollo 13, Apr 1970, aborted Moonlanding mission; pilot on Gemini 7, Dec 4 1965; Cdr, Gemini 12, on Nov 11 1966; CMP Apollo 8, 1st circumlunar flight, Dec 21 1968. Dep Dir Science & Applications, MSC, Houston, 1971. Pres Bay-Houston Towing Co. M; 2 sons, 2 dtrs. INACTIVE.

Mattingly, Thomas K. (b. Mar 17 1936): Cdr USN; pilot-astronaut, selected Apr 1966. Assigned as CMP to Apollo 13, but replaced following contact with German measles. CMP Apollo 16, Apr 1972. Working on Shuttle. M; 1 son. ACTIVE.

McCandless, Bruce (b. Jun 8 1937): Cdr USN; pilot-astronaut, selected Apr 1966. Support crew member, Apollo 14. Back-up pilot, Skylab 2, May 1973. Working on Shuttle. M; 1 son, 1 dtr. ACTIVE.

McDivitt, James (b. Jun 10 1929): Brig-Gen USAF (Ret); pilot-astronaut, selected Sep 1962. As Cdr Apollo 9, achieved 1st LM/CSM docking. Cdr, Gemini 4, Jun 3 1965, when White made 1st US spacewalk. Mgr Apollo Spacecraft Programme 1969. Res; Pres Pullman Co, Chicago Illinois. M; 2 sons, 2 dtrs (only astronaut to have child after flight). INACTIVE.

Michel, F. Curtis (b. Jun 5 1934): Civilian; scientist-astronaut, selected Jun 1965, resigned Aug 1969, to return to scientific research, Rice U. Houston. M; 1 son, 1 dtr. INACTIVE.

Mitchell, Edgar D. (b. Sep 17 1930): Capt USN (Ret), pilot-astronaut, selected Apr 1966. As LMP on Apollo 14, Feb 1971, was 6th man on Moon. Back-up LMP, Apollo 16. Ret from USN & NASA, 1972. Chairman Inst of Noetu Sciences in Calif, then Pres, Ed D Mitchell & Assoc. Inc, Palm Beach, Fla. M; 2 dtrs. INACTIVE.

Musgrave, Story (b. Aug 19 1935): Doc; scientist-astronaut selected Aug 1967. Back-up science-pilot, Skylab 2. M; 3 sons, 2 dtrs. ACTIVE.

O'Leary, Brian T. (b. Jan 27 1940): Civilian: selected Aug 1967. Res Apr 1968. Prof at Hampshire Coll. Amherst, Mass. M. INACTIVE.

Overmeyer, Robert F. (b. Jul 14 1936): Lt-Col USMC; pilot-astronaut transferred to NASA when MOL programme cancelled Aug 1969. ASTP Support. M; 2 dtrs. ACTIVE.

Parker, Robert A. (b. Dec 14 1936): Civilian, scientist-astronaut, selected Aug 1967. Support crew Apollo 15. M; 1 son, 1 dtr. ACTIVE.

Peterson, Donald H. (b. Oct 22 1933): Col USAF; pilot-astronaut transferred to NASA when MOL programme cancelled Aug 1969. Working on Shuttle. M; 1 son, 2 dtrs. ACTIVE.

Pogue, William R. (b. Jan 23 1930): Col USAF; pilot-astronaut selected Apr 1966. Support crew member Apollo 7, 11, 14. As pilot, Skylab 4, Nov 1973, is joint holder of record for longest spaceflight of 84 days. Ret Sep 75; Vice-Pres High Flt (Evangelist) Foundn. Then returned NASA as Special Asst, Earth Resources Payloads, Shuttle at JSC 1976. M; 2 sons, 1 dtr. ACTIVE.

Roosa, Stuart A. (b. Aug 16 1933): Col USAF (Ret); pilot-astronaut selected Apr 1966. CMP on Apollo 14 Jan 1971. Ret Feb 1976; Vice-Pres International Affairs, US Industries Inc; lives in Athens. M; 3 sons, 1 dtr. INACTIVE.

Schirra, Walter M. (b. Mar 12 1923). Capt USN (Ret); Mercury-astronaut, selected Apr 1959. Cdr, Apollo 7, Oct 1968. Cdr Gemini 6 Dec 15 1965 and achieved 1st RV with Gemini 7. Only man to fly Mercury, Gemini and Apollo missions. Ret Jul 1969; Chairman Environmental Control Co. Denver. Then Schirra Enterprises, Denver, Col. M; 1 son, 1 dtr. INACTIVE.

Schmitt, Harrison H. (b. Jul 3 1935): Civilian; scientist-astronaut, selected Jun 1965. Back-up LMP Apollo 15. As LMP, Apollo 17, Dec 1972, was 1st scientist and 12th man on Moon. Then Chmn of Astronauts' Office. Spec Assist on Energy Research to NASA Administrator, 1974. Res Aug 1975. Elected Sen (Dem, New Mexico) 76. Unmarried. INACTIVE.

Schweickart, Russell L. (b. Oct 25 1935): Civilian; pilot-astronaut, selected Oct 1963. As LMP, Apollo 9, in Mar 1969, carried out 1st Apollo EVA. Back-up Cdr, Skylab 2, in 1973. NASA Hq Payload Planning Assist May 74. M; 2 sons, 3 dtrs. ACTIVE.

Scott, David R. (b. Jun 6 1932): Col USAF; Pilot-astronaut, selected Oct 1963. As Cdr, Apollo 15, Jul 1971, was 7th man on Moon. Pilot Gemini 8 on Mar 16, 1966 for world's 1st docking. CMP Apollo 9, Mar 1969 for 1st docking of CSM and LM. Nominated Back-up Cdr Apollo 17, Dec 1972 but later replaced. Dep Dir NASA Flt Research Center, Edwards, Calif 1973. Dir 1975. M; 1 son, 1 dtr. INACTIVE.

See, Elliott M. (b. Jul 23 1927): Civilian; pilot-astronaut, selected Sep 1962. Was training as Cdr Gemini 9, when he and Charles Bassett were killed in T-38 crash at St Louis in Feb 1966.

Shepard, Alan B. (b. Nov 18 1923): Rear Ad USN (Ret). Mercury astronaut, selected Apr 1959. 1st American in space, on Mercury 3 suborbital flight, May 5 1961. Lost flight status owing to ear trouble, and became Chief of Astronaut Office, 1963. Restored flight status 1969; as Cdr Apollo 14 Feb 1971 was 5th man on Moon, and only Mercury astronaut to get there. Resumed duties as Chief of Astronaut Office. Ret Jun 1974. Pres Windward Co Deer Pk, Texas. M; 2 dtrs. INACTIVE.

Slayton, Donald K. (b. Mar 1 1924): Civilian; Mercury astronaut, selected Apr 1959. Lost flight status owing to heart condition. Co-ordinator, Astronaut Activities, Sep 1962. From Nov 1963, Dir of Flight Crew Operations. Returned to flight status 1972; DMP, ASTP Jul 1975. Manager, Shuttle Approach & Landing Tests. M; 1 son. ACTIVE.

Stafford, Thomas (b. Sep 17 1930): Maj-Gen USAF; test-pilot astronaut selected Sep 1962. Cdr Apollo 10, which in May 1969, flew within 14·5 km of Moon in final landing rehearsal. Pilot, Gemini 6, on Dec 15 1965, on world's first RV with Gemini 7. Cdr Gemini 9, on Jun 3 1966. Cdr ASTP Jul 1975. Cdr USAF Edwards AF Base, Calif 1975. M; 2 dtrs. INACTIVE.

Swigert, John L. (b. Aug 30 1931): Civilian; pilot-astronaut, selected Apr 1966. CMP on aborted Apollo 13 Moonlanding mission, Apr 1970. Staff Exec Dir, Committee on Science & Astronautics, House of Reps, Apr 1973. Available for Shuttle. Unmarried. ACTIVE.

Thornton, William P. (b. Apr 14 1929): Doctor; scientist-astronaut, selected Aug 1967. M; 2 sons. ACTIVE.

Truly, Richard H. (b. Nov 12 1937): Cdr USN; pilot-astronaut, transferred to NASA when MOL programme cancelled, Aug 1969. ASTP Support. 2nd pilot Shuttle ALT tests 1977. M; 2 sons, 1 dtr. ACTIVE.

Weitz, Paul L. (b. Jul 25 1932): Capt USN; pilot-astronaut, selected Apr 1966. Support crew member, Apollo 12. Pilot, Skylab 2, May 1973. Working on Shuttle. M; 1 son, 1 dtr. ACTIVE.

White, Edward H. (b. Nov 14 1930): Lt-Col USAF; pilot-astronaut, selected Sep 1962. As pilot, Gemini 4 on Jun 3 1965, made first US spacewalk. Died during Apollo launch-pad rehearsal fire, Jan 27 1967.

Williams, Clifton C. (b. Sep 26 1932): Maj USMC; pilot-astronaut, selected Oct 1963. Killed in T-38 jet crash, Oct 1967.

Worden, Alfred M. (b. Feb 7 1932): Col USAF (Ret); pilot-astronaut, selected Apr 1966. CMP, Apollo 15, 4th success-ful Moonlanding, Jul 1971. Back-up CMP, Apollo 17, Dec 1972. Airborne Science Officer, NASA Ames Research Center, 1972. Vice-Pres, High Flt Foundn; Colorado Springs, 1975. Divorced; 2 dtrs. INACTIVE.

Young, John W. (b. Sep 14 1930): Capt USN; test-pilot-astronaut selected Sep 1962. CMP, Apollo 10, final Moon-landing rehearsal. Pilot, Gemini 3, first US 2-man flight, Mar 23 1965; Cdr, Gemini 10, Jul 18 1966. Cdr Apollo 16, 5th Moonlanding mission, Apr 16 1972. Replaced Scott as Back-up Cdr, Apollo 17, Dec 1972. Chief, Astronaut Office, 1975. M; 2 sons. ACTIVE.

SOVIET COSMONAUTS

COSMONAUTS' COMPARATIVE FLIGHT TIME TO SOYUZ 24

Cosmonauts	Flights	Hr., Min.	Mission
Sevastyanov	2	1936:19	Soyuz 9, 18
Klimuk	2	1700:15	Soyuz 13, 18
Volynov	2	1254:20	Soyuz 5, 21
Zholobov	1	1181:24	Soyuz 21
Gubarev	1	709:20	Soyuz 17
Grechko	1	709:20	Soyuz 17
Volkov	2	689:03	Soyuz 7, 11
Dobrovolsky	1	570:22	Soyuz 11
Patsayev	1	570:22	Soyuz 11
Gorbatko	2	543:21	Soyuz 7, 24
Nikolayev	2	519:22	Vostok 3, Soyuz 9
Popovich	2	448:27	Vostok 4, Soyuz 14
Glazkov	1	424:08	Soyuz 24
Artyukhin	1	377:30	Soyuz 14
Bykovsky	2	309:00	Vostok 5, Soyuz 22
Kubasov	2	261:13	Soyuz 6, 19
Filipchenko	2	261:05	Soyuz 7, 16
Shatalov	3	237:59	Soyuz 4, 8, 10
Yeliseyev	3	214:24	Soyuz 5, 8, 10
Rukavishnikov	2	190:10	Soyuz 10, 16
Aksenov	1	189:54	Soyuz 22
Lebedev	1	188:55	Soyuz 13
Leonov	2	168:33	Voskhod 2, Soyuz 19
Shonin	1	118:42	Soyuz 6
Beregovoy	1	94:51	Soyuz 3
Tereshkova*	1	70:50	Vostok 6
Komarov	2	50:54	Voskhod 1, Soyuz 1
Sarafanov	1	48:12	Soyuz 15
Demin	1	48:12	Soyuz 15

Cosmonauts	Flights	Hr., Min.	Mission
Zudov	1	48:06	Soyuz 23
Rozhdestvensky	1	48:06	Soyuz 23
Khrunov	1	47:49	Soyuz 5
Lazarev	2	47:36	Soyuz 12, Apr 5 Anomaly†
Makarov	2	47:36	Soyuz 12, Apr 5 Anomaly†
Belyayev	1	26:02	Voskhod 2
Titov	1	25:18	Vostok 2
Yegorov	1	24:17	Voskhod 1
Feoktistov	1	24:17	Voskhod 1
Gagarin	1	1:48	Vostok 1
Total	58	14,426:50	31 Manned Missions

EXTRA VEHICULAR ACTIVITY (EVA) RECORD

Mission	Launch Date	Inflight	
		Cosmonauts	EVA Hr., Min.
Voskhod 2	18 Mar 65	Leonov	00:12
Soyuz 4	14 Jan 69	Yeliseyev	00:15
Soyuz 5	15 Jan 69	Khrunov	00:15
	Total		00:42

SOVIET UNION FLIGHT SUMMARY

	Flights
2 men (3 flights)	6
15 men (2 flights)	30
22 men (1 flight)	22
Total 39 men	58

*First female in space—Valentina V. Tereshkova
†Mission aborted before reaching orbit (20 min duration)

At the conclusion of Soyuz 24 in 1977, 39 cosmonauts had flown on Russia's 31 manned missions, including 1 woman. Of those, 6 are dead, 4 having been killed in re-entry accidents; of the other 2 Yuri Gagarin, first man in space, was killed in a military aircraft accident, and the other died from natural causes. Following the agreement with the US to fly the joint ASTP mission in 1975, Russia broke her strict policy of not giving cosmonauts' names until they had flown; as a result, 2 cosmonauts and 2 trainee cosmonauts who have not yet flown are included in the total of 43 named below, together with a 44th cosmonaut who has not yet flown but took part in the ASTP preparations. Gen Beregovoi recently disclosed that 50 Soviet cosmonauts had been trained. In addition, 16 cosmonauts from 8 other Communist countries are being trained at Baikonur. In the list below, full details are not always available.

Aksenov, Vladimir (b. 1935): Flt Eng Soyuz 22, Sep 1976. Involved in developing and testing new spacecraft systems and crews. Cosmonaut No. 36, joined group 1973. M; 2 sons.

Andreyev, Boris (b. 1940): Named in 1973 as 2nd member of 1st Back-up crew for ASTP, Jul 1975. Worked in a design bureau from 1965, and joined cosmonauts' training programme, 1970.

Artyuhkin, Yuri (b. 1930): Lt-Col Soviet AF; graduate of AF Engineering Academy. Flt-Eng Soyuz 14, Jul 4 1974. Joined Cosmonauts' team, 1963. Cosmonaut No. 30. M; 2 sons.

Belyayev, Pavel I. (b. 1932): Cdr Voskhod 2, from which Leonov made world's first spacewalk, Mar 18 1965. Cos-

Nikolayev in Vostok 3   Valentina Tereshkova   Pavel Popovich in Vostok 4

monaut No. 10, he was also a poet and painter. Died from internal trouble, Jan 11 1970, aged 37.

Beregovoi, Georgi T. (b. 1921): Maj-Gen Cdr Soyuz 3, Oct 1968, in RV flight with Soyuz 2. Head, Cosmonaut Training Centre 1975. Cosmonaut No. 12, joined cosmonaut team 1964.

Bykovsky, Valeri F. (b. 1934): Pilot, Vostok 5, Jun 1963, 2nd group flight, and was joined by Valentina Tereshkova, first woman in space, in Vostok 6. Training Officer for ASTP, Jul 1975. Cdr Soyuz 22, Sep 1976. Cosmonaut No. 5.

Demin, Lev (b. 1926): Col. Flt Eng Soyuz 15, Aug 1974. Wartime student with Komarov at special AF school, Moscow. First grandfather in space. Cosmonaut No. 32, joined team in 1963. M; 1 son, 1 dtr.

Dobrovolsky, Georgi (b. Jun 1 1928): Set new endurance record with 24-day flight in Soyuz 11 (23 days docked with Salyut) in Jun 1971. Killed following pressurization failure during re-entry. Cosmonaut No. 24, joined cosmonauts' team 1961. M; 2 dtrs.

Dzhanibekov, Vladimir (b. 1942): Maj. Named in 1973 as 1st member of 1st Back-up crew for ASTP, Jul 1975. Graduated from aviation school in 1965 with pilot-eng diploma. Joined cosmonauts' team 1970.

Feoktistov, Konstantin P. (b. 1926): Scientist-cosmonaut on Voskhod 1, first 3-man craft, Oct 1964. Cosmonaut No. 8, accepted for cosmonauts' group despite wartime injuries (shot and left for dead when acting as scout in German-occupied territory).

Filipchenko, Anatoli (b. Feb 26 1928): Lt-Col. Cdr Soyuz 7, in record 3-craft groupflight with Soyuz 6 & 8, Oct 1969. Cdr Soyuz 16, ASTP rehearsal, Dec 2 1974. Nominated Cdr, 2nd Prime crew, for ASTP, Jul 1975. Cosmonaut No. 19. M; 2 sons.

Gagarin, Yuri A. (b. Mar 9 1934): Col. World's first man in space, in Vostok 1, Apr 12 1961. Subsequently became Cdr of Soviet Cosmonauts' Detachment, and made world-wide public appearances. Killed with Col Seryogin during test flight of military jet, Mar 27 1968. Ashes buried in Kremlin Wall; crater on lunar farside named after him. M; 2 dtrs.

Glazkov, Yuri (b. 1939) Lt-Col. Flt Eng Soyuz 24, Feb 1977, which docked with Salyut 5. Specialist in 'man's activity in open space'. Joined cosmonauts' team 1965. Cosmonaut No. 39.

Gorbatko, Viktor (b. Dec 3 1934): Lt-Col. Research eng., Soyuz 7, group-flight with Soyuz 6 & 8, Oct 1969. Cdr Soyuz 24, docked with Salyut 5, Feb 1977. Joined Cosmonaut's group with Gagarin in 1960. Cosmonaut No. 21. M; 2 dtrs.

Grechko, Georgi (b. 1931): Flt Eng Soyuz 17 Jan 1975. Worked in Korolev's design team, and helped design automatic lunar landing systems. Joined cosmonauts' team 1966. Cosmonaut No. 36. M; 2 sons.

Gubarev, Alexei (b. 1931): Cdr/Lt-Col Soyuz 17 Jan 1975. Entered the 'Gagarin' Air Force Academy in 1957 and joined cosmonauts' detachment in 1963. Had to wait 12 yr for 1st flt. Wife works at Stellar Townlet. Cosmonaut No. 35. M; 1 son, 1 dtr.

Illavionov, Valeti: No details known. He was one of 9 Soviet specialists on Soyuz spacecraft who were at Mission Control, Houston for the ASTP mission.

Ivanchenko, Alexander (b. 1940): Named in 1973 as 2nd member of 2nd Back-up crew for ASTP, Jul 1975. Graduated from Moscow Aviation Inst and worked in design bureau from 1964. Joined cosmonauts' training programme, 1970.

Khrunov, Yevgeny V. (b. 1933): Lt-Col. Research eng., Soyuz 5, Jan 15 1969; first docking of 2-manned craft, space-walked with Yeliseyev to return in Soyuz 4. Cdr Soyuz 18, 63-day duration record. Cosmonaut No. 15, joined cosmonauts' team 1960. M; 1 son.

Klimuk, Pyotr (b. 1942): Maj AF. Cdr Soyuz 13, Dec 18 1973. Cosmonaut No. 28, joined cosmonauts' team 1965. M; 1 son.

Komarov, Vladimir M. (b. Mar 16 1927): Col. Cdr Voskhod 1, first 3-man craft, on Oct 12 1964. Flew twice despite heart condition similar to Slayton's. First man to be killed in space, when re-entry parachute snarled at end of Soyuz 1, Apr 23 1967. Cosmonaut No. 7.

Kubasov, Valeri (b. Jan 7 1935): Flt Eng Soyuz 6, on Oct 11 1969, involving manoeuvres with Soyuz 7 & 8. Kubasov carried out 1st space-welding experiments. Flt Eng Prime Crew ASTP, Jul 1975. Cosmonaut No. 18, joined cosmonauts' team 1966. M; 1 dtr.

Lazarev, Vasily (b. 1928): Lt-Col. Test pilot and physician. Cdr, Soyuz 12, Sep 1973. Stand-in Cdr, Soyuz 9. Cdr, Soyuz 00, aborted during launch Apr 1975. Cosmonaut No. 26, probably joined cosmonauts' team 1966. M; 1 son.

Lebedev, Valentin (b. 1942): Flt Eng Soyuz 13, Dec 1973. Cosmonaut No. 29, joined cosmonauts' team 1972. M; 1 son.

Leonov, Alexei A. (b. May 30 1934): Col. First man to walk in space on Voskhod 2, Mar 18 1965. Afterwards made vivid

Vladimir Komarov   Pavel Belyayev   Valery Kubasov

impressionistic paintings. Cdr Prime Crew, ASTP, Jul 1975. Cosmonaut No. 11.

Makarov, Oleg (b. 1933): Flt Eng Soyuz 12, Sep 1973. Flt Eng Soyuz 00, aborted during launch, Apr 1975. Cosmonaut No. 27, joined cosmonauts' team 1966. M; 1 son.

Nikolayev, Andrian G. (b. 1929): Col. Pilot, Vostok 3, in 1st double flight with Vostok 4, Aug 1962. Cdr Soyuz 9, which in Jun 1970, broke US endurance record with flight of nearly 18 days. Cosmonaut No. 3, joined team in 1960. Married V. Tereshkova, 1st space woman, in 1963; 1 dtr.

Patsayev, Viktor (b. Jun 19 1933): Test Eng Soyuz 11, on 24-

day flight in Soyuz 11 & Salyut space station, Jun 1971. Killed following pressurization failure during re-entry. Cosmonaut No. 25. M; 1 son, 1 dtr.

Popovich, Pavel R. (b. 1930): Col. Pilot, Vostok 4, in 1st double flight with Vostok 3, Aug 1962. Cdr, Soyuz 14, Jul 4 1974. Cosmonaut No. 4, believed to be first appointment to cosmonauts' team. Married to Marina, Test-pilot Col in Soviet AF; 2 dtrs.

Romanenko, Yuri (b. 1944): Capt. Named in 1973 as 1st member of 2nd Back-up crew for ASTP in Jul 1975. Graduated with distinction from higher aviation school in

1966 with pilot-eng. diploma. Young cosmonaut; joined team 1970.

Rozhdestvensky, Valeri (b. 1939) Lt-Col. Flt Eng, Soyuz 23 which made 1st Soviet splashdown in emergency return. Naval engineer and former Cdr of group of deep-sea divers in Baltic Rescue Service. Cosmonaut No. 38, joined group in 1965. M; 1 dtr.

Rukavishnikov, Nikolai (b. Sep 18 1932): Test Eng Soyuz 10, which docked briefly with Salyut, Apr 1971. Nominated in 1973 2nd crew member, 2nd Prime Crew for ASTP. Flt Eng Soyuz 16 ASTP rehearsal Dec 1974. Worked as space-craft designer under Korolev. Cosmonaut No. 23, joined cosmonauts' team Jan 1967. M; 1 son.

Sarafanov, Gennady (b. Jan 1 1942): Lt-Col. Cdr. Soyuz 15, Aug 1974. Cosmonaut No. 31, joined team 1965. M; 1 son, 1 dtr.

Sevastyanov, Vitali (b. Jul 8 1935): Flt Eng Soyuz 9, on 18-day endurance flight. Flt Eng Soyuz 18, 63-day duration record. Cosmonaut No. 22, joined team 1967. M; 1 dtr.

Shatalov, Vladimir A. (b. 1927): Maj Gen. Cdr, Soyuz 4, Jan 1969, for first docking, with Soyuz 5, of 2-manned craft. In Soyuz 8, Oct 1969, was Overall Cdr of Soyuz 6, 7 & 8 groupflight with 7 cosmonauts. Cdr, Soyuz 10, for docking with Salyut. Now Dir Cosmonauts' Training. Cosmonaut No. 13, joined group 1963. M; 1 son, 1 dtr.

Shonin, Georgi (b. 1935): Lt-Col. Cdr, Soyuz 6, on group-flight with Soyuz 7 & 8. Cosmonaut No. 17. M; 1 son, 1 dtr.

Tereshkova, Valentina V. (b. 1937). As pilot, Vostok 6, on Jun 16 1963, became first woman in space, in group flight with Vostok 5. Cosmonaut No. 6, she said that when Titov flew, she was 'not even dreaming of becoming a cosmonaut', but was in orbit 21 months after Titov. Married Nikolayev, 1963; 1 dtr.

Titov, Herman S. (b. 1935): Col. As pilot Vostok 2, was 2nd man in orbit, and 1st to spend one day there, on Aug 6 1961. Cosmonaut No. 2.

Volkov, Vladislav (b. Nov 23 1935): Flt Eng Soyuz 7, on group flight with Soyuz 6 & 8, Oct 1969. Flt Eng on 24-day flight in Soyuz 11 and Salyut space station, Jun 1971. Killed following pressurization failure during re-entry. Holds Soviet duration record (28 days, 16 hr, 21 min). Cosmonaut No. 20, joined group 1966. M; 1 son.

Volynov, Boris V. (b. 1934): Col. Cdr, Soyuz 5, for first docking of 2-manned craft with Soyuz 4, Jan 1969. Cdr Soyuz 21, Jul 6 1976; spent 48 days in Salyut 5. Cosmonaut No. 14, joined group 1960. M; 1 son, 1 dtr.

Yegorov, Boris B. (b. 1937). 1st space doctor, on 3-man Voskhod 1, Oct 1964. Cosmonaut No. 9.

Yeliseyev, Alexei S. (b. 1934): Flt Eng Soyuz 5, for first docking of 2-manned craft with Soyuz 4, Jan 1969. Space-walked from Soyuz 5, and returned in Soyuz 4. Flt Eng on Soyuz 8, for group flight with Soyuz 6 & 7, Oct 1969. Flt Eng Soyuz 10 Apr 1971. Flt Director, ASTP. Cosmonaut No. 16, joined group 1966. M; 1 dtr.

Zholobov, Vitaly (b. 1937) Lt-Col Soviet Army. Backup Flt Eng, Salyut 3. Flt eng Soyuz 21/Salyut 5, Jul 6 1976. Cosmonaut No. 35, joined Jan 1963. M; 1 dtr.

Zudov, Vyacheslav (b. 1942) Lt-Col. Cdr Soyuz 23 which made 1st Soviet splashdown in emergency return. Cosmonaut No. 37, joined group 1965. M; 2 dtrs.

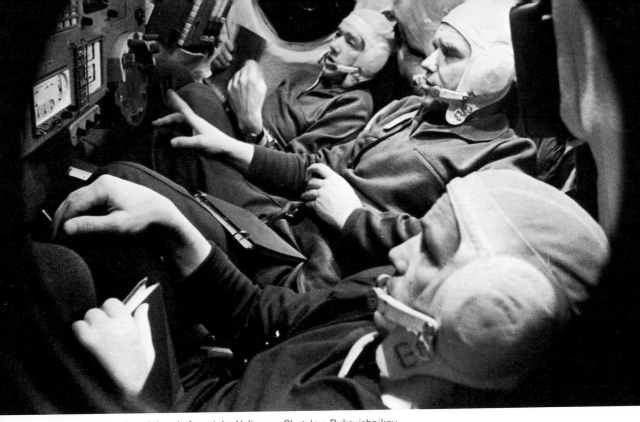

Soyuz 10 crew in training. *Left to right*: Yeliseyev, Shatalov, Rukavishnikov

INDEX

Main entries in **heavy** type; colour pictures in *italic* type.

VOYAGER MISSION HIGHLIGHTS

VOYAGER 2		Date	VOYAGER 1	
Event	Approximate range in kilometres		Event	Approximate range in kilometres
Launch		Aug 20, 1977		
Near Earth Science Seq		Aug 20–29, 1977		
TCM-1*		Aug 25–Sep 4, 1977		
		Sep 1, 1977	Launch	
		Sep 1–16, 1977	Near Earth Science Seq	
		Sep 6–16, 1977	TCM-1*	
		Dec 15, 1978	Start Jupiter Imaging	80,000,000
		Mar 4, 1979	Amalthea	415,000
		Mar 5, 1979	Jupiter Closest Approach	278,000
		Mar 5, 1979	Io	22,000
		Mar 5, 1979	Europa	733,000
		Mar 5, 1979	Ganymede	115,000
		Mar 6, 1979	Callisto	124,000
		Apr 1979	Conclude Jupiter Imaging	
Start Jupiter Imaging	75,000,000	Apr 20, 1979		
Callisto	220,000	Jul 8, 1979		
Ganymede	55,000	Jul 9, 1979		
Europa	201,000	Jul 9, 1979		
Amalthea	550,000	Jul 9, 1979		
Jupiter Closest Approach	643,000	Jul 10, 1979		
Conclude Jupiter Imaging		Aug 1979		
		Aug 1980	Start Saturn Imaging	100,000,000
		Nov 11, 1980	Titan	4000
		Nov 12, 1980	Saturn Closest Approach**	138,000
		Jan 1981	Conclude Saturn Imaging	
Start Saturn Imaging	100,000,000	Jun 1981		
Saturn Closest Approach***	102,000	Aug 27, 1981		
Saturn Rings Edge	38,000	Aug 27, 1981		
Conclude Saturn Imaging		Oct 1981		
Uranus Encounter		Jan 1986		

*Trajectory Correction Manoeuvre-1 may be conducted in two or three parts at about 24-hr intervals because of thermal constraints at this spacecraft–Sun distance. Each spacecraft will execute at least 7 additional TCMS prior to Saturn encounter

**Also encounter imaging of satellites Tethys, Enceladus, and Rhea

***Also encounter imaging of satellites Titan, Rhea, Tethys, and Enceladus